Q_3	third quartile
ρ	population correlation coefficient
r	Pearson product-moment correlation coefficient
R	Spearman rank-difference correlation coefficient
r^2	coefficient of determination
σ	population standard deviation; standard deviation of a probability distribution
σ^2	population variance; variance of a probability distribution
σ_d	standard error of the difference
σ_{dp}	standard error of the difference for proportions
σ_p	standard error of proportion
$\sigma_{\bar{x}}$	standard error of the mean
s	sample standard deviation
s^2	sample variance
s_d	standard error of the difference estimated from sample standard deviations
s_{dp}	standard error of the difference for proportion estimated from sample proportions
s_p	standard error of proportion estimated from sample proportion
$s_{\bar{x}}$	standard error of the mean estimated from sample standard deviation
SE	standard error for mean differences
$\sum$	summation sign
SS	sum of squares
SS_{BETWEEN}	sum of squares between samples
SS_{WITHIN}	sum of squares within samples
SS_{TOTAL}	total sum of squares
t	Student's t-distribution point
T	Wilcoxon T statistic
U	Mann-Whitney U statistic
x	data point of a random variable
x_d	difference score
$\bar{x}$	sample mean
x_d	mean difference
$\hat{x}$	assumed mean
x'	predicted value of x from regression equation
χ^2	chi-square distribution point
y	data point of a random variable
y'	predicted value of y from a regression equation
z	standard score
$\doteq$	is approximately equal to
$<$	is less than
$\leq$	is less than or equal to
$>$	is more than
$\geq$	is more than or equal to
$\binom{n}{r}$	binomial coefficient; combinations of n objects, r at a time
$!$	factorial

Elements of Statistics

ELEMENTS of

An Introduction to Probability and

D. Van Nostrand Company
New York • Cincinnati • Toronto • London • Melbourne

STATISTICS

Statistical Inference, 2nd edition

Donald R. Byrkit
The University of West Florida

To Will—
Who suggested this book,
and
To Marie—
Who put up with it

D. Van Nostrand Company Regional Offices:
New York Cincinnati Millbrae

D. Van Nostrand Company International Offices:
London Toronto Melbourne

Library of Congress Catalog Card Number: 74-25312
ISBN: 0-442-21413-8

Published by D. Van Nostrand Company
450 West 33rd Street, New York, N.Y. 10001

Published simultaneously in Canada by
Van Nostrand Reinhold Ltd.

10 9 8 7 6 5 4 3

Title page Photo: Allan D. Cruickshank from National Audubon Society

Cover Photo: Verna R. Johnston from National Audubon Society

Preface

This text is the outgrowth of the author's experience in teaching a beginning course in statistics to nonmathematically oriented students whose primary area of study is the biological or social sciences. Mathematical exposition, therefore, is kept to a minimum, and the major emphasis is on reasonable demonstration by example.

The second edition was prepared in response to requests to include additional topics and because a relative shift of emphasis in introductory statistics courses has occurred since the first edition was published. The emphasis in the second edition is on the use of statistical inference in decision making—the topic most pertinent to students who wish to use statistical techniques in their study of the biological or social sciences.

Part I covers the organization and presentation of data (Chapter 1) and measures of location and dispersion (Chapter 2); it is an essential introduction to the material in the remainder of the book. In Part II, probability (Chapter 3) and probability distributions (Chapter 4) are discussed. An optional discussion of conditional probability is included in Chapter 3. The binomial distribution (Chapter 5) and the normal distribution (Chapter 6) are the topics covered in Part III. Hypothesis testing is introduced for the first time in Sections 5.4 and 6.4, but the instructor may omit these sections and spend more time on Chapter 9, where this topic is discussed in detail. Part IV covers sampling distributions (Chapter 7) and confidence intervals (Chapter 8). In Part V, the student is introduced to hypothesis testing (Chapter 9), tests for sample means and proportions (Chapter 10), the Chi-Square distribution (Chapter 11), analysis of variance (Chapter 12), and correlation and regression (Chapter 13). A review of mathematics is included in Appendix A, and Appendix B has been expanded into a full chapter on nonparametric methods.

Changes in the second edition include an expansion of the material on descriptive statistics and a reduction of the emphasis on probability. Hypothesis testing is introduced early in the text (Chapter 5) as part of the expanded treatment of the binomial probability distribution. Widespread use of hand calculators has made the use of raw-data formulas more feasible, and these formulas are emphasized. Many sections have been expanded or rewritten for greater clarity. A large number of problems drawn from actual research studies have been added throughout the book. Tables of Poisson probabilities, binomial probabilities, and random numbers have been added. Scheffe's *post hoc* test has been incorporated into the discussion of the analysis of variance. The overall effect of these changes gives the instructor more flexibility in choosing or emphasizing selected topics to meet individual course requirements, while preserving the nonmathematical treatment of the material.

Each chapter section is followed by a set of problems which provide practice in the statistical techniques presented in the section and examples of the application of these techniques to research problems the students may encounter. Answers to almost all problems are provided. Detailed solutions are given to key problems to show the student the logic behind the solution.

The front endpaper contains a glossary of the symbols used throughout the text. The back endpaper contains a chart to help the student determine the appropriate statistical test to be used with a problem encountered in the research literature or in the student's own work.

The length of the course is approximately four semester-hours or six quarter-hours. A standard three semester-hour course might cover Chapters 1–10 and a few of the later topics if the instructor chooses to omit some of the earlier material.

Input from many sources helped in this revision, and the author is especially grateful to Jerry Oglesby, Nick Moore, Betsy Hill, William Price, and Robert Knapp for their helpful comments and suggestions.

I am indebted to the Biometrika Trustees for permission to use Tables 8, 12, and 18 from *Biometrika Tables for Statisticians*, Volume 1, Third Edition, by Pearson and Hartley. I am further indebted to the Literary Executor of the late Sir Ronald A. Fisher, F.R.S., to Dr. Frank Yates, F.R.S., and to Oliver & Boyd, Edinburgh, for permission to reprint Table VI from their book *Statistical Tables for Biological, Agricultural, and Medical Research.* Thanks go to Dr. R. A. Wilcox and Lederle Laboratories for permission to reprint data from *Some Rapid Approximate Statistical Procedures*, by Wilcoxon and Wilcox; to the authors and publishers of *Probability: A First Course*, Second Edition, 1970, by Mosteller, Rourke, and Thomas (Addison-Wesley), for permission to reprint Table IV, Part B; and to the authors and publishers of *Statistical Tables*, by R. R. Sokal and F. J. Rohlf (W. H. Freeman and Company), for permission to reprint part of Table O.

Donald R. Byrkit

Contents

PART I DESCRIPTIVE METHODS 1

Chapter 1 Organization and Presentation of Data 3

1.1 Frequency Distributions 3
1.2 Graphical Methods 11
1.3 Summary 19

Chapter 2 Measures of Location and Dispersion 22

2.1 Measures of Central Tendency 22
2.2 Measures of Dispersion 31
2.3 Other Measures of Location and Comparison 42
2.4 Summary 45

PART II PROBABILITY AND PROBABILITY DISTRIBUTIONS 49

Chapter 3 Probability 51

3.1 Probabilities 51
3.2 Permutations and Combinations 56
3.3 Rules of Probability 61
3.4 Independent Events and Conditional Probability 69
3.5 Summary 78

Chapter 4 Probability Distributions 84

4.1 The General Probability Distribution 84
4.2 Expected Value 89
4.3 Mean and Variance of a Probability Distribution 94
4.4 Summary 100

PART III THE BINOMIAL AND NORMAL DISTRIBUTIONS 103

Chapter 5 The Binomial Distribution 105

5.1 The Binomial Distribution 105
5.2 Applications of the Binomial Distribution 112
5.3 Using Binomial Tables (optional) 115
5.4 The Binomial Model in Hypothesis Testing 119
5.5 Other Important Distributions (optional) 123
5.6 Summary 130

Chapter 6 The Normal Distribution 134

6.1 Continuous and Discrete Variables 134
6.2 Normal Distributions 137
6.3 Applications of the Normal Distribution 149
6.4 The Normal Approximation to the Binomial 154
6.5 More on Hypothesis Testing 159
6.6 Summary 161

PART IV SAMPLES 165

Chapter 7 Sampling and Sampling Distributions 167

7.1 Samples 167
7.2 Parameters and Statistics 171
7.3 Sampling Distributions 174
7.4 Summary 178

Chapter 8 Estimation of Parameters 181

8.1 Point Estimation 181
8.2 Confidence Intervals for the Population Mean 186

8.3 Confidence Intervals Obtained from Small Samples 190
8.4 Confidence Intervals for Proportions 194
8.5 Summary 198

PART V DECISIONS 203

Chapter 9 Hypothesis Testing 205

9.1 Experimental Error 205
9.2 Formulating Hypotheses to Control Error 207
9.3 Tests of Significance 213
9.4 Summary 216

Chapter 10 Tests Concerning Means and Proportions 219

10.1 Means of One Large Sample 219
10.2 Means of One Small Sample 227
10.3 Proportions in a Sample 232
10.4 Differences between Means 234
10.5 Differences between Proportions 242
10.6 Summary 245

Chapter 11 The Chi-Square Distribution 251

11.1 Proportions 251
11.2 Contingency Tables 256
11.3 Curve Fitting 262
11.4 Summary 267

Chapter 12 Analysis of Variance 270

12.1 Basic Assumptions for Analysis of Variance 270
12.2 Estimating Population Variance 271
12.3 Other Methods for Analysis of Variance 278
12.4 Summary 283

Chapter 13 Correlation and Regression 287

13.1 The Coefficient of Correlation 287
13.2 Significance of the Correlation Coefficient 296
13.3 Linear Prediction—Regression 301
13.4 Summary 306

Appendix A Mathematics Review 311

Decimals	311
Fractions	313
Exponents	315
Signed Numbers	315
Order Relations	316
Square Roots	316
Linear Equations	318

Appendix B Nonparametric Tests 323

The Signs Test	324
The Wilcoxon *T*-Test	328
The Mann-Whitney *U*-Test	331
The Kruskal-Wallis *H*-Test	334
Summary	336

Tables 338

Answers 377

Index 427

Elements of Statistics

PART I

DESCRIPTIVE METHODS

Part I Photo: Courtesy of Chicago's Museum of Science and Industry

Chapter I

Organization and Presentation of Data

1.1 FREQUENCY DISTRIBUTIONS

Statistics can be defined as the science of classifying and manipulating data in order to draw inferences. An important aspect of classifying data is the efficient and effective organization and presentation of data. An unorganized mass of figures is often more confusing than clarifying. This chapter is concerned with the methods of deriving meaning from numerical data.

The term **data**, as used in the preceding paragraph, refers to the set of observations, values, elements, or objects under consideration. Each of these elements is called a **data point**, or **piece of data**. A score of 83 on a test made by one individual, for example, is a piece of data, while the collection of all scores made by individuals on the test would comprise the complete set of data.

Data are of two types, qualitative and quantitative. **Qualitative data**, or **attributes**, results from information that has been sorted into categories. Each piece of data belongs clearly to exactly one classification or category. Automobiles on a parking lot classified by make give one example of attribute data. **Quantitative**, or **variable data**, is data which is a result of counting or measuring. We might count the number of nicks and scratches in the paint of each car, then give the number of cars with 0 scratches, 1 scratch, 2 scratches, and so forth. A car has an integral number of scratches (it cannot have 3.7 scratches, for instance), so there are clear divisions between the values. This type of data is called **discrete**. If we weighed the cars, we could get the weight to the nearest pound, but it would be possible for a car to weigh slightly more or less than we reported. A weight of 3,456 pounds, for example, would indicate a weight somewhere between 3,455.5 and 3,456.5 pounds. This is an example of a **continuous** variable. Generally, data arising from measurement, distance, weight, duration, and so forth, are continuous, while data arising from counting or arbitrary classification are discrete. An example of the latter is a grading system where the grades are 4.0, 3.5, 3.0, 2.5, 2.0, 1.5, 1.0, and 0 (the familiar A, B+, B, C+, C, D+, D, F system).

Suppose you asked someone for data on the height of adult males in a city of some 25,000 population, and he responded by giving you a list of 8,968 heights. Unless the data were organized in some fashion, this list in its raw form would not be too usable. One of the means employed to organize data is known as the **frequency distribution**. In its simplest form, the frequency distribution consists of listing each possible value the data could have and enumerating the total number, called the **frequency**, for each such value. In the height example, such a frequency distribution might appear as follows:

Height (inches)	*Frequency*
83	11
82	44
81	116
80	132
79	157
78	284
77	316
76	388
75	547
74	731
73	783
72	808
71	817
70	931
69	848
68	712
67	604
66	411
65	206
64	94
63	28
Total	8,968

Such a table tells you at a glance that most of the values are from 67 to 75, inclusive, that the number of values above 78 and below 65 is relatively quite small, and, in short, gives you a very good and accurate profile of this set of data.

It is often useful to determine the **proportion** of cases for each value. This is also called the **relative frequency**, which is the number of cases for a given value divided by the total number of cases. In the preceding example 931 men were 70 inches tall out of the total of 8,968, so the relative frequency for 70 inches was 931/8,968 or about 0.104.

It is helpful to have a procedure to follow when constructing a frequency table. For small amounts of data, we can rewrite the data given into ascending (or descending) order. We then have the data ranked in order and the construction of the table is relatively simple. Another plan, more useful in cases where a great deal of data is involved, is to find the highest and lowest values, then list these values and all cases between them. In a second column we tally the number of cases, then summarize the results in a frequency table. The relative frequency can be included if desired. The table should be complete, including the title, with enough information to make the table completely self-explanatory if not accompanied with explanatory material.

EXAMPLE 1 A survey of prices of a plain loaf of bakery bread at each of 24 bakeries in the city yielded the data given below. Construct a table showing frequency and relative frequency for each price. Prices are given in cents.

37	39	38	41	36	37
43	37	49	39	39	41
37	43	52	39	42	38
40	39	43	39	45	39

SOLUTION There are just a few pieces of data here, so rewriting these in order would be logical. An example of the tally method might be helpful at this point, however, so it is presented here.

Cost (cents)	*Tally*
52	/
51	
50	
49	/
48	
47	
46	
45	/

—*continued*

Cost (cents)	*Tally*
44	
43	///
42	/
41	//
40	/
39	//// //
38	//
37	////
36	/

Using the tally chart we can construct the table below.

Cost of a plain loaf of bread[a]

Cost (cents)	*Frequency*	*Relative Frequency*
52	1	1/24
51	0	0
50	0	0
49	1	1/24
48	0	0
47	0	0
46	0	0
45	1	1/24
44	0	0
43	3	3/24
42	1	1/24
41	2	2/24
40	1	1/24
39	7	7/24
38	2	2/24
37	4	4/24
36	1	1/24
Total	24	

[a] Pottsville, June 7, 1974; survey of local bakeries.

COMMENT When a large number of classes (or values) contain no entries or only a few entries it is often convenient to combine them into one class. This is often done when these classes fall at one end of the distribution. Here we could write

44–52   3   3/24

This does not significantly impair our understanding of the table, but may make it difficult to perform arithmetic on the data. For example, we cannot compute the range of the grouped data.

DISCUSSION It generally does not matter whether tables are arranged in ascending or descending order. The guiding principles should be ease and clarity of presentation.

For some purposes a **cumulative frequency table** is useful. Generally the data are arranged in ascending order with the cumulative frequency listed. The cumulative frequency is the number of cases at or below the listed value. For the bread data, such a table could be as follows:

Cost (cents)	*Frequency*	*Cumulative Frequency*
36	1	1
37	4	5
38	2	7
39	7	14
40	1	15
41	2	17
42	1	18
43	3	21
44–52	3	24

A frequency table may be used to classify **attributes**. Attributes are a type of data that cannot be measured, but can only be described. For example, a table may classify the automobiles in a parking lot by make.

Make	*Number of Cars*
Chevrolet	33
Ford	28
Pontiac	11
Volkswagen	9
Plymouth	7
Other makes	4
Total	92

Another example of attributes is the division of data into **strata**, such as low, middle, and high income groups. This type of data is also called categorical data, since the material is placed into categories.

Now suppose that a set of data gives the incomes of families in a city. It is obvious that a listing of all possible incomes (even to the nearest dollar) would involve a listing of hundreds, even thousands of numbers. One way out of such a difficulty is the method of **grouping data**. Thus, we may consider all incomes, say, from 4,000 to 7,999 dollars, as constituting one **class**.

A few general observations govern the grouping of data into classes. First, we do not want too many or too few classes, since this might result in a distortion of the picture we wish to convey. Second, the classes should be of the same size. Third, the classes should be convenient to handle. To achieve these aims, it has been found that eight to fifteen classes are a reasonable number. To determine class size, the **range**—highest measure minus lowest measure—is divided by the approximate number of classes desired, then the number is rounded out to the nearest convenient division.

The height data presented earlier is an example of **continuous** data which has been grouped into separate, or **discrete**, classes. In this kind of arrangement, a height of 72 inches represents all heights from 71.5 to 72.5 inches. It is convenient and useful to accept this representation, since otherwise we may run into difficulties when we classify the data into larger classes. For example, if we decide to combine our data into classes covering four inches and we start with 63 inches, our classes become 63–67, 67–71, 71–75, 75–79, and 79–83 inches. Into which class does a height of 71 inches fall? To avoid this problem with the endpoints we let the endpoints of each class fall *between* the integral values of the class. Each integer becomes the midpoint of an interval extending one-half unit in each direction. Thus the lowest height represented is 62.5 inches and the highest is 83.5 inches. The range, then, is 21 inches, $83.5 - 62.5$. If the data is divided into ten classes, each class has a size of 2.1 inches, while eight classes produce a class size of 2.63 inches. Generally we wish the size of the classes, or **class interval**, to be an integer. There are some advantages to having an odd-size class interval. This is because the center of each class, called the **class mark**, is then an integer. Suppose in this example we let the class interval be 3 inches. Since the lowest represented height is 62.5 inches, we can start there, if we wish, and the lowest interval would be 62.5 inches to 65.5 inches. The lowest and highest values of each class are called the **class limits**. Here they are 62.5 and 65.5. The class mark is the center of each interval and can be found by adding the class limits and dividing by 2. For example, $(62.5 + 65.5)/2 = 64$. To check our work, observe that the difference between successive lower class limits, successive upper class limits, and successive class marks all equal the class interval.

Many experts prefer to present a table in terms of the actual limits for each class, but others feel that if our understanding of the table is not hampered, we can use the *apparent* limits, generally listed as integers. A class whose actual limits are, say, 71.5 and 74.5 may be thought of as having apparent limits of 72 and 74 since these are the integral values placed in the class. By convention, a value such as 71.5, midway between two classes, is counted in the higher class; that is, it is put at the *beginning* of the class.

Using these general guides, the following frequency table could be constructed for the height data. In this example the apparent limits are used since heights are generally measured to the nearest inch.

Height (inches)	*Frequency*	*Class Mark*
63–65	328	64
66–68	1,727	67
69–71	2,596	70
72–74	2,322	73
75–77	1,251	76
78–80	573	79
81–83	171	82
Total	8,968	

EXAMPLE 2 Suppose that income data for another city shows the lowest family income to be $343 per year, and the highest $43,764. Construct a frequency distribution with approximately fifteen classes for these data.

SOLUTION The range is $43,764 – $343, or $43,421. Dividing this by 15, we obtain $2,895. A convenient class interval is $3,000. We could start with $343, but it is more usual to start with some " nice " number, such as zero or $1,000. Since 343 is less than 1,000, we might arrange the data like this.

Income (dollars)	*Number of Families*
42,000–44,999	3
39,000–41,999	7
36,000–38,999	12
33,000–35,999	17
30,000–32,999	28
27,000–29,999	60
24,000–26,999	216
21,000–23,999	268
18,000–20,999	448
15,000–17,999	621
12,000–14,999	949
9,000–11,999	1,421
6,000– 8,999	2,844
3,000– 5,999	2,123
0– 2,999	294
Total	9,311

DISCUSSION A value such as \$29,999.37 does not fit into this table, but recall that each class is considered to extend down one-half unit from the lower value and up one-half unit from the higher value. Thus, the class given by 27,000 to 29,999 is actually considered to be the interval 26,999.50 to 29,999.50, while 30,000 to 32,999 would represent the interval 29,999.50 to 32,999.50.

It is tempting, and sometimes convenient, to represent attenuated cases such as this one by lumping everything higher than, say, \$30,000 into one class, such as

30,000 or above

This does not hinder our understanding of the table, but it may make it impossible to perform arithmetic on the data since an **open class**, one which lacks either an upper or lower class limit, has *no* class mark.

Problems

1. Several thousand IQ test scores were found to fall between a high of 168 and a low of 72.

   (a) What is the range of the scores?
   (b) What class interval should be selected to divide the data into approximately twelve classes?
   (c) Make two lists of classes, one using the value from (b), one using ten as class interval.

2. The following are the weekly sales, in dollars, of Rogers' Cat Food at each of 80 retail outlets.

14	27	81	36	92	60	17	34	83	54
37	40	27	30	26	36	29	71	23	37
31	37	12	36	61	17	70	36	35	77
83	13	39	61	48	54	23	37	35	23
61	97	31	46	13	24	30	19	26	73
70	17	23	10	38	11	65	67	14	45
70	55	24	27	45	64	24	86	28	16
27	25	11	15	53	65	12	58	62	53

   (a) Group these numbers into a frequency table with 8 class intervals.
   (b) Use the results of (a) to make a relative frequency table.
   (c) Use the results of (a) to make a cumulative frequency table.

3. The following are lengths, in centimeters, of mullet taken from the bay.

31	29	35	37	39	35	40	32	37	35	36	41
42	33	39	35	41	36	37	37	34	41	37	31
38	43	37	38	37	34	36	35	36	32	39	37
37	38	36	37	35	37	35	33	36	30	34	33
34	36	37	36	36	33	40	39	38	37	37	37
35	37	40	33	37	38	38	36	37	35	33	36

(a) Determine the proper class interval if we wish to divide the data into seven to twelve classes.
(b) Make frequency, relative frequency, and cumulative frequency tables for the data.

4. Pollution indices for a city for 120 consecutive days are given below.

47	70	84	46	29	64	43	61	46	40	41	59
58	72	88	57	39	60	47	62	58	38	33	54
67	59	81	63	44	57	54	54	60	47	42	63
72	54	77	69	57	51	59	57	52	62	48	60
74	50	70	72	60	58	62	44	48	54	41	66
88	61	54	61	61	61	67	30	54	70	52	69
67	73	42	52	53	59	68	33	58	83	48	58
73	68	41	44	50	54	70	41	54	88	44	42
64	79	43	37	44	60	74	49	67	69	42	47
61	82	37	33	58	48	66	52	49	57	45	48

(a) Group the numbers into a frequency table with a class interval of 5.
(b) Make cumulative frequency and relative frequency tables for this data.

1.2 GRAPHICAL METHODS

One disadvantage of frequency distributions is that their presentation lacks visual appeal. They do not call attention to the outstanding or important features of the distribution as a graphical presentation does. Graphs appear in many forms, only a few of which will be discussed here. Certain types of data lend themselves well to certain types of graphs.

Data which are classified naturally into distinct groups are usually represented well by means of **bar graphs**. Examples of bar graphs are given here.

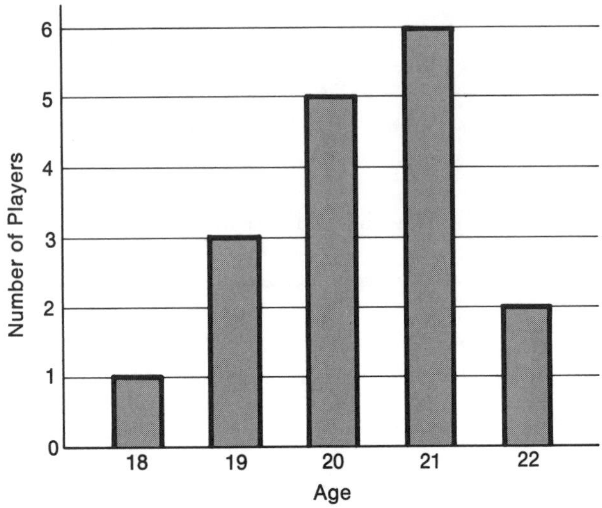

Distribution of Ages of Players on the Meridian College Basketball Team

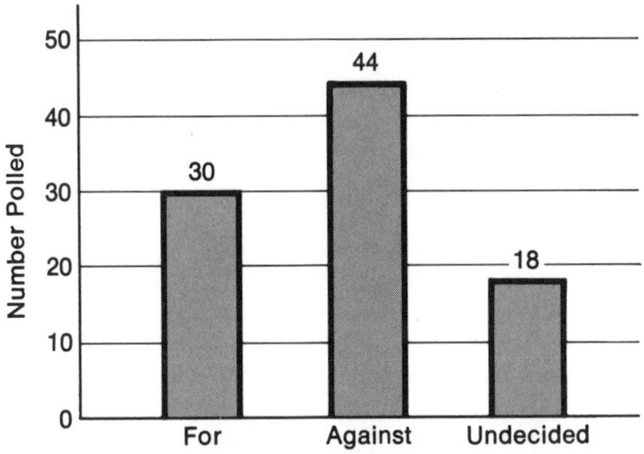

Results of a Poll of Voters Stand on an Upcoming Bond Election

It is not necessary to list the number of cases at the top of the bar, but it sometimes helps.

The width of the bar has no significance on a bar graph. In fact, a single line will work as well. It can be seen that a bar graph is particularly suitable for attribute data. A bar graph can also be used for discrete data or for any data reported in distinct groups, but for continuous data it is more usual to use a graph called a **histogram**. In a histogram, the class intervals are generally equal since the *areas*, not simply the heights, of the rectangles represent the frequencies. The assumption in a histogram is that the scale is continuous and the data are spread evenly throughout the class. Histograms are discussed further in section 4.1. An example of a histogram is given here.

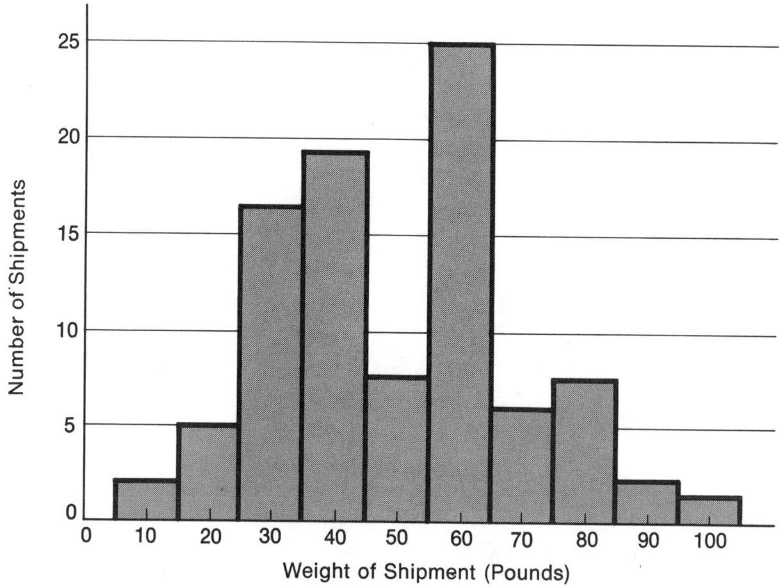

Distribution of Diamond Shipment Weight from the Abercrombie Mine for 1971

In this example one can see that most shipments were about 30, 40, or 60 pounds. The classes are ten pounds, with multiples of ten falling at or near the center of the classes at the class mark.

If the vertical scale represents *relative frequency*, the histogram is called a **relative frequency histogram**. The only difference is the vertical scale, as shown below.

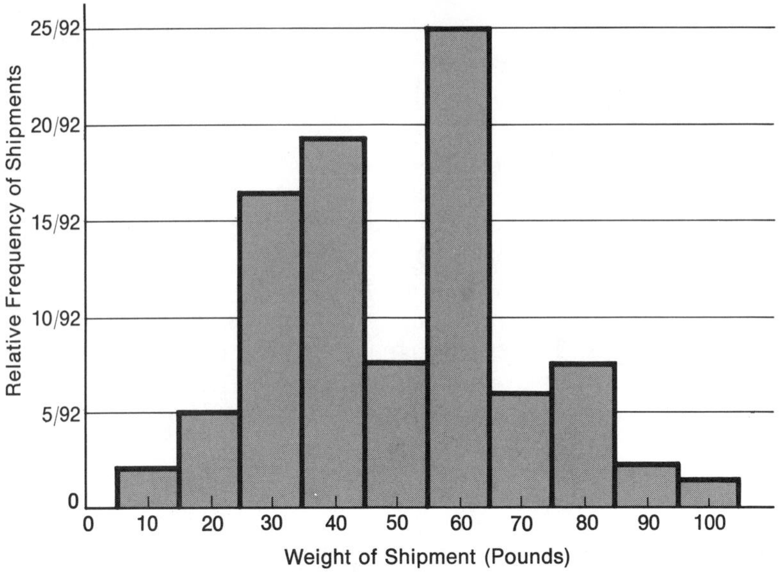

Distribution of Diamond Shipment Weight from the Abercrombie Mine
for 1971

The **frequency polygon** or **line graph** is often used as a substitute for a histogram. For a frequency polygon, the class mark, or center, of the uppermost edge of the rectangle is marked with a dot, which represents the class mark, and the dots are connected in order. The height of the dot corresponds to the frequency of the class. A general practice is to add a class with zero frequency at each end to complete the polygon. An example of a frequency polygon based on the diamond shipment data follows:

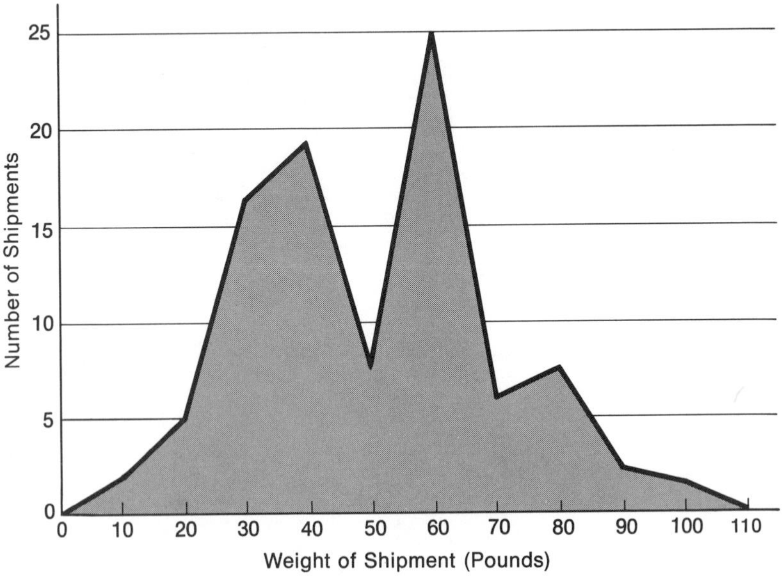

Distribution of Diamond Shipment Weight from the Abercrombie Mine
for 1971

If the vertical scale represents relative frequency, we have a relative frequency polygon.

Sometimes we wish to show cumulative data; that is, we want to represent the number of cases, or the proportion of cases, which fall below a certain figure. The graph used for this purpose is called an **ogive**, or **cumulative frequency polygon**. To construct an ogive, we use the upper class limits, since we have not accumulated all the data from a class until we reach the upper class limit. An ogive is also useful for reading percentiles. (See section 2.3.)

EXAMPLE 1 Suppose we collect heights of 50 incoming freshmen men and arrange the data in the following table.

Height (inches)	*Frequency*	*Relative Frequency*	*Cumulative Relative Frequency*
62.5–65.5	3	3/50	3/50
65.5–68.5	5	5/50	8/50
68.5–71.5	12	12/50	20/50
71.5–74.5	15	15/50	35/50
74.5–77.5	9	9/50	44/50
77.5–80.5	4	4/50	48/50
80.5–83.5	2	2/50	50/50
	50	50/50	

An ogive for this data looks like this.

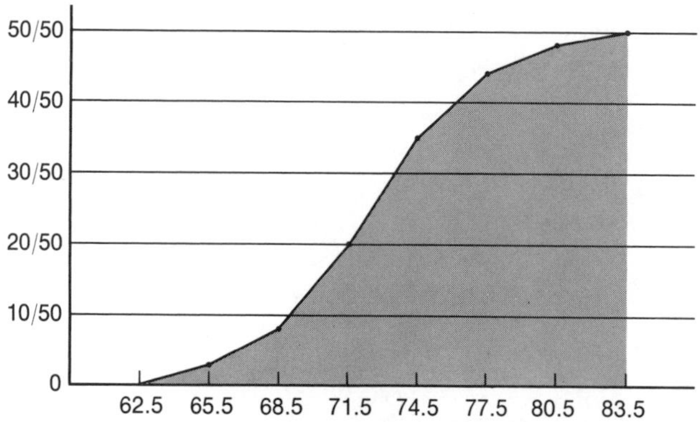

Other methods of graphical presentation include pictograms and pie charts. Pie charts are particularly useful in presenting percentages or proportions and are familiar to many because of their use in showing sources of dollar income and expenditure in the national government.

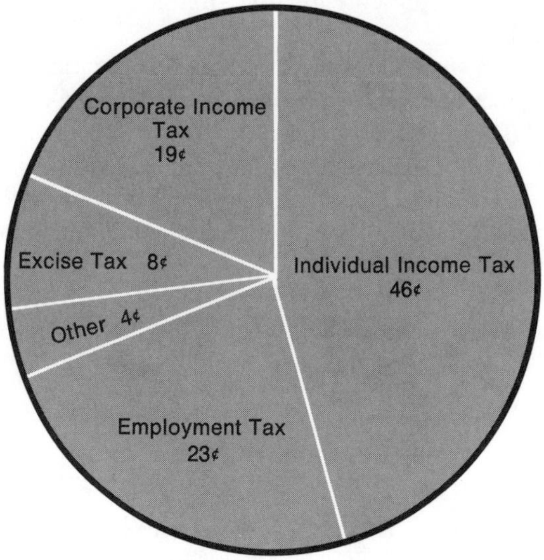

Federal Income 1970

A hazard of pictograms is the possibility of conveying a false impression.

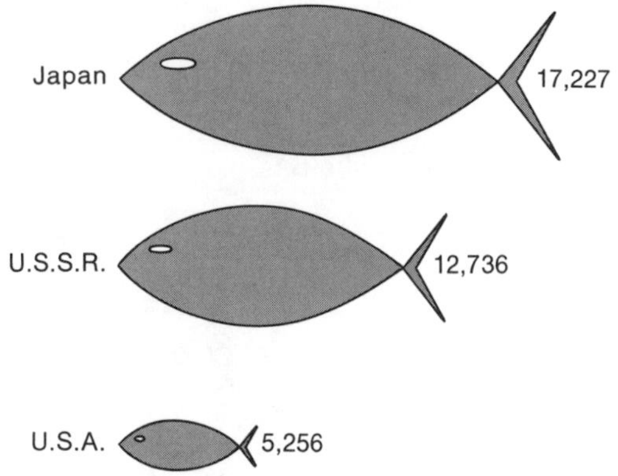

1969 Catch of Fish and Shellfish
(Millions of Pounds)

Although the fish on top is just about three times as long as the one on the bottom, its area is about ten times as great, and this is the impression the eye receives. A similar impression is given by adding the illusion of depth.

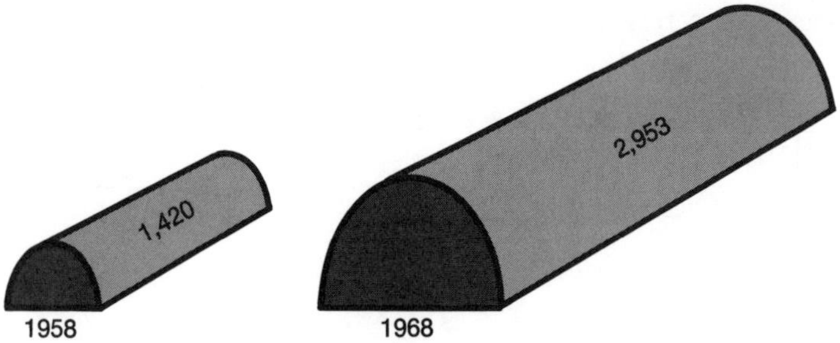

U.S.A. Aluminum Production (Thousands of Tons)

The graph which follows conveys a fairer impression.

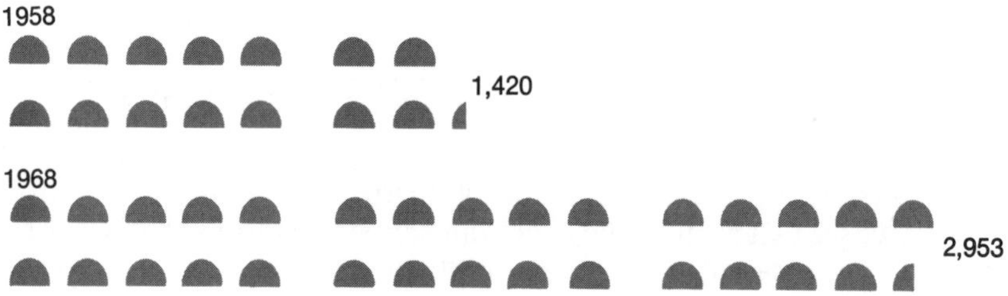

U.S.A. Aluminum Production (Thousands of Tons)
Each ingot represents 10,000 tons.

The problem of creating false impressions is not restricted to pictograms. The vertical scale can cause problems if it does not start at zero. Unions asking for a wage increase could use the following graph to support their claim that the company is "reaping in profits."

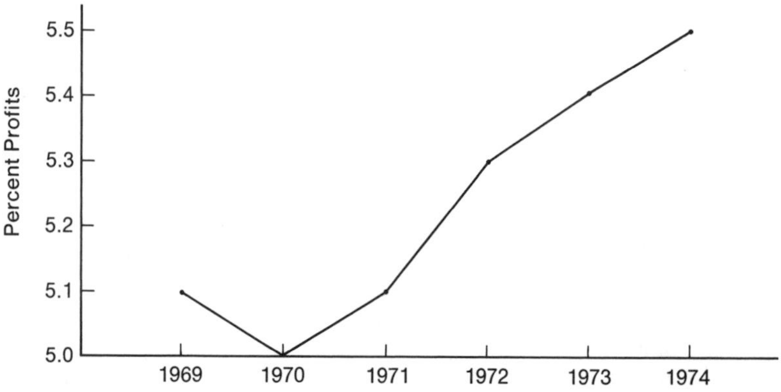

In proper perspective, the graph should look like this.

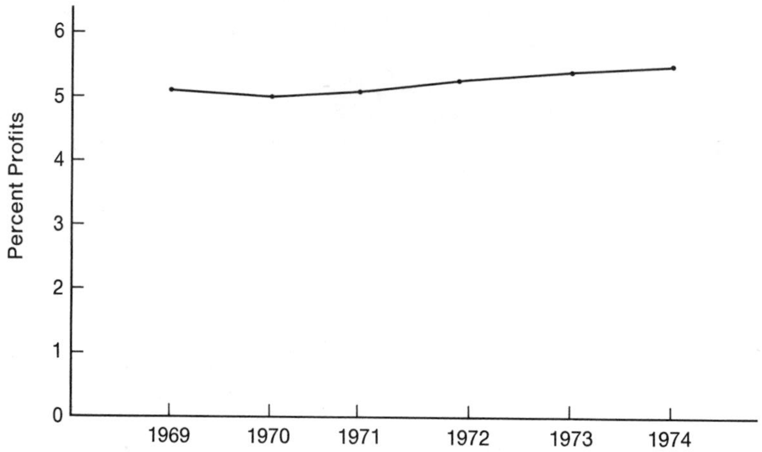

In the first graph we get a picture of sharply increasing profits. In the second, we see that profits were approximately stable throughout this period. A similarly misleading graph could probably be used by the company to show increases in worker income.

Problems

1. Construct a bar graph for the bread data of example 1, section 1.1.
2. Construct a histogram for the height data of section 1.1.
3. Draw a histogram and a relative frequency polygon for the data of problem 2, section 1.1.
4. Construct an ogive for the diamond shipment data of this section.
5. Draw a relative frequency histogram and a relative frequency polygon for the data of problem 4, section 1.1
6. Draw a histogram and an ogive for the data of problem 3, section 1.1.
7. Draw a pictogram for the basketball team data of this section.
8. Construct a pie chart, using proportions or percentages, of the results of the voter poll.
9. Using the income data of example 2, section 1.1, draw

   (a) a histogram;
   (b) a relative frequency histogram;

(c) a frequency polygon;
(d) a relative frequency polygon;
(e) an ogive.

1.3 SUMMARY

The organization and presentation of data is one of the primary purposes of descriptive statistics. Large amounts of data must be well organized in order to be understandable, and smaller amounts of data may also be more meaningful after organization.

One of the best ways to organize data is to use a frequency distribution. To organize the data into a frequency distribution, we first determine the desired number of classes which will cover the entire set of data. Each piece of data is then placed in a class and the results are presented in a table. In general, each class is the same size, although attenuated distributions may have data from one or both ends combined into a single, larger class. Related data distributions are the relative frequency and cumulative frequency or relative cumulative frequency distributions.

Bar graphs, histograms, frequency polygons, ogives, pictograms, and pie charts are used to present data in a visually appealing manner. From a statistical viewpoint, histograms are the most important because classes in a histogram can be compared by areas. Great care must be exercised when using graphs to avoid conveying a false impression.

Problems

1. A set of data has a high of 363 and a low of 107. Construct a table with eleven classes which contain the data. Give the actual class limits and the class marks.
2. Suppose that weights of shipments, to the nearest pound, are grouped into classes 0–75, 76–150, 151–225, 226–300, 301–375, 376–450, 451–525, 526–600, 601–675, 676–750, 751–825, 826–900, and 901 or more. Is it possible to determine the number of shipments

   (a) weighing 300 pounds or less?
   (b) weighing less than 300 pounds?
   (c) weighing more than 750 pounds?
   (d) weighing 750 pounds or more?
   (e) weighing more than 1,000 pounds?
   (f) weighing less than 700 pounds?

3. The class marks of a distribution of blood cholesterol counts are 125.5, 175.5, 225.5, 275.5, 325.5, 375.5, 425.5, 475.5, 525.5, 575.5. What are the apparent class limits? The actual class limits?

4. A frequency distribution is given below.

Class	*Frequency*
11.00–12.99	2
13.00–14.99	7
15.00–16.99	13
17.00–18.99	24
19.00–20.99	11
21.00–22.99	4

(a) What are the actual class limits implied for the class 17.00–18.99?
(b) What is the class mark for the class 13.00–14.99?
(c) What is the class interval?
(d) What is the range if the highest score was 22.68 and the lowest 11.74?

5. Sales of a certain type of candy bar over a period of sixty days are listed here.

48	34	36	40	26	29	25	30	22	32	34	41
47	24	42	37	37	30	26	37	42	41	41	23
31	22	34	41	24	42	35	41	40	29	32	32
22	31	28	41	27	34	37	24	33	33	24	28
29	43	46	32	32	28	34	34	28	27	43	34

Set up a frequency distribution and draw a bar graph for the data. Recall that these values are discrete.

6. During a period of one month, a total of 120 patients were admitted to a certain hospital. Their ages were as follows, in order of admittance.

37	54	81	64	11	34	80	71	51	12	52	13
62	8	1	2	39	22	4	84	28	76	30	73
47	33	16	24	16	24	35	8	56	6	38	55
54	7	3	10	4	6	58	10	67	29	3	4
9	68	59	43	11	64	60	23	88	37	56	46
33	52	71	28	46	3	52	55	9	7	81	57
8	19	52	17	24	27	7	83	25	57	7	24
77	37	83	48	7	38	57	43	58	82	34	9
64	50	8	23	18	14	10	7	60	44	39	29
74	41	74	64	27	44	63	6	52	8	9	54

(a) Construct a frequency table, using a class interval of ten, assuming ages to nearest birthday rather than last birthday.
(b) Draw a histogram using the results of (a).
(c) Construct a relative frequency table and a relative frequency polygon for the data.
(d) Draw an ogive for the data.

7. A family's monthly expenditures were listed as follows: housing, \$380; utilities, \$60; medical expenses, \$15; food, \$110; transportation, \$135; clothing, \$25; savings, \$25; miscellaneous, \$50. Draw a pie chart showing this information in terms of percentages or proportions.

Chapter II

Measures of Location and Dispersion

2.1 MEASURES OF CENTRAL TENDENCY

Although organized data is more meaningful in presentation form than unorganized data, there are still many other things which can be learned by a closer examination. In this chapter we shall study a few of the ways to *describe* a set of data which can aid in our understanding.

When we deal with a set of data, we generally assume that the data all come from the same source; that is, we assume we have a representative *sample* of some *population*. A **population** is a set of data which consists of all possible or hypothetically possible values that a certain set of observations can take. A **sample** is a limited set of data drawn from the complete set of values, the population. There are many kinds of samples, some of which are not representative of the population from which the sample is drawn. Types of samples will be discussed in Chapter 7.

Since the characteristics of a complete population are rarely known, except, perhaps, theoretically, most information which a statistician or researcher uses comes from samples. We cannot know, for example, the actual distribution of weights of all the people in the world.

Quite often it is desirable to describe a set of data by using one number. The most useful numbers are **measures of central tendency** or **averages**. There are three such numbers, and each has its uses. Consider the following episode.

A campaigner knocks on a door, asking for contributions to the County Fund. He stresses that he expects a lot, since the average salary in this neighborhood is \$16,000. A few days later, a campaigner asks for contributions for the poor of the neighborhood. After all, he says, the average salary in the neighborhood is only \$5,000. Confused, the object of this attention asks a nearby university sociologist, who tells him that the average salary in the neighborhood is \$8,000! Who is correct, and who is lying, if anyone?

To sort out the truth, look at the actual yearly salaries of the thirteen families in the neighborhood.

Salary (dollars)	*Frequency*
\$100,000	1
34,000	1
10,000	2
8,000	3
5,000	6

A picture begins to unfold here. If we add up all 13 salaries and divide by 13, we obtain \$16,000, which is known as the **mean**. If we count down from the top until we find the middle (seventh) measure, we obtain \$8,000, which is known as the **median**. If we look at the most frequently occurring salary we obtain \$5,000, which is known as the **mode**.

The mode is the least stable and the least frequently used of these values. For instance, a raise of \$3,000 given to two of the lowest paid people would change the mode from \$5,000 to \$8,000; this change would increase the mean from \$16,000 to about \$16,462 and leave the median unchanged. Use of the mode is generally restricted to those cases in which it is desired to represent a set of data by its most typical value.

If no one value appears more often than the others, there is no mode. If, however, there are two or more classes containing the most frequently occurring measures and these classes are separated by other classes, the distribution is called **bimodal** or **multimodal**, as the case may be. A bimodal or multimodal distribution has two or more peaks when the distribution is graphed. A histogram of a bimodal distribution appears in the following figure.

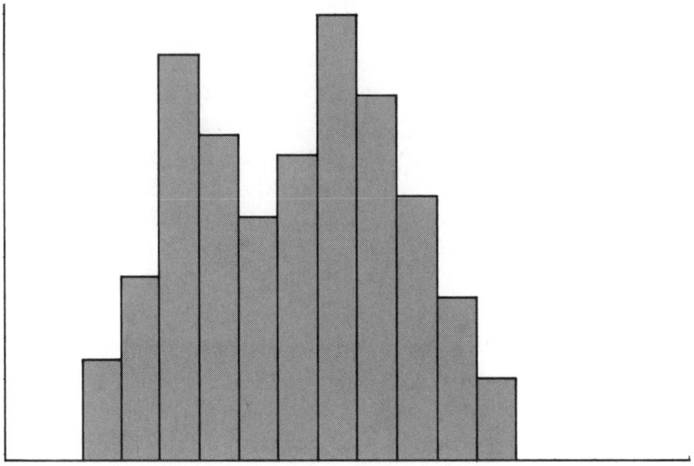

If a distribution is bimodal, this may mean that we have collected data from two different sources or populations. A height distribution of the general adult population might be bimodal with peaks at 5′ 5″ (the modal female height) and 5′ 10″ (the modal male height). When such bimodality occurs, it is a signal for us to examine the possibility that the data are drawn from two or more distinct populations.

The median, or central value of the data sample is most useful when there are extremes which influence the mean unduly. In the neighborhood case, the \$100,000 value is extreme, so the median is probably more representative than the mean.

To find out which number is at the median, simply take $(1/2)(n + 1)$, where n is the total number of cases. If n is odd, this will give us a specific number, as previously demonstrated. If n is even, it will not. For example, if $n = 30$, $(1/2)(30 + 1) = 15.5$. The median, then, is said to lie midway between the fifteenth and sixteenth values. It is usually helpful to arrange the values in order first, then simply count from the top or bottom.

EXAMPLE 1 Determine the median of the following measures:

47, 44, 41, 40, 39, 38, 37, 36, 36, 33, 32, 28, 28, 22, 20, 14, 6, 3

SOLUTION There are eighteen numbers, so the median is at (1/2)(18 + 1) or 9.5. The ninth and tenth numbers are 36 and 33, so the median (symbol Md) is (1/2)(36 + 33) or $Md = 34.5$.

The most stable, most frequently used measure of central tendency is the mean. The mean is particularly effective with large numbers of data. If exact figures are known, the mean can be found by the following formula.

MEAN OF A SAMPLE

The mean of a sample containing n pieces of data is given by

$$\bar{x} = \frac{\sum x}{n}$$

Here $\bar{x}$ represents the mean; $\sum$ means "the sum of"; x is each score; n is the total number in the sample.

EXAMPLE 2 A random sample of five accounts in a department store showed the following balances at the end of the month: \$67.32; \$108.97; \$17.64; \$412.11; \$81.96. Compute the mean balance.

SOLUTION Since the sum of the five amounts is \$688.00, the mean is given by $\bar{x}$ = \$688.00/5 = \$137.60.

EXAMPLE 3 The ages of 25 people in a certain income bracket are distributed as follows:

Age	29	33	37	38	39	40	42	43	45	47	50	59	66
Frequency	1	1	3	4	2	3	2	2	3	1	1	1	1

Estimate the mean age of this sample.

SOLUTION If there are several data points which have the same value, it is obvious that we simply multiply each value by its frequency rather than adding it in the required number of times. Thus, the mean is given by

$$\bar{x} = \frac{29 + 33 + 3 \cdot 37 + 4 \cdot 38 + 2 \cdot 39 + 3 \cdot 40 + 2 \cdot 42 + 2 \cdot 43 + 3 \cdot 45 + 47 + 50 + 59 + 66}{25}$$

$$\bar{x} = \frac{29 + 33 + 111 + 152 + 78 + 120 + 84 + 86 + 135 + 47 + 50 + 59 + 66}{25}$$

$$\bar{x} = \frac{1050}{25} = 42.$$

Here we have used an alternate formula for the mean,

$$\bar{x} = \frac{\sum x \cdot f}{n}$$

where $\bar{x}$ represents the mean, $\sum$ means "the sum of," $x \cdot f$ is each score times its frequency, and n is the total number of data points.

EXAMPLE 4 Find the mean of the thirteen salaries discussed at the beginning of the section.

SOLUTION

$$\bar{x} = \frac{(100{,}000)(1) + (34{,}000)(1) + (10{,}000)(2) + (8{,}000)(3) + (5{,}000)(6)}{13}$$

$$= \frac{208{,}000}{13}$$

$$= 16{,}000$$

DISCUSSION Note that the mean of the twelve salaries excluding \$100,000 is \$9,000. The median of these is \$6,500, while the mode is still \$5,000. In many cases involving small sets of data, the median is the best representative measure of the information because the mean is unduly influenced by extremes.

If the data are grouped, some accuracy is lost because we do not know exactly what each value represents. The ease of presentation of grouped data may compensate for the loss of accuracy. The *mode* is still easy to find with grouped data. The modal class is the class with the most values in it. The mode is the midpoint of the modal class.

To determine the *median* from grouped data, it is assumed that the data points are spread evenly over the interval. If an interval contains four data points, they divide the interval into five equal subintervals.

Suppose we want to locate the third number from the bottom, if there are four numbers in the interval and the apparent class limits are 102 and 107. The actual limits of the class interval are 101.5 and 107.5, and the third number from the bottom is above three of the subintervals. The distance of the third interval from the bottom interval is (3/5)(6) or 3.6. The median is then 101.5 + 3.6 = 105.1. In practice the median is often rounded off and given as an approximation. In this case, the median would be 105.

In calculating the *mean* from grouped data, it is assumed that each member of a class falls at the *center* (class mark) of the class interval. The results would be the same if the members of each class were considered to be spread out over the interval because of the way in which the mean is calculated. In the income data, the class interval 27,000 to 29,999 would be represented by 28,500.

NOTE: The center is actually 28,499.50, but minor discrepancies of this type are generally ignored since the result is only an estimate.

EXAMPLE 5 Find the median and mean of the data represented by the following frequency distribution.

Class	*Frequency*
81–83	3
78–80	4
75–77	8
72–74	7
69–71	3
66–68	4
63–65	0
60–62	1

SOLUTION The median is at (30 + 1)/2 = 15.5 and is located midway between the fifteenth and sixteenth pieces of data. These are at the top of the class interval 72–74 and the bottom of the class interval 75–77, so we can say that the median is approximately 74.5.

To determine the mean of grouped data, a table which summarizes the whole procedure can be constructed. Letting x represent the class mark and f, the frequency, we have

x	f	$x \cdot f$	
82	3	246	
79	4	316	
76	8	608	
73	7	511	$\bar{x} = 2{,}220/30 = 74$
70	3	210	
67	4	268	
64	0	0	
61	1	61	
Total	30	2,220	

DISCUSSION In the above table there is only one entry from 60 to 65, so it is tempting to combine these two classes to form a class interval from 60–65 with a frequency of one. This would not change the median, but it would increase the mean to 74.05, since the class mark would become 62.5. It is not good practice to have classes of different sizes or to combine attenuated values if we wish to perform arithmetic on the data.

If the data are numerically large, the best way to find the mean is to use a calculator, if one is available. Some calculators give the mean automatically, but the majority do not. For statistical use, a calculator should have at least one storage register, and preferably more, so that intermediate answers can be stored and recalled when needed. If a calculator is not available, a procedure called *coding* may be used. There are many variations of this technique, some of which are more difficult than they should be, but the basic principle is the use of an **assumed mean**, usually an integer $\hat{x}$ in the vicinity of the true mean. The mean difference between $\hat{x}$ and each value is calculated and added to $\hat{x}$ to obtain the true mean. The formula is as follows:

MEAN OF A SAMPLE FROM AN ASSUMED MEAN

$$\bar{x} = \frac{\sum (x - \hat{x})}{n} + \hat{x}$$

EXAMPLE 6 Given the following frequency distribution, estimate the mean and determine the median.

Class	*Frequency*
171–175	4
166–170	8
161–165	14
156–160	22
151–155	27
146–150	19
141–145	17
136–140	11
131–135	3

SOLUTION Using the class marks and an assumed mean of 153 (since this is quite close to the midpoint of the distribution), we have the following table.

x	f	$x - 153$	$(x - 153) \cdot f$
173	4	20	80
168	8	15	120
163	14	10	140
158	22	5	110
153	27	0	0
148	19	−5	−95
143	17	−10	−170
138	11	−15	−165
133	3	−20	−60
Total	125		−40

Then $\bar{x} = -40/125 + 153 = -0.32 + 153 = 152.68$.

To determine the median, we note that the median will be located at data point number $(125 + 1)/2$ or number 63. Counting from the bottom, we find that 50 pieces of data fall in classes through 146–150, while 77 are contained through 151–155. Thus the median lies in the interval with apparent class limits 151 and 155. For some purposes it is reasonable to use the midpoint of the interval, so that we may say that the median lies in the interval with class mark 153, or md $\doteq$ 153. The symbol $\doteq$ means "is approximately equal to" and is used whenever the number we give is only a good approximation to the actual value. Another approach is to assume that the data are distributed evenly over the interval. Thus data point number 63 in the distribution will be number 13 from the bottom of the interval containing the median. The actual class limits of that interval are 150.5 and 155.5, so that the 27 pieces of data are distributed evenly over five units. This means that there are actually 28 intervals. For example, if there were only one data point in the class, it would divide the class interval into two parts, one above and one below it. Thus the thirteenth data point in the interval lies above 13/28 of the interval. The interval is five units, so we have

$$Md = 150.5 + (13/28)(5) \doteq 152.82 \text{ or about } 152.8.$$

This procedure is of particular value when the median lies toward one end of the interval.

NOTE In the remainder of the text, most problems assume that a calculator is available to readers, although it may not be necessary. In the event that it is not, the reader may wish to check the examples by using an alternate procedure. Problems marked **c** are recommended only if a calculator is available.

Problems

Using the best technique, determine the mean, median, and mode of each of the samples in problems 1–4.

1. 20, 22, 23, 26, 29, 30.
2. 28, 25, 20, 33, 27, 29, 23, 21, 24, 18, 30, 25.
3. 1.34, 1.69, 1.78, 1.89, 2.03, 2.27, 2.39, 2.88, 3.16, 3.34, 4.92, 5.57, 6.83, 7.44, 9.63, 11.82.
4.

x	*Frequency*	x	*Frequency*
134	1	171	12
137	1	173	10
143	1	174	13
150	2	178	8
152	3	181	6
153	1	186	3
161	3	193	2
164	2	201	1
166	8	217	1
169	7	234	1
170	13	246	1

Determine the mean, median, and mode of each of the following sets of data:

5. The bread data of example 1, section 1.1.
6. The data of problem 4, section 1.3. Use the actual class marks, then round upward to two decimal places (for the mean) and see if the difference to two decimal places is worth the extra arithmetic.
c7. The height data of section 1.1. Use the raw scores, then the scores as given in the frequency table and compare the results.
c8. The cat food data of problem 2, section 1.1.
c9. The lengths of mullet in problem 3, section 1.1.
c10. The pollution indices of problem 4, section 1.1.
c11. The candy bar sales of problem 5, section 1.3.
c12. The hospital admissions of problem 6, section 1.3.
c13. The income data given below. Use class marks to the nearest dollar and a reasonable assumed mean.

Income (dollars)	*Number of Families*
42,000–44,999	3
39,000–41,999	7
36,000–38,999	12
33,000–35,999	17
30,000–32,999	28
27,000–29,999	60
24,000–26,999	216
21,000–23,999	268
18,000–20,999	448
15,000–17,999	621
12,000–14,999	949
9,000–11,999	1,421
6,000– 8,999	2,844
3,000– 5,999	2,123
0– 2,999	294

2.2 MEASURES OF DISPERSION

When we determine the approximate center of a distribution, we obtain one description of it. We can describe it even better by obtaining a measure which tells us something about the spread of the distribution. This kind of measure tells us whether the values in the distribution cluster closely about the mean or stretch out in both directions. A measure of dispersion which is large compared to the mean indicates that the distribution is spread out while a small number for this measure shows that the data points cluster closely about the mean.

The simplest measure of dispersion, which was discussed in Chapter 1, is the range, which gives the distance between the highest and lowest values in the data set. The range is obtained by subtracting the lowest value in the data set from the highest. If the data set contains a few extreme scores, or even one, the range will give a misleading impression of the spread of the distribution, so its use is limited.

The range has definite uses, however, in circumstances in which the data set does not contain extreme values. It is a quick, easy measure of dispersion, although it may not be very representative. A position-vacant advertisement may give a salary range of \$600–\$765 per month, depending on qualifications. The use of the term "range" is rather loose in this case, however, since a statistical range would be \$165. In addition to this sort of use, the range can be used to estimate the standard deviation, another measure of distribution dispersion, which is discussed next in this section. The standard deviation is approximately one-fourth the range, by rule-of-thumb.

For distributions that are relatively symmetrical about the mean and taper off in both directions, the best measure of dispersion about the mean is probably the *standard deviation*, which will be used throughout the rest of the book.

The difference between each data point in the distribution and the mean is the dispersion of that data point. We might think that the average of all dispersions would be a good indicator of the spread of the data. Recall, however, that the measures above the mean balance those below, so the sum of all the dispersions is zero, no matter what the distribution looks like. If we take only the absolute values of the dispersions, which ignore the plus and minus signs, we obtain a perfectly good measure of spread, called the **mean deviation**. A formula for the mean deviation is this one.

MEAN DEVIATION OF A SAMPLE

$$\text{mean deviation} = \frac{\sum |x - \bar{x}|}{n}$$

The mean deviation of a sample containing the data points 1, 3, 4, 7, 9, 12, is found easily. The mean is $(1 + 3 + 4 + 7 + 9 + 12)/6$ or 6. The absolute values of the differences between each data point and the mean are found and added.

	x	$x - \bar{x}$	$\lvert x - \bar{x} \rvert$
	1	−5	5
	3	−3	3
	4	−2	2
	7	1	1
	9	3	3
	12	6	6
Total	36	0	20

Then the mean deviation is $20/6 \doteq 3.33$. Note in the above table that the sum of all $x - \bar{x}$ terms is zero. This is always true and can be used to check your arithmetic.

The mean deviation is a valid measure of dispersion, but its applicability is limited. The cancelling effect of adding the $x - \bar{x}$ terms can also be eliminated by squaring the terms before adding. Thus the sum is positive but the units are squared. We get back to the original units by taking the square root. This principle is utilized to find the **standard deviation.**

STANDARD DEVIATION OF A SAMPLE

The standard deviation of a sample with mean $\bar{x}$ and containing n pieces of data is given by

$$s = \sqrt{\frac{\sum (x - \bar{x})^2}{n - 1}}$$

Here s is used to indicate that it is the standard deviation of a *sample*, $\bar{x}$ is the mean of the sample, and n is the number of data points in the sample. The square of the standard deviation is also a measure of dispersion. It is called the **variance**, s^2, which is defined as

$$s^2 = \frac{\sum (x - \bar{x})^2}{n - 1}$$

It might seem more natural to divide by n, but statisticians have found that division by n yields a value which is, on the average, too small to provide what is technically known as an "unbiased" estimate of the variance of the population. The term "unbiased" means that if we take a great number of samples from a population, the mean of the variances of these samples will be a better estimate of the population variance than the variance of each sample, by itself. We arrive at an unbiased estimate if we divide by $n - 1$ rather than n. All other uses for s^2 are basically unaffected by this change.

One reason for using the standard deviation as our primary measure of dispersion is that it has been investigated extensively. One investigation which produced *Chebyshev's theorem* found that for *any* set of data, 3/4 of all pieces of data lie within two standard deviations of the mean, 8/9 lie within three standard deviations of the mean, and, in general, $1 - 1/k^2$ of the values lie within k standard deviations of the mean. If a distribution has a mean of 75 and a standard deviation of 10, then 3/4 of all data lie between 55 and 95, and 8/9 of them lie between 45 and 105. For distributions which more closely approximate the so-called "normal" distribution, which is bell-shaped, symmetrical about the mean, and trails off sharply in both directions, about 68% of the data lie within one standard deviation of the mean, 95% within two, and more than 99% within three standard deviations of the mean. Features of the normal distribution will be discussed at great length in Chapter 6 and later in the text.

EXAMPLE 1 Find the standard deviation of 1, 3, 4, 7, 9, 12.

SOLUTION We set up a table as given below.

x	$x - \bar{x}$	$(x - \bar{x})^2$
1	−5	25
3	−3	9
4	−2	4
7	1	1
9	3	9
12	6	36
36	0	84

Then $s = \sqrt{84/5} = \sqrt{16.8} \doteq 4.099$. A table of square roots is given in Table 1 in the Tables section at the end of the book. Instructions for the use of Table 1 are given in Appendix A.

EXAMPLE 2 Using the data of example 2 of the previous section, determine the standard deviation of the balances.

SOLUTION Again a table is helpful.

	x	$x - \bar{x}$	$(x - \bar{x})^2$	
	67.32	−70.28	4,939.2784	
	108.97	−28.63	819.6769	$s^2 = 98{,}600.9066/4$
	17.64	−119.96	14,390.4016	$s^2 = 24{,}650.2665$
	412.11	274.51	75,355.7401	$s \doteq 157.00$
	81.96	−55.64	3,095.8096	
Total	688.00	0	98,600.9066	

DISCUSSION The extreme tedium of working with numbers as cumbersome as these points out the desirability of having the use of a calculator. If a calculator is available, there is no reason to find the mean before beginning the arithmetic for the standard deviation or variance. There are three formulas available which are used more often than the defining formula. They are presented here without their derivation from the original formula.

ALTERNATE FORMULAS FOR THE STANDARD DEVIATION

$$s = \sqrt{\frac{\sum x^2 - n(\bar{x})^2}{n - 1}}$$

$$s = \sqrt{\frac{n \sum x^2 - (\sum x)^2}{n(n - 1)}}$$

$$s = \sqrt{\frac{\sum x^2 - \frac{(\sum x)^2}{n}}{n - 1}}$$

The first of these is most useful if the mean is known, while the second has been the most widely used raw data formula prior to the advent of the electronic calculator. The standard deviation of the set of data: 1, 3, 4, 7, 9, 12 can be found using any one of these formulas.

x	x^2
1	1
3	9
4	16
7	49
9	81
12	144
36	300

Then

$$s = \sqrt{\frac{300 - 6(6)^2}{5}} = \sqrt{\frac{300 - 216}{5}} = \sqrt{\frac{84}{5}} = \sqrt{16.8} \doteq 4.1;$$

or

$$s = \sqrt{\frac{6(300) - (36)^2}{6(5)}} = \sqrt{\frac{1800 - 1296}{30}} = \sqrt{\frac{504}{30}} = \sqrt{16.8} \doteq 4.1;$$

or

$$s = \sqrt{\frac{300 - \frac{(36)^2}{6}}{5}} = \sqrt{\frac{300 - 216}{5}} = \sqrt{\frac{84}{5}} = \sqrt{16.8} \doteq 4.1.$$

The second formula has been used to avoid rounding errors which usually occur with other formulas. Modern electronic calculators, even the hand-held variety, are seldom limited in this way. If a calculator which can sum x and x^2 in storage registers (such as the Wang 300 series) is available, the third formula is the most useful. Some relatively inexpensive calculators, such as the Hewlett-Packard HP45 and HP80, and the Monroe 1930, calculate the mean and standard deviation for you.

EXAMPLE 3 Estimate the mean and standard deviation of the following sample of Nielsen ratings: 20.6; 11.3; 13.7; 9.2; 18.1; 7.2.

SOLUTION Using an alternate formula we have:

	x	x^2
	20.6	424.36
	11.3	127.69
	13.7	187.69
	9.2	84.64
	18.1	327.61
	7.2	51.84
Total	80.1	1,203.83

Then, $\bar{x} = 80.1/6 = 13.35$, and

$$s^2 = \frac{1{,}203.83 - \dfrac{(80.1)^2}{6}}{5} = \frac{1{,}203.83 - \dfrac{6{,}416.01}{6}}{5} = \frac{1{,}203.83 - 1{,}069.335}{5}$$

$$= \frac{134.495}{5} \doteq 26.90$$

and

$$s \doteq \sqrt{26.90} \doteq 5.2.$$

If the data are grouped into classes by means of a frequency distribution, both the mean and standard deviation are found by using the class marks or centers of the classes.

EXAMPLE 4 Using the age data of example 3, section 2.1, calculate the standard deviation of the ages.

SOLUTION The standard deviation can be calculated in the same way as in the preceding example. Note, however, that introduction of frequencies necessitates three additional columns in the table if the calculations are done by hand. If a calculator is used, no such problems are encountered since each number is entered the appropriate number of times.

	x	f	$x \cdot f$	x^2	$x^2 \cdot f$
	29	1	29	841	841
	33	1	33	1,089	1,089
	37	3	111	1,369	4,107
	38	4	152	1,444	5,776
	39	2	78	1,521	3,042
	40	3	120	1,600	4,800
	42	2	84	1,764	3,528
	43	2	86	1,849	3,698
	45	3	135	2,025	6,075
	47	1	47	2,209	2,209
	50	1	50	2,500	2,500
	59	1	59	3,481	3,481
	66	1	66	4,356	4,356
Total		25	1,050		45,502

Then

$$s = \sqrt{\frac{45{,}502 - \dfrac{(1{,}050)^2}{25}}{24}} = \sqrt{\frac{1{,}402}{24}} = \sqrt{58.4167} \doteq 7.64.$$

In the event that a calculator is not available and the raw data are messy to work with, we can use the assumed mean mentioned in connection with *coding*, which was described in the previous section. As mentioned there, the mean difference between $\hat{x}$ and the data points is calculated and added to $\hat{x}$ to obtain the true mean. The standard deviation is calculated as usual, using $\hat{x}$ in place of $\bar{x}$, then *before* dividing by $n - 1$, a correction factor is subtracted to compensate for the fact that $\hat{x}$ is not the true mean. The formulas used are the following:

MEAN AND STANDARD DEVIATION FROM AN ASSUMED MEAN

$$\bar{x} = \frac{\sum (x - \hat{x})}{n} + \hat{x}$$

$$s = \sqrt{\frac{\sum (x - \hat{x})^2 - nc^2}{n - 1}} \qquad \text{where } c = \bar{x} - \hat{x}$$

EXAMPLE 6 Given the following frequency distribution, estimate the mean and the standard deviation.

Class	*Frequency*
171–175	4
166–170	8
161–165	14
156–160	22
151–155	27
146–150	19
141–145	17
136–140	11
131–135	3

SOLUTION Using the class marks and an assumed mean of 153 (since this is quite close to the midpoint or median of the distribution), we have the following table.

x	f	$x - 153$	$(x - 153) \cdot f$	$(x - 153)^2$	$(x - 153)^2 \cdot f$
173	4	20	80	400	1,600
168	8	15	120	225	1,800
163	14	10	140	100	1,400
158	22	5	110	25	550
153	27	0	0	0	0
148	19	−5	−95	25	475
143	17	−10	−170	100	1,700
138	11	−15	−165	225	2,475
133	3	−20	−60	400	1,200
	125		−40		11,200

Then $\bar{x} = -40/125 + 153 = -0.32 + 153 = 152.68$; and

$$s^2 = \frac{11{,}200 - 125(-0.32)^2}{124} = \frac{11{,}200 - 12.8}{124} \doteq 90.22$$

and

$$s = \sqrt{90.22} \doteq 9.50.$$

If a distribution is highly assymetric, or has a few extreme scores, counting methods should be used. These distributions are those for which the median is the best measure of central tendency. First we must determine some measures of *location*. Measures of location are discussed in general in the next section, but it is appropriate to discuss the **quartiles** at this time. The first quartile, $\boldsymbol{Q_1}$, is the value which exceeds one-fourth of the data, the second quartile, $\boldsymbol{Q_2}$, is the value which exceeds one-half of the data, and the third quartile, $\boldsymbol{Q_3}$, is the value which exceeds three-fourths of the data. It can be seen that Q_2 is the median; thus we are concerned basically with only Q_1 and Q_3 as quartiles.

To locate Q_1 and Q_3, we proceed as with the median: Q_1 is at $\frac{1}{4}(n + 1)$ and Q_3 is at $\frac{3}{4}(n + 1)$ from the bottom of the distribution.

The quartiles themselves are not measures of dispersion, but their placement says something about the dispersion of the distribution. The distance between Q_3 and Q_1 is called the **interquartile range**. A more widely used measure is the **semi-interquartile range, $\boldsymbol{Q}$**, which is $\frac{1}{2}(Q_3 - Q_1)$.

EXAMPLE 7 Find Q for the age data of example 4.

SOLUTION $n = 25$, so that Q_1 is at $\frac{1}{4}(26) = 6.5$ from the bottom, and Q_3 is at $\frac{3}{4}(26) = 19.5$. Both the sixth and seventh pieces of data are 38, so $Q_1 = 38$. The nineteenth and twentieth pieces of data are both 45, so $Q_3 = 45$. Then $Q = \frac{1}{2}(45 - 38) = 3.5$. This is a moderate spread.

EXAMPLE 8 Find the median, Q_1, Q_3, and Q, of the following data on ages of football players.

Age	*Number*
23	1
22	13
21	16
20	18
19	6
18	2

SOLUTION There are 56 pieces of data, so that Q_1 is piece of data numbered 14.25, the median is number 28.5, and Q_3 is piece of data numbered 42.75, all from the bottom. Both the fourteenth and fifteenth piece of data from the bottom are age 20, so $Q_1 = 20$. Both the twenty-eighth and twenty-ninth pieces of data are 21, so $Md = 21$.

Now the forty-second measure from the bottom is 21, and the forty-third measure is 22, so measure number 42.75 is $\frac{3}{4}$ of the way from 21 to 22; that is, $Q_3 = 21 + \frac{3}{4}(22 - 21) = 21.75$.

Finally, $Q = \frac{1}{2}(21.75 - 20) = \frac{1}{2}(1.75) \doteq 0.88$. When compared with the median, 21, Q is quite small, so the distribution has very little spread.

Problems

Using the best technique, determine the variance and standard deviation of the samples in problems 1-3.

1. 20, 22, 23, 26, 29, 30.
2. 28, 25, 20, 33, 27, 29, 23, 21, 24, 18, 30, 25.
3. 1.34, 1.69, 1.78, 1.89, 2.03, 2.27, 2.39, 2.88, 3.16, 3.34, 4.92, 5.57, 6.83, 7.44, 9.63, 11.82.

4. Find the first and third quartiles and the semi-interquartile range for the data of problem 3.

5. Determine the standard deviation and semi-interquartile range for the following distribution.

x	*Frequency*
134	1
137	1
143	1
150	2
152	3
153	1
161	3
164	2
166	8
169	7
170	13
171	12
173	10
174	13
178	8
181	6
186	3
193	2
201	1
217	1
234	1
246	1

In problems 6–13 determine the standard deviation of each set of data.

6. The bread data of example 1, section 1.1.

7. The data of problem 4, section 1.3. Use the actual class marks, then round upward to two decimal places and see if the difference to two decimal places is worth the extra arithmetic.

8. The height data of section 1.1. Use the raw scores, then the scores as given in the frequency table and compare the results.

c9. The cat food data of problem 2, section 1.1.

c10. The lengths of mullet in problem 3, section 1.1.

c11. The pollution indices of problem 4, section 1.1.

c12. The candy bar sales of problem 5, section 1.3.

c13. The hospital admissions of problem 6, section 1.3.

c14. Using the income data given below, determine the standard deviation and the semi-interquartile range.

Income (dollars)	*Number of Families*
42,000–44,999	3
39,000–41,999	7
36,000–38,999	12
33,000–35,999	17
30,000–32,999	28
27,000–29,999	60
24,000–26,999	216
21,000–23,999	268
18,000–20,999	448
15,000–17,999	621
12,000–14,999	949
9,000–11,999	1,421
6,000– 8,999	2,844
3,000– 5,999	2,123
0– 2,999	294

2.3 OTHER MEASURES OF LOCATION AND COMPARISON

Pieces of data have values which are generally called "raw scores." Raw scores are basically meaningless unless they are compared with something. A man's blood test shows a cholesterol count of 320. Is that good or bad? Average is said to be 200. He is above the average, but how much above? Suppose we also know that the standard deviation of such data is 60. He is then two standard deviations above the mean. If the cholesterol counts are "normally distributed," this means that about 97.5% of all individuals have cholesterol counts lower than his and that, probably, his count is too high.

The number of standard deviations a given value is above or below the mean is called the **standard score** or **z-score** and is determined by the formula

STANDARD SCORE OF A SAMPLE

$$z = \frac{x - \bar{x}}{s}$$

where x is the value under consideration, $\bar{x}$ is the mean of the sample, and s is the standard deviation. The value of z locates the approximate place of the value in the set of data in relation to the spread of the distribution. The z-score also gives us a way to compare scores measuring similar quantities which were obtained on different instruments or from different samples. If two students take different IQ tests, for example, their scores are not directly comparable. Suppose John scores 112 on IQ test A while Fred scores 118 on IQ test B. Which was really the better score? If IQ test A had a mean of 98 and a standard deviation of 16, John's standard score was $(112 - 98)/16 = 14/16 \doteq 0.88$. If IQ test B had a mean of 104 and a standard deviation of 18, Fred's standard score was $(118 - 104)/18 = 14/18 \doteq 0.78$. John did better than Fred in this comparison. This does not mean, however, that John's IQ is higher than Fred's. On a retest it is possible that the results would be reversed. The difference in scores on the first test might be due to chance. All we can do, at this point, is to compare their scores on the tests taken.

Another measure of location is the **percentile**. In this case the data must be arranged in order preferably from low to high. The p-th percentile is the value which exceeds p percent of the scores. In the case of 100 pieces of ranked data, the lowest score would be at the zero percentile since it exceeds none of the scores, while the highest score would be at the ninty-ninth percentile since it exceeds 99% of the scores. There is, of course, no hundredth percentile since this would indicate that a score exceeded all of the scores, including itself.

To determine the percentile of a particular score, count the scores below it and divide by the total number of scores. Thus if r is the rank of a score (counted from the bottom), its percentile would be given by $p = \dfrac{r - 1}{n} \times 100$ where n is the number of scores.

To determine the value at a particular percentile, several methods are currently used, each having certain advantages. One method which works well is to obtain the location of the upper limit of p percent of the scores. Then the pth percentile is the next highest score. To obtain the thirty-fourth percentile of a set of 73 scores, for example, we take $(0.34) \times (73) = 24.82$. The lowest score above score number 24.82, then, is the twenty-fifth score from the bottom. This score is the thirty-fourth percentile. The fifty-sixth percentile of 50 scores would be the score above (0.56)(50). This number = 28 and so the twenty-ninth score from the bottom would be the fifty-sixth percentile. Percentiles are used to analyze great masses of scores.

For grouped data certain conventions are followed. We define the score at percentile p to be the value of the score $(p/100)(n + 1)$ from the bottom. This definition is also consistent with the accepted definition of Q_1, Md, and Q_3 as, respectively, the twenty-fifth, fiftieth, and seventy-fifth percentiles.

EXAMPLE 1 Determine the fifth, ninetieth, and ninety-ninth percentiles for the following frequency distribution.

	Class	*Frequency*
	171–175	4
	166–170	8
	161–165	14
	156–160	22
	151–155	27
	146–150	19
	141–145	17
	136–140	11
	131–135	3
Total		125

SOLUTION Using the definition, the fifth percentile is number $\frac{5}{100}(126)$ or number 6.3 from the bottom. We round off to the sixth measure and note that this is in the interval 136–140. It is the third value in this interval since there are three in the interval below it. The class size is 5 and the lower class limit is 135.5. There are 11 scores in the interval and if they are considered to be evenly spaced, there are 12 spaces in the 5 unit interval; the sixth is at the third space, hence above 3/12 of the interval. Then its value is $135.5 + (3/12)(5) = 136.75$. Similarly the ninetieth percentile is at $\frac{90}{100}(126) = 113.4$, or about 113 from the bottom. This is the top measure in the interval 161–165. There are 14 scores, hence 15 spaces in the interval, so the top score is 14/15 of the 5 unit interval above the lower class limit, 160.5. Thus it is $160.5 + (14/15)(5)$ or about 165.17. The ninety-ninth percentile is at $\frac{99}{100}(126) = 124.74$, thus at the top (125th) value. Its value is then estimated at $170.5 + (4/5)(5)$ or 174.5.

NOTE Some users of percentiles prefer to keep the fractions while performing the arithmetic. Using this method we would have

$$p = 5; \qquad x = 135.5 + \frac{3.3}{12}(5) \doteq 136.9;$$

$$p = 90; \qquad x = 160.5 + \frac{14.4}{15}(5) = 165.3;$$

$$p = 99; \qquad x = 170.5 + \frac{3.74}{5}(5) = 174.24.$$

It can be noted that the differences are slight and either method will yield an acceptable result.

Comparisons using percentiles are quite simple. If two raw scores are converted into percentiles, the score with the higher percentile is better than the other score in relation to the sample from which each is taken.

To make score comparison easy, it is a quite usual practice to *standardize* an evaluation instrument. This is done by applying it to a large group; perhaps giving a new IQ test to a, hopefully, representative sample of people around the country. Standard scores and percentiles are then computed based on this reference sample, and anyone subsequently taking the test can be compared to this reference sample.

Problems

1. A distribution has a mean of 47 and a standard deviation of 12. Determine standard scores for data points 17, 24, 33, 44, 53, 67, and 81.
2. For the frequency distribution of problem 5, section 2.2, determine the tenth, twenty-second, forty-third, seventy-first, ninety-third, and ninety-sixth percentiles.
3. A distribution has a mean of 156 with a standard deviation of 34. What value has a standard score of -1.63? -0.44? 0.76? 2.38?
4. For the data of problem 3, section 2.2, determine the percentiles of 2.03, 3.16, 4.92, and 9.63.
5. A subject was given three tests for mental dexterity. The means and standard deviations of scores for each test, determined by the reference sample, were as follows: Test *A*, $\bar{x} = 4.20$, $s = 0.40$; Test *B*, $\bar{x} = 160$, $s = 20$; Test *C*, $\bar{x} = 36.2$, $s = 5.6$. The subject's scores for the three tests were as follows: Test *A*, 4.38; Test *B*, 168; Test *C*, 37.6. Comparing his results to the reference samples, on which test did he do best? worst?
6. Referring to the income data given in problem 14, section 2.2, determine the fifth, twenty-seventh, forty-fourth, sixty-third, eighty-second, and ninety-fourth percentiles.

2.4 SUMMARY

In order to characterize a distribution we obtain a **measure of central tendency**—a value which is approximately at the center of the distribution—and a **measure of dispersion**—a number which tells us something about the spread of the distribution. One useful measure of central tendency is called the mode. The **mode** is the value or class with the greatest number of data points. If two or more values or classes have the greatest number of data points and are separated by other values or classes, the distribution is bi- or multimodal and may represent more than one population.

The **median** of a set of data is the value which separates the data into two equal sets: one with values greater than the median, the other with values less than the median. The median is used to describe data which are scattered or bunched at one end, or which contain a few extreme scores.

For most sets of data, the primary measure of central tendency is the **mean**.

MEAN OF A SAMPLE

The mean of a sample containing n pieces of data is given by

$$\bar{x} = \frac{\sum x}{n}$$

For sets of data for which the mean is appropriate, the best measure of dispersion is the **standard deviation**. The most important formulas for finding the standard deviation are given here.

ALTERNATE FORMULAS FOR THE STANDARD DEVIATION

$$s = \sqrt{\frac{\sum x^2 - n(\bar{x})^2}{n-1}}$$

$$s = \sqrt{\frac{n \sum x^2 - (\sum x)^2}{n(n-1)}}$$

$$s = \sqrt{\frac{\sum x^2 - \frac{(\sum x)^2}{n}}{n-1}}$$

Important measures of location are **standard scores** (**z-scores**) and **percentiles**. These measures are also useful for comparing scores obtained by different tests which evaluate the same quantity.

Problems

1. A total of 25 patients admitted to a hospital are tested for levels of blood sugar, with the following results.

87	71	83	67	78
77	69	76	68	85
84	85	70	68	80
74	79	66	85	73
81	78	81	77	75

(a) Find the mean, median, and mode for the data.
(b) Determine the standard deviation.
(c) What is Q?
(d) Determine the standard score of the highest and lowest values.
(e) Find the eleventh, fortieth, and fifty-ninth percentiles.

2. A store manager wishes to determine if a product sells well enough to warrant carrying it on the shelves. The number of units sold in each of the last twelve weeks is as follows:

61	44	51	32	76	44	38	52	43	56	18	67

(a) Determine the mean and median of the data.
(b) Determine the standard deviation.
(c) What is Q?
(d) Find the standard score for the highest and lowest values.
(e) Find the eleventh, fortieth, and fifty-ninth percentiles.

3. A hundred rats are fed a special diet and their weight gain is recorded after four weeks. The following represents the gain of weight (in grams) for the rats over the four weeks.

15	7	11	17	9	23	13	6	9	15
11	6	18	5	14	11	22	9	15	8
3	11	14	18	17	21	19	2	17	3
17	8	14	8	18	9	17	24	13	9
19	15	11	17	20	7	11	14	18	12
8	16	8	10	31	11	17	13	7	13
27	19	9	2	5	12	7	11	9	10
8	14	8	11	6	3	19	22	7	11
11	5	3	18	22	16	8	14	7	17
4	10	7	11	8	17	13	9	11	4

(a) Determine the mean, median, and mode for this set of data.
(b) Calculate the standard deviation.
(c) Find Q.
(d) Determine the fifth, twenty-eighth, fifty-seventh, and seventy-eighth percentiles.

4. A restaurant serves chicken and seafood dinners. The number of seafood dinners served each day for 60 days is given here.

56	43	44	58	39	41	54	61	39	36
48	41	47	51	57	46	31	48	39	49
52	63	51	28	48	33	44	48	57	37
40	45	38	44	58	54	37	41	51	36
47	44	53	37	33	44	37	46	48	66
44	38	44	39	46	52	55	38	40	46

(a) Find the mean, median, and mode of the data.
(b) Determine the standard deviation.
(c) Calculate the value of Q.
(d) Determine the fifteenth, forty-fourth, fifty-eighth, and eighty-third percentiles.

5. The **coefficient of variation** is defined as $V = (s/\bar{x}) \cdot 100$ for a sample. This gives the standard deviation of a set of data as a percentage of the mean. This characteristic is useful in comparing different sets of measures. Determine the coefficients of variation of the data in problems 1–4.

6. A child needed an operation (myringotomy and adenoidectomy) and the father surveyed a sample of doctors in two cities to arrive at the mean cost of the operation, with the following results:
CITY A: \$150, \$175, \$150, \$200, \$175.
CITY B: \$250, \$200, \$175, \$225, \$200, \$250, \$200, \$220.
Figure out the mean and standard deviation of each sample.

7. The highest charge for the operation in city A was \$200; in city B it was \$250. Which of these fees was most expensive *relative* to its group? (Hint: find the standard score for each value.)

8. Mr. Jones belongs to an age group for which the systolic blood pressure should be 130 with a standard deviation of 10; for Mr. Adams' group, the systolic blood pressure reading should be 140 with a standard deviation of 12. If Mr. Jones' blood pressure is 142 and Mr. Adams' blood pressure is 152, which is higher compared to their respective groups?

PART II

PROBABILITY & PROBABILITY DISTRIBUTIONS

Part II Photo: Dunes Hotel & Country Club, Las Vegas

Chapter III

Probability

3.1 PROBABILITIES

Although the primary emphasis of this text is statistics and use of statistical tools, a knowledge of basic probability is necessary to be able to make proper use of appropriate statistical techniques and to understand the results of their use. Probability deals with the determination of the chances that a particular outcome will occur in a known population of possible outcomes. For example, if a particular pool contains 50 black fish and 30 yellow fish, we can determine the chance of a sample of three fish consisting of exactly two black and one yellow.

Statistics, on the other hand, deals with the attempt to determine the makeup of a population by using results taken from samples belonging to that population. If we caught two yellow and one black fish from a pool whose contents are unknown, statistics will enable us to make educated guesses about this pool. Thus probability and statistics are complementary.

Let us first try to determine what is meant by the term *probability*. Some years ago, in Monte Carlo, a certain roulette wheel paid off on red for more than twenty consecutive times. Each time during the streak onlookers made their bets on red or black. Some reasoned that since red and black are equally likely, each should come up an equal number of times. Thus, they thought each time red came up made black more likely on the next turn of the wheel. This, they said, is what is meant by the *law of averages*. They also lost a lot of money.

The *law of averages* is a name given to what may be the most widely misinterpreted mathematical statement in existence. In point of fact, no law of averages exists. The popular conception is a misunderstood interpretation of a mathematical statement called the **law of large numbers**. Simply stated, this law says that if an act is repeated a large number of times, the proportion of those times a particular thing happens tends to a fixed number. For instance, if you tossed a coin several million times, you would expect the coin to fall heads up approximately one-half of the tosses. This means that if a coin is tossed a large number of times, the proportion of heads tends to one-half of the total number of tosses. The number one-half is called the **probability** that heads will occur on one toss of the coin.

If four cards—an ace, a king, a queen, and a jack—are thoroughly shuffled and the top card turned over, you would expect it to be the ace about one-fourth of the time. Thus, the probability that the ace will be the top card in this case is one-fourth.

Now suppose that you performed this experiment twenty times and no ace appeared. What then of the probability that the ace will appear on the twenty-first drawing? The cards have no memory—each repetition of the experiment has no effect on any other repetition. The repetitions are **independent**. The cards are no more—and no less—likely to produce an ace on the next repetition than on any other.

In the case of the roulette wheel, there was no connection between any two of the repetitions and no greater likelihood of black on the *next* roll than on any other. The law of large numbers *guarantees* that after an extensive number of trials it is very likely that the proportions of red and black will be approximately equal on an honest roulette wheel; heads and tails will each turn up approximately half the time in coin flipping; and a six will occur on approximately one-sixth of the rolls of a balanced die. It says nothing, however, about any particular set of occurrences. The law of large numbers does not work by compensation; instead, any unusual deviation is simply buried over the long haul. If a fair coin turns up heads ten times in a row, this occurrence will have less effect upon the proportion of heads after one hundred tosses, appreciably none after one thousand.

The law of large numbers is not necessary, of course, to determine all probabilities. It is not even usable in many cases, such as the determination of probability of rain on a particular day. The principle, however, can be abstracted to many cases. If we wish to determine the probability that a coin will fall heads up when tossed, we can toss it a few million times, determine the proportion of heads, and use this to estimate the probability of heads on a single toss. A more practical method is to observe that there are exactly two possibilities, a head and a tail, and that they are equally likely. The probability is therefore one-half (1/2 or 0.50). In one throw of a die, there are six equally likely possibilities, so that the probability that a particular number, say six, will appear on top on a single throw is one-sixth. If one person is to be selected by lot from among five—say two men and three women—the probability that the person selected will be a man is 2/5 since two of the five possibilities are men and all the possibilities are equally likely.

Now suppose that a small boy knocks five books off a shelf. These books are volumes 1–5 of an encyclopedia. He replaces them randomly and hurriedly on the shelf. What is the probability that they are in the correct order? Since there are five possible volumes to put in the first place, four volumes to put in the second place after the first place is filled, three subsequent choices for the third place, two for the fourth, and only one for the fifth and last place, there are exactly $5 \cdot 4 \cdot 3 \cdot 2 \cdot 1$ or 120 different ways to place the five books back on the shelf, only one of which is correct. If it is assumed that all the possibilities are equally likely, the probability that the books have been replaced in the correct order is 1/120.

It is apparent that in order to determine probability it is often necessary to be able to count the number of ways something might occur. If, for instance, three roads lead from town A to town B, and two roads lead from town B to town C, the number of different paths from town A to town C, using these roads, is six. For each of the three roads from A to B, a traveler has a choice of two roads from B to C, so there are $3 \cdot 2$ or 6 different choices. If the number of roads from C to D is four, our traveler has a choice of four roads for each of the six choices from A to C, so that the number of paths he may travel from A to D, using these roads, is $3 \cdot 2 \cdot 4$ or 24. This example illustrates a basic principle: if an act is performed in several steps and each step can be performed in several different ways, the total number of ways the act can be performed is the product of the individual numbers of ways each step can be performed. This can be stated in mathematical language as follows:

MULTIPLICATION RULE	Suppose that an act requires n steps to complete. If the steps can be performed successively in $m_1, m_2, m_3, \ldots, m_n$ ways, then the act can be performed in the order stated in $m_1 \cdot m_2 \cdot m_3 \cdots m_n$ different ways.

If the act requires two steps, the first step can be done in m_1 different ways and the second in m_2 different ways, then the act can be performed in $m_1 m_2$ different ways. Each of the different ways the act can be performed is called an **outcome**. A set which contains one or more outcomes is called an **event**.

(Actually, an empty set can also be considered to be an event. See section 3.3.)

One way to determine the number of possible outcomes if the multiplication rule is used is a tree diagram. A tree diagram shows each step and all the possible outcomes at each step. The total number of branches gives the number of different ways the act can be performed. For example, suppose that we toss a coin until a tail appears or three times, whichever occurs first. A tree diagram illustrating this experiment is given here:

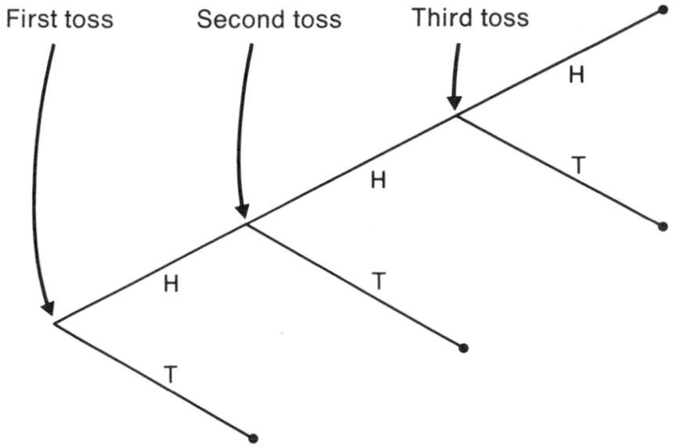

Inspection of this figure shows that four different outcomes are possible. Tree diagrams are useful in counting possible outcomes for acts requiring small numbers of steps where there are few choices at each step. For larger numbers of steps or choices, tree diagrams become unwieldly.

Intuitively, we can see that if an act can be performed in s equally likely ways, r number of which constitute an event, the probability of that event occurring on a particular repetition of the act, or **experiment**, is r/s. This is the *classical* definition of probability.

CLASSICAL PROBABILITY

If an experiment has s equally likely outcomes, r of which constitute an event, then the probability of the event is r/s.

EXAMPLE What is the probability of obtaining two heads on two tosses of a fair coin?

SOLUTION This experiment consists of two steps, each of which can be performed in two ways, so there are four equally likely outcomes. One of these outcomes, head followed by head, makes up the desired event. The probability of this outcome is 1/4 or 0.25.

DISCUSSION It may seem that there are only three possible outcomes: two heads, two tails, or one of each. It is possible to look at this experiment in this way (see section 3.3), but you should note that these three outcomes are not equally likely. A head and a tail can be obtained in two ways—head followed by a tail, or tail followed by a head. This outcome is then twice as likely as each of the other two and has, in fact, a probability of 2/4 or 0.50. The student should try it for himself if he is skeptical.

Problems

1. Toss a coin twenty times and record the number of heads. Perform the experiment twenty times and add the totals. Do you notice a leveling effect? If you do not obtain a ratio of approximately 1/2 heads, repeat an additional ten times. The probability of obtaining less than 268 or more than 332 heads in the 30 experiments is only 1/100!
2. Select a page from your telephone book and record the frequency of occurrence of last digits of telephone numbers. Combine your results in class and determine the proportions for each digit. The proportions should be approximately equal.
3. Shuffle a deck of cards and deal them out, one at a time, until the first heart appears. Record the number of cards dealt, including the heart. Perform the experiment one hundred times or combine results of five to ten experiments per student in class. Compare your average number of cards dealt with the theoretical number, approximately 3.79 per deal. (See solution for explanation.)
4. Joe's Pizza Parlor is offering a special. For a given price one can purchase a large pizza and a drink or else a small pizza, a burger, and a drink. Pizzas come plain or with a choice of either of two toppings. Joe serves hamburgers, cheeseburgers, garlicburgers, and onionburgers. For drinks, the choice is cola, root beer, orange, grape, and cherry. How many different dinners can be purchased on the special?
5. A sociologist is making a survey of apartment dwellers. He wants to choose one family from each floor of a building. If a particular apartment building has eight families on the first floor, four on the second, and three on the third, in how many ways can he choose his sample of three? What is the probability that Mr. White's family, which lives on the first floor, is chosen as one of the three?
6. One of a group of twelve students is to be chosen class president. If eight are men and four are women, what is the probability that the one chosen is a man? What is

the probability that Fred Jones, one of the men, will be chosen? Assume the outcomes are equally likely, although this may not be the case.

7. Three varieties of fruit flies are kept in a container. There are 150 of variety *A*, 200 of variety *B*, and 100 of variety *C*. A small hole appears in the container and a fruit fly escapes. What is the probability that it is of variety *B*? What is the probability that it is not of variety *A*?
8. A rat is allowed to proceed along the paths of a maze which has no dead ends. After 10 centimeters of straight path, the path splits into two parts. This procedure is repeated so that there are a total of five splits per path. Tabulations are kept about which of the exits the rat uses, and a variety of different rewards/punishments are used for incentive. How many different exits are there?
9. Two groups were given therapy, and the improvement, if any, in their condition noted. It was found that a particular *order* of treatments was better than a particular set of treatments. Three treatments were isolated as being most beneficial and it was decided to try all possible orders of these three treatments. How many different experimental groups should be selected in order to try a different order of the three treatments with each group?

3.2 PERMUTATIONS AND COMBINATIONS

To help us to determine the number of possible outcomes, we must study the mathematical concepts of permutation and combination. Consider a set of three letters, A, B, C. In how many ways can these three letters be arranged as a set of three? This question can be answered by listing the arrangements as follows: {ABC, ACB, BAC, BCA, CAB, CBA}. Counting these arrangements, we find that there are six different ones. Instead of writing down these combinations, we could have applied our basic multiplication rule. There are three choices for the first place, two for the second and one for the third, so by the multiplication rule we have $3 \cdot 2 \cdot 1$ different choices. When we multiply this out, we get six different choices. If we wrote down any three different letters of the alphabet, there would be $26 \cdot 25 \cdot 24$ or 15,600 different possible arrangements or **permutations**. This principle can be formulated as follows:

PERMUTATION RULE A	If r objects are to be selected from a set of n different objects in such a way that the order of selection is important, the number of permutations is given by $n \cdot (n-1) \cdot (n-2) \cdots (n-r+1)$.

This rule follows directly from the multiplication rule where $m_1 = n$, $m_2 = n - 1$, $m_3 = n - 2, \ldots, m_r = [n - (r - 1)] = n - r + 1$. The number of permutations of n objects taken r at a time is often symbolized by ${}_nP_r$ or P^n_r.

EXAMPLE 1 How many ways can a president, vice-president, secretary, and treasurer be chosen from an organization containing 30 members?

SOLUTION Since the order is important, we have 4 positions to be selected from 30 persons, so the number of permutations is $30 \cdot 29 \cdot 28 \cdot 27$ or 657,720.

As a consequence of this rule, it can be seen that the number of permutations of n objects taken all together, is $n(n - 1)(n - 2) \cdots 3 \cdot 2 \cdot 1$. This expression is generally written $\boldsymbol{n!}$ (read "n-factorial") to facilitate writing it down.

It follows then that the number given in permutation rule A can be written as $n!/(n - r)!$ since

$$\frac{n!}{(n-r)!} = \frac{n(n-1)(n-2)\cdots(n-r+1)(n-r)(n-r-1)\cdots 3\cdot 2\cdot 1}{(n-r)(n-r-1)\cdots 3\cdot 2\cdot 1}$$

$$= n(n-1)(n-2)\cdots(n-r+1)\frac{(n-r)!}{(n-r)!}$$

$$= n(n-1)(n-2)\cdots(n-r+1).$$

Note that for this to be valid in all cases, we must define $0! = 1$. For example, ${}_nP_n$ should be $n!$, but

$${}_nP_n = \frac{n!}{(n-n!)} = \frac{n!}{0!} = n!$$

only when $0! = 1$.

EXAMPLE 2 In how many ways may eleven people be seated in a row containing seats numbered from one to eleven?

SOLUTION Since there are exactly eleven "objects" (the people), all of whom will be seated, there are 11! or 39,916,800 different ways in which they can be seated. Note that the numbering or the presence of the seats is immaterial. If the persons were seated in a row on the ground, the answer would be the same.

EXAMPLE 3 In how many ways may eleven people be seated in a circle?

SOLUTION Unless there is some specified "first" position or numbered seats, it doesn't matter where the first person sits. After he sits, all the others may be seated in reference to him. Thus after one person is situated anywhere, the remaining ten persons may be seated in any of 10! or 3,628,800 different ways.

Thus far we have discussed permutations in which all the elements are different. If some of the items are repetitious, as in the letters of the word "tree," some modification in our procedures must be made. If the two letters, "e," were distinguishable, there would be 4! or 24 different permutations. Exchanging e's, however, does not produce a different permutation. Thus, we must divide the 4! by 2!, which is the number of ways the repeat letters may be switched without changing the combination. Similarly, the word "bubble" can be rearranged in 6!/3! or 120 different ways. If more than one element is repeated, as in the word "classification," each such repetition is incorporated into the denominator. For "classification," the number of permutations of all the letters is (14!/2! 2! 2! 3!). This can be stated formally as follows:

PERMUTATION RULE B

If a set contains n elements of which r_1 are of one kind, r_2 are of a second kind, r_3 are of a third kind, and so on through r_k, the number of permutations of all n objects is given by

$$\frac{n!}{r_1!\,r_2!\,r_3!\cdots r_k!}$$

If the order is not significant, the formulas for permutations do not apply. For instance, if three letters are to be selected from the set {a, e, i, o, u}, the number of permutations is given by $5 \cdot 4 \cdot 3$ or 60. If, however, the set {a, e, o} and the set {e, o, a} are considered to be the same, that is, the order is immaterial, again some modifications must be made. Sets of this type are called **combinations**. Of each distinct set of three elements forming a combination there are 3! or 6 different permutations. Thus, of the total of 60 permutations, sets of six form just one combination. There are, therefore, 60/6 or 10 distinctly different combinations of five different letters, taken three at a time. If we want the number of combinations of n things taken r at a time, we know that there are

$$n(n-1)(n-2)\cdots(n-r+1)\left(\text{or } \frac{n!}{(n-r)!}\right)$$

different permutations. Of each set of r objects there are $r!$ permutations of those r objects. These $r!$ permutations yield exactly one combination. Thus, we have the following rule:

COMBINATION RULE

The number of combinations of n objects taken r at a time is given by

$$\frac{n(n-1)(n-2)\cdots(n-r+1)}{r(r-1)(r-2)\cdots 3\cdot 2\cdot 1} \quad \text{or} \quad \frac{n!}{r!\,(n-r)!}$$

This number is often symbolized ${}_nC_r$ or C_r^n, and most often $\binom{n}{r}$. Note that $\binom{n}{r}$ is a symbol for a positive integer and not a fraction such as n/r.

EXAMPLE 4 How many committees of 4 can be chosen from an organization containing 30 members?

SOLUTION In contrast to the previous example, the order of selection is unimportant. Thus the number of possible committees is

$$\binom{30}{4} \quad \text{or} \quad \frac{30\cdot 29\cdot 28\cdot 27}{4\cdot 3\cdot 2\cdot 1} \quad \text{or} \quad 27{,}405.$$

Problems

1. How many 4-letter "words" can be made from 8 different letters if (a) unlimited repetitions of a letter are allowed? (b) repetitions of a letter are not allowed?
2. How many different poker hands of 5 cards can be dealt from an ordinary deck of 52 different cards?
3. A perfect bridge hand consists either of 13 cards of the same suit, or of 4 aces, 4 kings, 4 queens, and a jack. What is the probability of being dealt a perfect hand?
4. Jones Stables has entered three horses in a race, and Foster Stables has entered two horses in the same race. After the race, Mrs. Jones and Mrs. Foster talk about the race, which had only the five horses entered. Mrs. Jones notes that one of her horses finished first but Mrs. Foster recalls that one of Mrs. Jones horses finished last as well. If the Foster horses did not finish consecutively, how many different orders are possible, based on this information?
5. A committee of 3 is to be selected from a group of 5 men and 3 women. In how many ways can the committee be selected so that

   (a) it consists of 2 men and 1 woman?
   (b) each sex is represented?
   (c) at least 1 man is on the committee?
   (d) at least 1 woman is on the committee?
   (e) there is no discrimination (that is, no restrictions)?
6. A student is writing an examination consisting of two parts. Part A contains 6 problems, part B, 5 problems. If he is to omit two problems from each part, how many essentially different examinations can he write?
7. How many different permutations are there of the letters in each of the following words?

   (a) syzygy
   (b) bookkeeper
   (c) calamity
   (d) gorgeous
   (e) Mississippi
8. Sixteen astronauts are bound for Mars, and four of them will actually set foot on the planet. If the captain is an automatic choice and the remaining positions are selected by lot, what is the probability that Tom, Dick and Harry, who are buddies, will be selected to go along with the captain?
9. Three Ferrari's, two Porsche's, and five different-make American cars are in a race. By make, how many possible ways can the cars finish?
10. Six friends arrive at a movie house and wish to sit together. The only row with six seats vacant, however, has people sitting in seats 1, 5, 9 and 10. In how many different ways can the six friends be seated in this row, assuming a row has only

ten seats? If all ten seats were vacant, in how many different ways could the six friends be seated?

11. Show that $\binom{n}{r} = \binom{n}{n-r}$.

12. Show that $\binom{n+1}{r} = \binom{n}{r-1} + \binom{n}{r}$ for $1 \leq r \leq n$.

    This is called *Pascal's rule*.

3.3 RULES OF PROBABILITY

If an experiment is performed, the set consisting of all possible outcomes is called the **sample space** for the experiment. The sample space is usually denoted by S. Each outcome is called a **point** of the sample space or a **sample point**. A sample space can best be obtained by listing every possible outcome in its simplest form. If a coin is to be tossed three times, for example, the sample space of possible outcomes of heads and tails is given by

$$S = \{(H, H, H), (H, H, T), (H, T, H), (H, T, T), (T, H, H), (T, H, T), (T, T, H), (T, T, T)\}$$

Each of the eight possible outcomes thus becomes a point of the sample space.

EXAMPLE 1 An experiment consists of tossing a coin and rolling a die. Give the sample space for the experiment.

SOLUTION The sample space consists of twelve outcomes since there are two possible outcomes for the coin and six for the die. Applying the multiplication rule, $2 \cdot 6 = 12$. These outcomes are as follows:

$$S = \{(H, 1), (H, 2), (H, 3), (H, 4), (H, 5), (H, 6), (T, 1), (T, 2), (T, 3), (T, 4), (T, 5), (T, 6)\}$$

An alternate representation is given below in which each dot represents the point of the sample space corresponding to the outcome on the coin (to the left) and on the die (above).

coin \ *die*	*1*	*2*	*3*	*4*	*5*	*6*
H	•	•	•	•	•	•
T	•	•	•	•	•	•

Any collection of points in the sample space is called an **event** or **simple event**. The set of no points is the **empty set**, denoted by $\varnothing$, and is included for completeness. In the experiment in which a coin is tossed three times, the event "heads appear twice" consists of the points (H, H, T), (H, T, H), and (T, H, H).

EXAMPLE 2 Three dice are rolled. List the points of the sample space corresponding to the event "a total of six is obtained on the upper faces."

SOLUTION It may help if we imagine the dice to be of three different colors, say red, yellow, green. We let the ordered triple (1, 2, 3) represent the numbers on, respectively, the red, yellow, and green die. There are six such orders (by Permutation Rule A). Other possibilities are (1, 1, 4) and its permutations, three in all, and (2, 2, 2). No other possible outcomes add up to six. The ten points which correspond to the event: "a total of six on the three dice" are as follows:

$$\text{"6"} = \{(1, 2, 3), (2, 1, 3), (1, 3, 2), (3, 1, 2), (3, 2, 1),$$
$$(2, 3, 1), (1, 1, 4), (1, 4, 1), (4, 1, 1), (2, 2, 2)\}$$

NOTE A listing of all sample points for the experiment and extraction of those which correspond to "6" is possible, but cumbersome since there are $6 \cdot 6 \cdot 6$ or 216 points in the sample space for the experiment.

The classical definition of probability can be used to determine the probability of an event, if each of the sample points is equally likely. In this case the probability of an event is equal to the number of sample points in the event divided by the number of points in the sample space. In example 2, the probability of obtaining a six on one roll of three dice is 10/216 (or 5/108 or about 0.046) since the points are equally likely. There are ten points in the event and 216 in the sample space. We symbolize this by writing

$$P(6) = 10/216$$

The symbol **P(6)** is read "P of 6" or "probability of 6." It does not refer to multiplication, but is a way of symbolizing a probability even if we do not know its value.

EXAMPLE 3 Suppose that a drawing of one marble is to be made from a box containing three white marbles, four red marbles, and one clear marble. Determine the following probabilities: P(the marble is red), P(the marble is white), P(the marble is not red).

SOLUTION The sample space may be illustrated schematically by using a **Venn diagram** as illustrated below.

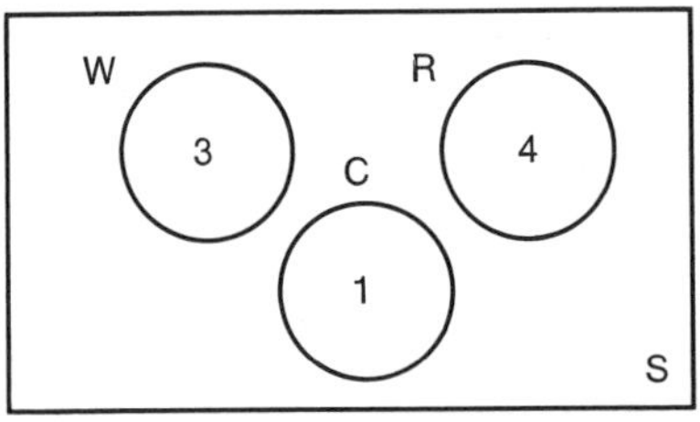

In this diagram, W represents the set of white marbles, R represents the set of red marbles, C represents the set of clear marbles, and S represents the sample space. The number of marbles in each set is marked within the set. If there were marbles which did not belong to any of the sets, that number would appear outside the circles, but in S. There are 4 red marbles, of the total of 8, so $P(R) = 4/8$, or 0.5. Similarly there are three white marbles, so $P(W) = 3/8$. Five marbles are not white, so this probability is 5/8. To symbolize this event, not white, we write **W′**, read W-prime, so that $P(W') = 5/8$. An event of this type, consisting of sample points not in another event, is called a *complement*. The **complement** of any event A, denoted A', consists of all sample points not in A.

In the preceding example, the events did not have any points in common. Such events are said to be **mutually exclusive**. If two events are mutually exclusive, they cannot both occur at the same time. The three events, R, W, C, accounted for all points of the sample space. Such events are said to be **exhaustive**. (Exhaustive events need not be mutually exclusive.) Complementary events are any two events which are *both* mutually exclusive and exhaustive.

EXAMPLE 4 A die is tossed once. Determine the probability of obtaining (a) at least a 3; (b) at most a 4; (c) more than 4; (d) 5 or less; (e) 2 or more; and (f) less than 5.

SOLUTION The sample space is $\{1, 2, 3, 4, 5, 6\}$, each point with probability 1/6. The set corresponding to each event with its probability follows:

	Set	*Symbol*	*Value*
(a)	$\{3, 4, 5, 6\}$	$P(x \geq 3)$	4/6 or 2/3
(b)	$\{1, 2, 3, 4\}$	$P(x \leq 4)$	4/6 or 2/3
(c)	$\{5, 6\}$	$P(x > 4)$	2/6 or 1/3
(d)	$\{1, 2, 3, 4, 5\}$	$P(x \leq 5)$	5/6
(e)	$\{2, 3, 4, 5, 6\}$	$P(x \geq 2)$	5/6
(f)	$\{1, 2, 3, 4\}$	$P(x < 5)$	4/6 or 2/3

DISCUSSION Many students have difficulty with the concepts "greater than," "less than," "at least," and "at most." Note that "at least k" means either "k" or "greater than k," while "greater than k" excludes k. Note also that there is a basic difference in cases where integers only are involved. In the case of integers, the events "$x > 4$" and "$x \geq 5$" are identical. Both cases include all integers from 5 on. Where real or rational numbers are involved, however, these events are different since, for instance, 4.3 is greater than 4 and hence satisfies $x > 4$, but it is less than 5 and hence does not satisfy $x \geq 5$.

In the preceding examples, some of the rules by which probability is calculated are seen. First, we observe that we are working only with one sample space for each of the probability models we have examined. Then it is obvious that the probability of a particular event occurring must be positive or zero, that is, $P(A) \geq 0$ for any event A. The probability of any event, however, cannot exceed one, so $P(A) \leq 1$. Combining these two statements, we obtain the **first probability rule**: The probability of an event is expressed as a real number from zero to one, inclusive, or $0 \leq P(A) \leq 1$. Further, the sum of the probabilities of the sample points which make up the sample space is one. Thus the probability of *some* event in the sample space occurring is one, and the probability of no event in the sample space occurring is zero; that is, if S represents the whole sample space, $P(S) = 1$, and if $\varnothing$ represents the empty set, $P(\varnothing) = 0$.

In example 4, note that the events given in (b) and (c) are complementary and the sum of their probabilities is equal to one. This is true for any complementary events, so we here summarize the probability rules for simple events.

PROBABILITY RULES FOR SIMPLE EVENTS	1. For any event A, $0 \leq P(A) \leq 1$ 2. For any sample space S, $P(S) = 1$; $P(\varnothing) = 0$ 3. For any event A, $P(A') = 1 - P(A)$

Probabilities can also be calculated for combinations of simple events. Such combinations are called **compound events**. There are two types of compound events. The probability that *either* event A *or* event B (or both) will occur is symbolized by $P(A \text{ or } B)$; the probability that *both* event A *and* event B will occur is symbolized by $P(A \text{ and } B)$. Rules governing compound events can best be illustrated by an example.

Suppose that one marble is to be drawn from a box containing three solid white marbles, four solid red marbles, two with red and white stripes, and one clear marble. We let W represent the event "white marble" and R represent "red marble." A marble with red and white stripes is considered to belong to *both* W and R. A Venn diagram for the sample space is given here.

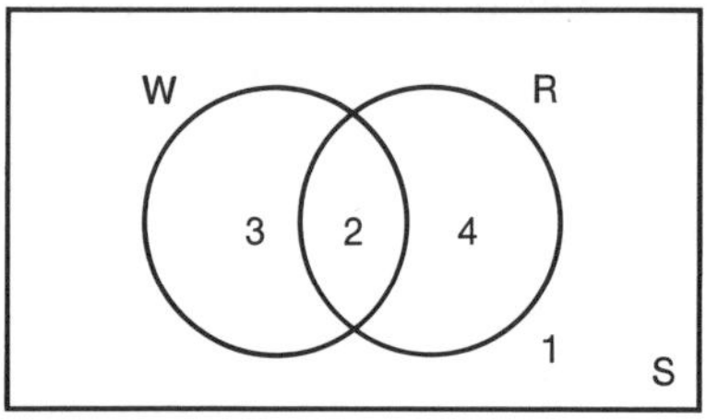

For the simple events, W and R, we note that $P(W) = 0.5$ since 5 of the ten marbles have white markings, and $P(R) = 0.6$, since 6 of the ten marbles have red markings. There are a total of 9 marbles which have either red or white markings (or both), so we have $P(W \text{ or } R) = 0.9$. Similarly, exactly two of the marbles have *both* red *and* white markings, so $P(W \text{ and } R) = 0.2$. Now note that $P(W) + P(R) = 0.5 + 0.6 = 1.1$. This is not equal to $P(W \text{ or } R)$ since the overlapping portion is counted twice, so $P(W \text{ and } R)$ must be subtracted from 1.1 to obtain the correct value for $P(W \text{ or } R)$. This leads to the **general rule of addition**: If A and B are events in the same sample space, the probability that at least one will occur is the sum of the individual probabilities minus the probability that both will happen.

ADDITION RULE FOR PROBABILITY	For any events, A and B, in a sample space $P(A \text{ or } B) = P(A) + P(B) - P(A \text{ and } B)$

In the marble case we observe that $P(A \text{ or } R) = 0.9$ and $P(W) + P(R) - P(W \text{ and } R) = 0.5 + 0.6 - 0.2 = 0.9$. A little arithmetic gives us a useful consequence of the addition rule, namely $P(A \text{ and } B) = P(A) + P(B) - P(A \text{ or } B)$.

Now if two events, A and B, are mutually exclusive, they cannot both occur, so $P(A \text{ and } B) = 0$. We then have the special case of the addition rule, for mutually exclusive events. If A and B are mutually exclusive events, the probability that one (or both) of them will occur is the sum of their probabilities.

ADDITION RULE FOR PROBABILITY (Special Case)	If A and B are mutually exclusive events, $P(A \text{ or } B) = P(A) + P(B)$

Applications of the probability rules for simple and compound events will enable us to determine probabilities for many different events. In the example with ten marbles, some of the events are given here, together with their probabilities. The student should take the time to verify each one before going on.

EVENT	PROBABILITY	
The marble drawn has markings which are	*Symbol*	*Value*
White	$P(W)$	0.5
Red	$P(R)$	0.6
White and Red	$P(W \text{ and } R)$	0.2
White or Red	$P(W \text{ or } R)$	0.9
Not White	$P(W')$	0.5
Not Red	$P(R')$	0.4
Neither white nor red	$P(W' \text{ and } R')$	0.1
Not white and not red	$P(W' \text{ and } R')$	0.1
Not both white and red	$P(W \text{ and } R)'$	0.8
Not white or not red	$P(W' \text{ or } R')$	0.8
White but not red	$P(W \text{ and } R')$	0.3
Red but not white	$P(W' \text{and } R)$	0.4

Note that the events "white and red" and "white but not red" are mutually exclusive, and their union is simply the event "white." In terms of probability, $P(W) = P(W \text{ and } R) + P(W \text{ and } R')$. This special rule has many applications and can be written as follows: If A and B are any events, $P(A) = P(A \text{ and } B) + P(A \text{ and } B')$.

A summary of the **probability rules** presented in this section is now appropriate:

PROBABILITY RULES	
	1. For any event, A, $0 \leq P(A) \leq 1$
	2. For any sample space S, $P(S) = 1$; $P(\varnothing) = 0$
	3. For any event A, $P(A') = 1 - P(A)$
	4. If A and B are any events, $P(A \text{ or } B) = P(A) + P(B) - P(A \text{ and } B)$
	5. If A and B are mutually exclusive, $P(A \text{ or } B) = P(A) + P(B)$
	6. If A and B are any events, $P(A) = P(A \text{ and } B) + P(A \text{ and } B')$

Problems

1. If $P(A) = 0.6$ and $P(B) = 0.7$, under what conditions may A and B be mutually exclusive?
2. If A and B are mutually exclusive and $P(A) = 0.4$ and $P(B) = 0.5$, what is $P(A \text{ or } B)$?
3. A psychologist states that if a rat enters a maze, the probability that he will emerge at point A is 0.44, at point B is 0.29, at neither is 0.17. Do you agree or disagree with his assertion? Why?
4. An enthusiastic football fan claims that there are only 2 chances in 15 that his team will lose while the probability that it will win is 0.9. A second fan, more conservative, agrees that there are only 2 chances in 15 of losing, but that the probability of winning is only 0.8. Comment on these two claims.
5. A family is considering buying a dog. If the probability that they will buy a small dog is 0.1, that they will buy a medium-sized dog is 0.3, that they will buy a large dog is 0.2, and that they will buy a very large dog is 0.1, what is the probability that the family will buy a dog?
6. An urn contains 3 white, 2 red, 1 blue, and 4 black balls. If one ball is drawn at random, what is the probability that it is

   (a) black?
   (b) either black or white?
   (c) not white?
   (d) both red and blue?

7. Of 10 students, 3 are mathematics majors, one of whom is from New York. Of the 10 students, 4 are from New York. If a student is chosen at random from among these 10, what is the probability that he is

   (a) a math major from New York?
   (b) a New Yorker who is not a math major?
   (c) either from New York or a math major?
   (d) neither a math major nor from New York?
   (e) not from New York?
   (f) either not a math major or not from New York?
   (g) not a math major?
   (h) either a math major or else not from New York?

8. From experience, the toll taker feels that the probability that an automobile paying a toll on the Sunshine Skyway is a Chevrolet is 0.21, that it is a Ford is 0.17, that it is a Pontiac is 0.11. Assuming he is correct, if an automobile pulls up to pay a toll, what is the probability that it is

   (a) a Chevrolet or a Ford?
   (b) a Ford or a Pontiac?
   (c) neither a Chevrolet nor a Pontiac?
   (d) none of these three makes?

9. Let R be the event "the stock market rises," K the event "Amalgamated Checkers stock rises," and T the event "the stock market falls." Suppose that $P(R) = 0.3$, $P(T) = 0.4$, and $P(K) = 0.2$. Assuming that K is independent of both R and T, symbolize and determine each of the following probabilities.

   (a) The market rises and Amalgamated rises.
   (b) The market falls and Amalgamated rises.
   (c) The market does not rise, but Amalgamated rises.
   (d) The market remains steady.
   (e) Amalgamated does not rise, or the market falls.
   (f) Neither the market nor Amalgamated rises.

10. The probability that Fred will watch television on a Friday night is 0.32; the probability that he will go to a movie is 0.24; the probability that he will go to a friend's house is 0.28; the probability that he will play a game with his family is 0.12; the probability that he will stay home and read a magazine is 0.04.

    (a) What is the probability that he will watch television, go to a movie, or read a magazine?
    (b) What is the probability that he will go to a friend's house, play a game with his family, or read a magazine?
    (c) What is the probability that he will stay home?
    (d) What is the probability that he will neither watch television nor go to a movie?

11. If a patient is admitted to a hospital with a particular disease, the probability that he will die of the disease is 0.12, and the probability that he will recover quickly is 0.24. The probability of recovering after a long convalescence without developing complications is 0.22, and the probability that he will develop complications is 0.42. Half the patients developing complications die, and half recover.
    (a) What is the probability that he will recover from the disease without developing complications?
    (b) What is the probability that a patient will recover from the disease?
12. Among a group of children, 35% suffer from cognitive dissonance as a result of school experiences and 40% have distorted senses of reality as measured by a standard measuring device. In a group of 200 children for which the above percentages hold, a total of 32 have both cognitive dissonance and a distorted sense of reality. How many of these 200 children are free from both problems?
13. A sociogram shows that 12% of the members of a group are isolates. One-third of those are from low-income groups. If a member of this group is selected at random, what is the probability that this person is an isolate not from a low-income group?
14. For one variety of mosquito, 30% are resistant to repellent A, 25% are resistant to repellent B, and 50% are resistant to neither repellent. A subject walks into a room after having used both repellents. There is one mosquito of this variety in the room. Assuming nothing else is done to deter the mosquito, what is the probability that the subject will be bitten?

3.4 INDEPENDENT EVENTS AND CONDITIONAL PROBABILITY

If a coin is tossed twice in succession, the probability that the first toss comes up a head is 1/2. If the first toss is a head, the probability that the second toss is a head is 1/2. If the first toss is a tail, the probability that the second toss is a head is still 1/2. In other words, the probability of a head on the second toss does not depend upon the outcome of the first toss. The results of the first toss have absolutely no bearing on the outcome of the second toss. The two outcomes are *independent*. Two events are said to be **independent** if neither outcome affects the probability of occurrence of the other. The probability of two heads in succession on two successive tosses of a coin can be determined with reference to the sample space. There are four equally likely points since $S = \{(H, H), (H, T), (T, H), (T, T)\}$ and exactly one of them is the event "two heads," so the probability is 1/4. We can also reason that 1/2 of the tosses will be heads the first time, and 1/2 of those will be heads the second time, so 1/4 will be heads both times. The reasoning is correct and leads to the special case of the **multiplication rule for probability**.

MULTIPLICATION RULE FOR PROBABILITY (Special Case)	If A and B are independent events, then $P(A \text{ and } B) = P(A) \cdot P(B)$

The concept of independence is central to probability and needs more foundation than the intuitive approach given above. A sounder understanding of this concept can be obtained by examining **conditional probability**.

Consider two sets of playing cards—set X consisting of two aces and three kings, and set Y consisting of three aces and one king. If an experiment consists of tossing a coin to decide which set to pick (heads, X, tails, Y), and then we shuffle the set and look at the top card, it is apparent that the probability of getting a king is much greater if set X is selected than if set Y is selected. Three-fifths of the cards in set X are kings while only 1/4 of those in set Y are kings. Now, if the event K means that a king is on top, while X means that set X is selected, $P(K \mid X)$ (read "probability of K, given X") is the probability that if set X is selected, a king will be on top. Similarly, $P(K \mid Y)$ is the probability that if Y is selected, a king will be on top. Then $P(A \mid X)$ and $P(A \mid Y)$ are the probabilities that an ace is on top if X and Y are selected, respectively. From the information given previously,

$$P(K \mid X) = 3/5,\ P(K \mid Y) = 1/4,\ P(A \mid X) = 2/5,\ P(A \mid Y) = 3/4$$

The sum of these probabilities is greater than one, so that these events cannot be considered all a part of the same sample space.

Now the sample space corresponding to this experiment has four points whose probabilities can be determined intuitively as follows: X will be selected half of the time; 3/5 of those times a king will be selected, so that X will be selected *and* a king will turn up about $(3/5) \cdot (1/2)$ or 3/10 of the time. Hence $P(K \text{ and } X) = 3/10$. Similarly $P(A \text{ and } X) = (2/5) \cdot (1/2) = 2/10$, $P(K \text{ and } Y) = (1/4) \cdot (1/2) = 1/8$. $P(A \text{ and } Y) = (3/4) \cdot (1/2) = 3/8$. We note that in each case $P(A \text{ and } B) = P(A \mid B) \cdot P(B)$ so that

$$\frac{P(A \text{ and } B)}{P(B)} = P(A \mid B)$$

This gives us the formal definition of conditional probability.

CONDITIONAL PROBABILITY If an event B is given ($P(B) \neq 0$), the conditional probability of A is defined by

$$P(A \mid B) = \frac{P(A \text{ and } B)}{P(B)}$$

The event B is called a restricted sample space.

The definition of conditional probability gives us the general form of the multiplication rule for probability since if $P(A \mid B) = P(A \text{ and } B)/P(B)$, then $P(A \text{ and } B) = P(A \mid B) \cdot P(B)$ and we have

MULTIPLICATION RULE FOR PROBABILITY For any events A and B,
$P(A \text{ and } B) = P(A \mid B) \cdot P(B)$

EXAMPLE 1 Consider a group of 40 students, 20 of whom take an English course and 16 take a business course. Of these two groups, 12 are taking both. Determine $P(B \mid E)$ and $P(E \mid B)$.

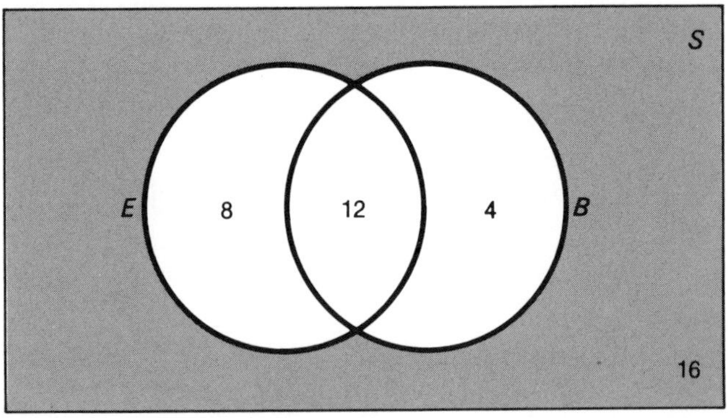

SOLUTION $P(E) = 20/40 = 0.50$, $P(B) = 16/40 = 0.40$, $P(E \text{ and } B) = 12/40 = 0.30$ where E and B constitute the obvious events. If one student is selected and it is known that he is taking an English course, then he is one

of 20 students in set E. Since 12 of the students taking English also take business, the probability that our English-taking student takes business is 12/20 or 0.60. By the rule for conditional probability,

$$P(B \mid E) = \frac{P(B \text{ and } E)}{P(E)} = \frac{0.30}{0.50} = 0.60$$

Thus, the probability of B for the restricted sample space E is, as expected, the conditional probability of B given E. Similarly, 12 of the 16 students taking business also take English. By the rule,

$$P(E \mid B) = \frac{P(E \text{ and } B)}{P(B)} = \frac{0.30}{0.40} = 0.75$$

Since $12/16 = 0.75$, we have affirmed the rule once again.

Returning to the example which opened the section, suppose we wish to know the probability before tossing the coin that the card we turn up will be a king. There are only two possibilities: we select set X by obtaining a head and draw a king, or we select set Y and draw a king. Since X and Y are complementary we know that

$$P(K) = P(K \text{ and } X) + P(K \text{ and } Y) = 3/10 + 1/8 = 34/80 = 17/40$$

Naturally, since K and A are complementary, $P(A) = 1 - 17/40 = 23/40$. Another way to state $P(K) = P(K \text{ and } X) + P(K \text{ and } Y)$ is

$$P(K) = P(K \mid X) \cdot P(X) + P(K \mid Y) \cdot P(Y)$$

Applying the same reasoning to the addition rule, we have the **rule of total probability**.

RULE OF TOTAL PROBABILITY	If $B_1, B_2, \ldots, B_n$ are mutually exclusive events which make up the sample space, then $$P(A) = P(A \mid B_1) \cdot P(B_1) + P(A \mid B_2) \cdot P(B_2) + \cdots + P(A \mid B_n) \cdot P(B_n).$$

EXAMPLE 2 A golfer has a probability of getting a good shot equal to 9/10 if he uses the correct club. If he uses an incorrect club, the probability drops to 7/10. If only one club is correct for a particular shot and he has 14 clubs in his bag, determine the probability that he gets a good shot in the unlikely event that he chooses one of the clubs at random.

SOLUTION Let G represent the event "he gets a good shot" and C represent the event "he uses the right club." Then $P(G|C) = 9/10$, $P(G|C') = 7/10$, $P(C) = 1/14$, and $P(C') = 13/14$. By the rule of total probability, then,

$$\begin{aligned} P(G) &= P(G \mid C) \cdot P(C) + P(G \mid C') \cdot P(C') \\ &= (9/10)(1/14) + (7/10)(13/14) \\ &= 9/140 + 91/140 \\ &= 100/140 \\ &= 5/7 \end{aligned}$$

A tree diagram approach to this problem may prove helpful. Let the first step represent selection of club, the second step represent the actual shot. Probabilities are multiplied on each branch by virtue of the multiplication rule.

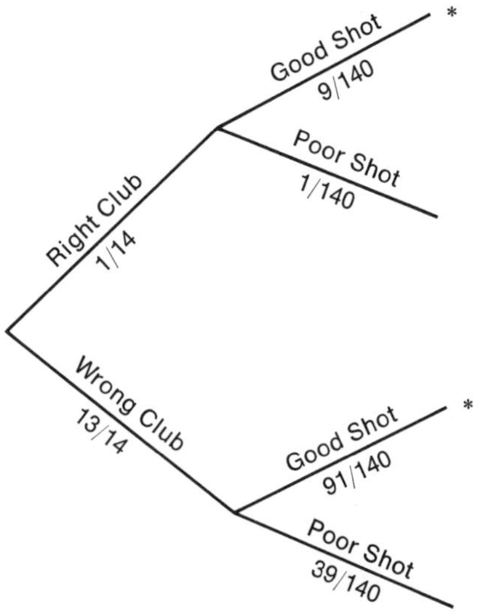

We find the outcomes which correspond to the good shot (marked above) and the probabilities of each. Then we simply add up the probabilities to obtain $P(G)$.

The similarities between the general and special multiplication rules for probability lead us to ask how independence and conditional probability are related, since this is the only difference between the events. To answer that question let us examine the group of 40 students of example 1. Of these, 20 take an English course, 16 take a business course. Twelve students take both. Suppose we bet, with a friend, that we can guess whether or not a particular student takes an English course. We can ask him one question and one question only, and it may not pertain to whether he takes English or not. Since $P(E|B) = 0.75$ and $P(E|B') = 0.33$, we can considerably improve our chances of winning by asking whether the student is taking a business course. If so, we should guess that he is taking an English course since 0.75 of the students taking a business course take an English course. That is, $P(E \mid B)$ and $P(E)$ are different so that occurrence or nonoccurrence of event B has an effect on the probability of event E.

Now the game has changed. Our friend has another group of 40 students. Twenty of these take accounting, 16 take math, but now only 8 of those take both. The rules remain the same, only now we must guess whether or not a student takes accounting. We confidently sit down to determine how to guess after we find out whether or not he takes math. We have $P(A) = 0.50$, $P(M) = 0.40$, $P(A \text{ and } B) = 0.20$. Then $P(A \mid M) = 0.20/0.40 = 0.50$. But then $P(A \mid M) = P(A)$; that is, the proportion of math students taking accounting is equal to the proportion of students taking accounting in the whole sample space. Further,

$$P(A \mid M') = \frac{P(A \text{ and } M')}{M'} = \frac{0.30}{0.60} = 0.50 = P(A)$$

again. Thus, whether a student takes math or not has no effect whatsoever on the probability of the student's taking accounting. The two events are truly **independent**.

INDEPENDENT EVENTS	Two events, A and B, are independent if and only if $P(A \mid B) = P(A)$.

Note that if both A and B are independent, then

$$P(A \mid B) = \frac{P(A \text{ and } B)}{P(B)} = P(A), \quad \text{so} \quad P(A \text{ and } B) = P(A) \cdot P(B),$$

which is the special case of the multiplication rule for probability. That is, the probability that both will occur is then the product of their individual probabilities (but only if the events are independent).
Moreover, if the events are independent,

$$P(B \mid A) = \frac{P(B \text{ and } A)}{P(A)} = \frac{P(A \text{ and } B)}{P(A)} = \frac{P(A) \cdot P(B)}{P(A)} = P(B).$$

Thus, if the probability that the home team will win any particular game of the World Series is 0.6, the probability of a clean sweep (winning the first four games) by one team is (0.6) (0.6) (0.4) (0.4) or 0.0576. This is assuming independence of outcomes which is not very likely. Recall that the first two games are played on one field, the next two games on the other team's field.

EXAMPLE 3 An urn contains 3 red balls and 2 white balls. A ball is drawn, then replaced. What is the probability that a white ball will be drawn three times in a row?

SOLUTION The probability of drawing a white ball on any particular drawing is 2/5. Successive drawings are independent, since the ball drawn is replaced. Therefore, the probability, $P(WWW)$, is $(2/5) \cdot (2/5) \cdot (2/5)$ or $(2/5)^3$ which is 8/125 or 0.064.

EXAMPLE 4 Two dice are thrown and the total number of dots observed. List a sample space for the experiment, together with associated probabilities.

SOLUTION We must consider the dice separately to arrive at an accurate space—for example, we could have one red, and one green. The probability of obtaining a "1" on each die, hence a total of two, is $(1/6) \cdot (1/6)$ or 1/36. Thus $P(2) = 1/36$. There are two ways to obtain three—"2" on the red die,

"1" on the green, or vice versa. Again each has a probability of 1/36. These two outcomes are mutually exclusive, so $P(3) = 2/36$ or 1/18. Continuing, we have the following table:

x	$P(x)$
2	1/36
3	2/36
4	3/36
5	4/36
6	5/36
7	6/36
8	5/36
9	4/36
10	3/36
11	2/36
12	1/36

CAUTION Many students confuse the terms "independent" and "mutually exclusive." These terms are different, and apply to different things. This can be shown easily by observing that if A and B are events with nonzero probabilities, if they are mutually exclusive, $P(A \text{ and } B) = 0$, while if they are independent, $P(A \text{ and } B) \neq 0$. Mutually exclusive events are events which cannot both occur at the same time, while independent events are events which do not influence each other.

A striking example of this can be given by football games. If Florida and Alabama play football games on a given day, the events "Florida wins" and "Alabama wins" are independent unless they play each other. In that case, the events are mutually exclusive.

Problems

1. Find $P(A \text{ and } B)$ if $P(A) = 2/3$, $P(B) = 1/5$ and

   (a) $P(A \mid B) = 1/10$
   (b) $P(B \mid A) = 1/10$
   (c) $P(A \mid B) = P(A)$
   (d) $P(B \mid A) = P(B)$

2. In a group of 100 students, it is found that 80 take an English course, 60 take a mathematics course, and 10 take neither. Are the two events independent? Why?

3. If $P(A) = 0.30$, $P(B) = 0.50$, and A and B are mutually exclusive events, find

   (a) $P(A \mid B)$
   (b) $P(B \mid A)$

4. If a coin is tossed five times, what is the probability of getting a head on the first and last tosses, and tails on the middle three?

5. Fifteen white marbles and 6 red marbles are placed in a box. What is the probability of drawing 3 white marbles

   (a) if the marbles are replaced after each drawing?
   (b) if the marbles are not replaced?

6. In a set of 60 teachers, 20 teach history, 6 are coaches. It is found that coaching and teaching history are independent events. How many of the coaches teach history?

7. The owner of a department store is giving away a door prize by randomly selecting one of 1,000 slips filled out by the customers (only one to a customer). The breakdown into those who are regular customers and those who are not and also into those who purchased a special sale item is as follows:

	Regular Customers	*Not Regular Customers*	*Totals*
Sale Item Purchasers	350	130	480
Nonpurchasers	210	310	520
Totals	560	440	1000

   If R represents the event "prize is won by a regular customer," P the event "prize is won by a sale item purchaser," find each of the following probabilities:

   (a) $P(R)$
   (b) $P(R \mid P)$
   (c) $P(P)$
   (d) $P(P \mid R)$
   (e) $P(R \mid P')$
   (f) $P(P \mid R')$
   (g) $P(R' \mid P')$
   (h) $P(P' \mid R')$
   (i) $P(R' \mid P)$
   (j) $P(P' \mid R)$

8. A die is cast three times. What is the probability

   (a) of obtaining three 6's?
   (b) of not obtaining three 3's?
   (c) of obtaining the same uppermost face on all three tosses?
   (d) of obtaining three different faces?
   (e) of obtaining a total of 15 for the three throws?

9. A beginning bowler has a probability of 1/5 of getting a strike if he uses the proper weight ball. If he uses the improper weight, the probability drops to 1/8. There are 6 balls in the rack, only 2 of which are the proper weight for him. If he chooses a ball at random, what is the probability that he will get a strike?

10. A machine produces parts and they are stored in a large bin. The parts are then boxed for shipment. On the average 1 box in 50 contains one or more defective parts. If a company orders 4 boxes of parts, what is the probability that all 4 boxes contain at least one defective part? What is the probability that no defective parts will be in any of the boxes? Assume independence of the outcomes.

11. A certain species of toad has two interesting traits: resistance to a certain type of fungus infection varies widely among individuals, and skin variations known as warts also vary widely among individuals. A biologist thinks that perhaps there is some relation between number of warts and resistance to the fungus. In a sample of toads, he classifies them into equal groups of high, average, and low incidence of warts. He then tests the toads for resistance to the fungus and finds 1/4 of them to have high resistance to the fungus. If number of warts and resistance to the fungus are actually independent, what proportion of the toads would be expected to have a large number of warts *and* high resistance to the fungus. (One means of testing such a situation will be given in Chapter 11).

12. Daily water samples are taken from an estuary to determine the proportion of plankton in bloom. It is not known if the water sample is acceptable until it is analyzed. To be acceptable it must contain a minimum number of plankton so that the proportion in bloom can be accurately estimated. The sample is to be taken at 6 A.M., but because of the possibility that a sample may not be acceptable, samples are also taken at 5 A.M., 5:30 A.M., 6:30 A.M., and 7:00 A.M. The analysis is begun at 8 A.M. The 6:00 analysis is performed first. If it is acceptable, no other analyses are performed. If not, the 5:30, 6:30, 5:00, and 7:00 analyses are performed in that order until two of them are acceptable. The probability that an analysis is acceptable has been found to be 0.7. If the acceptability of two samples selected a half-hour apart are independent events, determine the probability of being able to determine the proportion of plankton in bloom on a particular day.

13. A selection of two persons is to be made from a group of people which has been classified on the basis of social mobility and economic stability. The proportion of socially mobile people is 0.40, while that of economically stable people is 0.50. If the two factors are independent, what is the probability that at least one person selected from this group demonstrates both characteristics? What is the probability that at least one of those selected demonstrates at least one of the characteristics?

3.5 SUMMARY

In this chapter we have examined some of the basic principles of probability, including methods of counting outcomes. The most important among the latter are briefly restated below.

MULTIPLICATION RULE	Suppose that an act requires n steps to complete. If the steps can be performed successively in $m_1, m_2, m_3, \ldots, m_n$ ways, then the act can be performed in $m_1 \cdot m_2 \cdot m_3 \cdots m_n$ different ways.

PERMUTATION RULE A

If r objects are to be selected from a set of n different objects, the number of permutations is given by either $n!/(n-r)!$ or

$$n \cdot (n-1) \cdot (n-2) \cdots (n-r+1).$$

PERMUTATION RULE B

If a set contains n elements of which r_1 are of one kind, r_2 are of a second kind, r_3 are of a third kind, etc., through r_k, the number of permutations of all n objects is given by

$$\frac{n!}{r_1!\,r_2!\,r_3!\cdots r_k!}$$

COMBINATION RULE

The number of combinations of n objects taken r at a time is given by either

$$\binom{n}{r} = \frac{n!}{r!\,(n-r)!} \quad \text{or}$$

$$= \frac{n(n-1)(n-2)\cdots[n-(r+1)]}{r(r-1)(r-2)\cdots 3\cdot 2\cdot 1}$$

The basic concepts of classical probability, conditional probability, and independence are given once again.

CLASSICAL PROBABILITY

If an experiment has s equally likely outcomes, r of which constitute an event, then the probability of the event is r/s.

CONDITIONAL PROBABILITY

If $P(B) \neq 0$, then $P(A \mid B) = \dfrac{P(A \text{ and } B)}{P(B)}$

INDEPENDENT EVENTS	Two events, A and B, are independent if and only if $P(A \mid B) = P(A)$. Further, A and B are independent if and only if $P(A \text{ and } B) = P(A) \cdot P(B)$.

We have also examined sample spaces and some rules for obtaining probabilities. These rules are restated below.

PROBABILITY RULES

1. For any event A, $0 \leq P(A) \leq 1$.
2. For any sample space S, $P(S) = 1$, $P(\varnothing) = 0$.
3. For any event, A, $P(A') = 1 - P(A)$.
4. If A and B are any events,
   $P(A \text{ or } B) = P(A) + P(B) - P(A \text{ and } B)$.
5. If A and B are mutually exclusive,
   $P(A \text{ or } B) = P(A) + P(B)$.
6. If A and B are any events,
   $P(A) = P(A \text{ and } B) + P(A \text{ and } B')$.
7. If A and B are any events,
   $P(A \text{ and } B) = P(A|B) \cdot P(B)$.
8. If A and B are independent events,
   $P(A \text{ and } B) = P(A) \cdot P(B)$.
9. If $B_1, B_2, \ldots, B_n$ are mutually exclusive events which make up the sample space, then
   $P(A) = P(A|B_1) \cdot P(B_1) + \cdots + P(A|B_n) \cdot P(B_n)$.

Problems

1. Find $P(A \text{ and } B)$ if $P(A) = 1/2$, $P(B) = 1/3$ and
   (a) $P(A \text{ or } B) = 4/5$
   (b) $P(A \mid B) = 3/4$
   (c) $P(B \mid A) = 2/5$
   (d) A and B are independent
   (e) A and B are mutually exclusive
   (f) $P(A' \text{ and } B) = 1/10$

2. Determine $P(A)$ if $P(A \text{ or } B) = 0.7$, $P(B) = 0.3$ and
   (a) $P(A \text{ and } B) = 0.2$;
   (b) $P(A \mid B) = 1/3$;

(c) $P(A') = 0.4$;
(d) A and B are independent;
(e) A and B are mutually exclusive;
(f) $P(A \text{ or } B') = 0.9$.

3. How many ways can a cross-country match finish by school if three schools have entered, respectively, 6, 7, and 8 runners?

4. A 10-volume encyclopedia and a 3-volume edition of *Tristram Shandy* are knocked onto the floor. If they are replaced at random into the vacant spaces (13 adjacent), what is the probability that the volumes of *Tristram Shandy* are together? In the proper order?

5. Five different colored marbles are placed in a jar; a marble is drawn and replaced, then another marble is drawn. Determine the number of different points in the sample space and the probability that the 2 marbles drawn are the same color.

6. A pair of dice is rolled until a 7 appears. What is the probability that *more* than three rolls will be necessary?

7. The probability that a car parked at Oak Street Beach has Illinois license plates is 0.82; the probability that it is a convertible is 0.18; the probability that it is a convertible with Illinois license plates is 0.08. Determine the following probabilities:

   (a) that such a car is either a convertible *or* has Illinois license plates;
   (b) that such a car either does not have Illinois license plates or else is not a convertible;
   (c) that a car with Illinois license plates parked on the lot is a convertible;
   (d) that a convertible parked on the lot has Illinois license plates.

8. An experiment consists of putting a dime, a nickel, and a penny in a cigar box, closing the lid, and shaking vigorously. The box is then opened and each coin observed as to whether it shows a head or a tail. The results are recorded as an ordered triple, with the dime indicated first, then the nickel, and finally the penny. Thus, (h, t, h) would show that the dime and penny came up heads, the nickel was a tail; (t, h, h) shows that the dime was a tail, while the nickel and penny were heads.

   (a) How many points are there in the sample space?
   (b) List all points corresponding to the event where at least two coins showed heads.
   (c) Let D be the event in which the dime showed heads in all cases. List all sample points corresponding to D.
   (d) If N is the event corresponding to the outcomes in which the nickel showed heads, list all points in the event N *and* D'.

9. Among a group of 40 talented musicians, 4 play the piano but no other instrument. A total of 20 play the piano, 20 play the cello, and 32 play the violin, and of these, 4 play all three. Sixteen play both the violin and the piano, and 12 play the violin and the cello, but not the piano.

   (a) How many play none of the three instruments?
   (b) Are any of the two events independent? Which ones?
   (c) What is the probability that a pianist plays the violin? The cello? All three?

10. The probability of Mr. Jones being selected for jury duty is 0.40. The probability that Mr. Smith will be selected is 0.25. The probability that neither will be selected is 0.45. What is the probability that *both* will be selected?

11. The probability that Richard and his brother will go swimming this afternoon is 0.44. The probability that Richard will go swimming but his brother will not is 0.16. What is the probability that Richard will not go swimming this afternoon?

12. Two urns are placed on a shelf. Urn A contains 3 white balls and 2 red balls; urn B contains 3 white balls and 5 red balls. A die is cast. If 3 or 6 dots appear on the top face of the die, urn A is chosen, otherwise urn B. A ball is then drawn from the chosen urn. What is the probability that a white ball will be drawn? A red ball? If a red ball is drawn, can you determine the probability that it came from urn A?

13. An airplane can successfully complete its flight if at least half the engines are in working order. Suppose that the probability of engine failure is 0.01 and engine failures are independent events. Calculate the probability that an airplane will successfully complete its flight if it has two engines; three engines; four engines.

14. Each machine in an assembly line has a probability of failure during a single day of 0.01. There is a "back-up" machine for each machine that can be used in case of failure, and each of these also has a 0.01 probability of failure. The assembly line will be shut down only if both a machine and its "back-up" fail. There are five machines on the assembly line. If all are independent, what is the probability that the assembly line will be shut down on any particular day?

15. Suppose that a utility tray is assembled from three pieces, a tray, a leg, and a stand. These pieces are manufactured separately and randomly assembled into a finished product. One in twenty of the trays, one in one hundred of the legs, and one in fifty of the stands are flawed. The finished products are examined and rejected if and only if at least two of the three pieces are flawed. What is the probability that a finished utility tray, selected at random, will be rejected?

16. Heredity is governed by laws of probability. In a certain species of guinea pig, 3/4 of each generation is short-haired, the remainder long-haired. An offspring of two long-haired guinea pigs is always long-haired, while 8/9 of the offspring of two short-haired guinea pigs are short-haired. If the parents are of two different types, 2/3 of the offspring are short-haired. If a long-haired guinea pig is selected at random, what is the probability that both its parents were short-haired?

17. An ancient problem is the following: A traveler is beset by ferocious natives who immediately decide to burn him at the stake. Upon learning that he is a math-

ematician, however, the chief offers him a chance to save his life. Two containers will be set in front of him, together with ten black berries and ten red berries. He must use all the berries and distribute them as he pleases between the two containers, putting at least one in each container. The chief will then select a container at random, then draw one berry. If it is red, the traveler will go free; if black, the barbeque will go on as planned. How should the traveler distribute the berries to maximize his chances? If he does so, what is the probability that he will be allowed to go free?

18. Closely akin to probabilities are *odds*. Roughly speaking, odds are the ratio of favorable probabilities to unfavorable. For example, if two of twelve eggs are spoiled, the odds in favor of selecting a spoiled egg at random is 1 to 5 (since there is one spoiled egg for five good ones). If an event has probability 0.15, then the odds in favor of its occurrence are 3 to 17 (0.15 to 0.85) and the odds against its occurrence are 17 to 3.

    (a) If the probability of rain on a given day is said to be 0.4, what are the odds in favor of rain? against rain?
    (b) On a roulette wheel, described in problem 4, section 4.2, what are the odds against getting a selected number? against red? against odd?
    (c) A fuse box has 20 fuses in it, of which 4 are 20 ampere size. If a man takes a fuse out at random, what are the odds against his getting a 20 ampere fuse?

19. It is important that a research study on the latest variety of influenza virus be completed before November 1 so that a suitable vaccine can be developed and marketed. A vital piece of equipment breaks down on September 1. Before it broke down, the project was on schedule, but now every day without the equipment decreases the probability of completing the project on time by 0.05. The supplier of the equipment makes deliveries every 5 days. He will be unable to deliver the equipment on the first delivery date, but feels that the probability it can be delivered in 10 days is 0.50. The probability that it can be delivered in 15 days, if not in 10, is 0.65 by his estimate and he is sure that it will be delivered within 20 days. What is the probability that the project will be completed on time?

20. Approximately 1/15 of the male population is color blind. Color blindness is a sex-linked characteristic which means that a female must have two color blindness genes in order to be color blind. The probability that a color blindness gene will be passed on by any given parent is thus 1/15. What is the probability that a female will be color blind? Thus what proportion of the female population is color blind? If the proportion of male to female in the population is 103 to 105, what proportion of the entire population is color blind?

21. It is highly unlikely, in problem 9, section 3.3, that the performance of Amalgamated Checkers stock would be independent of market fluctuations. Suppose that the probability is 0.7 that Amalgamated Checkers stock will rise if the market rises, 0.4 that it will rise if the market remains steady, and 0.1 that it will rise if the market falls. Using this information, and assuming $P(R) = 0.3$ and $P(T) = 0.4$, as in problem 9, determine $P(K)$.

Chapter IV

Probability Distributions

4.1 THE GENERAL PROBABILITY DISTRIBUTION

One of the central ideas in probability, and in statistics, is that of a **probability distribution**. Intuitively, we might guess that a probability distribution is something which describes how the probabilities in a sample space are distributed. We can best understand the concept of probability distribution by first understanding the concept of a random variable.

A **random variable** is defined in relation to a sample space simply by assigning numbers to all the possible outcomes or events in the sample space generated by a particular probability experiment.

Let us look at an example of a random variable. If you roll two dice and observe the number of dots which come up, you can examine the sample space generated by the experiment. To do this you list an array of ordered pairs, 36 in all, which gives all possible outcomes as points of the sample space. To understand that the ordered pairs, (1, 2) and (2, 1), represent different points, suppose that one die is red, the other green. You then look at the number of dots on the red die first and then the number of dots on the green die. The first number in each ordered pair is the number of dots on the red die; the second number, the number of dots on the green die. The sample space is given here. For clarity, the braces are omitted.

(1, 1)	(2, 1)	(3, 1)	(4, 1)	(5, 1)	(6, 1)
(1, 2)	(2, 2)	(3, 2)	(4, 2)	(5, 2)	(6, 2)
(1, 3)	(2, 3)	(3, 3)	(4, 3)	(5, 3)	(6, 3)
(1, 4)	(2, 4)	(3, 4)	(4, 4)	(5, 4)	(6, 4)
(1, 5)	(2, 5)	(3, 5)	(4, 5)	(5, 5)	(6, 5)
(1, 6)	(2, 6)	(3, 6)	(4, 6)	(5, 6)	(6, 6)

A random variable can be defined on this sample space by letting x, the variable, represent the number of dots showing on the two dice. Then x can take on all integral values from 2 through 12.

If a coin is tossed four times and we let the random variable be the number of heads that come up, the variable can take on all integral values from 0 through 4.

It can thus be seen that a random variable can be used to assign numbers to mutually exclusive events which make up a sample space. The rules of probability described in the last chapter will enable us to determine the probability of each of the events designated by a particular value of the random variable. The entire collection of values the random variable can take on, together with the associated probabilities for each of these values, define a **probability distribution**.

In this chapter we shall be concerned primarily with probability distributions derived from **discrete random variables**, that is, those which can take on only a countable number of values.

EXAMPLE 1 Two dice are tossed. If x represents the total number of dots on the two faces, determine a probability distribution for this random variable.

SOLUTION Referring to the sample space given previously, we find that the sample points group themselves into distinct events as shown below.

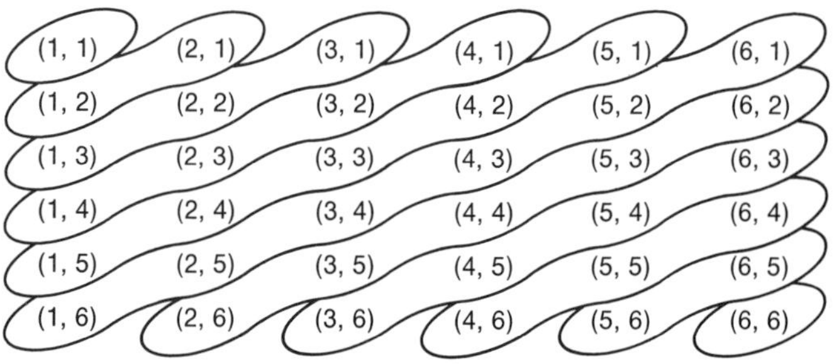

We see that there are 36 points in total, each equally likely, so the probability corresponding to each event is the number of points in the event divided by 36. As you can see from this illustration, the number of points which correspond to the value $x = 2$ is one, so $P(2) = 1/36$. We can summarize the results with a table such as the following.

x	$P(x)$
2	1/36
3	2/36
4	3/36
5	4/36
6	5/36
7	6/36
8	5/36
9	4/36
10	3/36
11	2/36
12	1/36

EXAMPLE 2 A jar contains 15 marbles; 1 is red, 2 are white, 3 yellow, 4 green, and 5 blue. Define a random variable for the experiment of drawing one marble from the jar and give a probability distribution for the variable.

SOLUTION We cannot define a random variable whose values are colors. We can define a sample space with associated probabilities, for this experiment, but there are advantages to using random variables which will become clear in succeeding sections. We could paint numbers on the marbles (or otherwise assign numbers). We will number the red marble "1," the white ones "2," the yellow ones "3," the green ones "4," and the blue ones "5." We can then give the following table for the probability distribution of the random variable x, where x represents the number on the marble drawn.

x	$P(x)$
1	1/15
2	2/15
3	3/15
4	4/15
5	5/15

It is also possible to define this same distribution by the rule $P(x) = x/15$ where $x = 1, 2, 3, 4, 5$. It is important to explicitly specify the set of values that x can take on. There is no place in this distribution for a "6" because the admission of any values other than 1, 2, 3, 4, or 5 would violate one of the rules of probability—that the probabilities of the events in a sample space must add up to one. If we exclude all values with zero probability we have a discrete probability distribution.

DISCRETE PROBABILITY DISTRIBUTION	For any discrete random variable x, the probability $P(x)$ is positive for each value of x, and the sum of the probabilities is one.

Discrete distributions can be graphed by various techniques, one of the most important of which is the **histogram** previously discussed in section 1.2. In the histogram, each value of the variable is represented by a unit of length on the base and a rectangle drawn above it to indicate its probability. The probability is equal to the area of the rectangle. In the marble example, we can represent the distribution as follows:

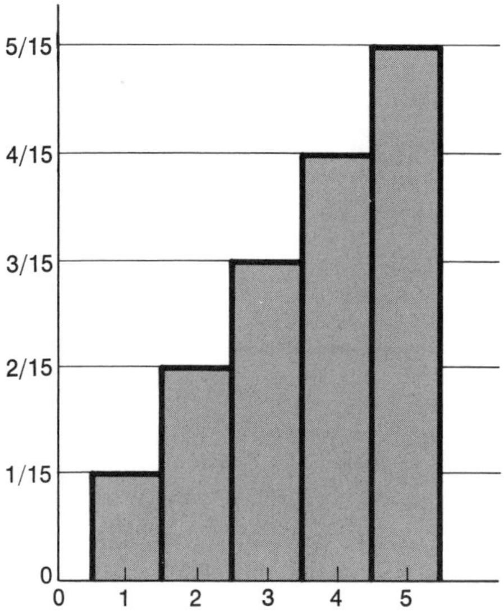

Note that the numbers 1, 2, 3, 4, 5 lie in the *center* of the rectangles. That is, we represent "1" by the interval from 0.5 to 1.5, "2" by the interval 1.5 to 2.5, and so forth. This is done because the probability of an event is equal to the area of the rectangle. For example, the probability of drawing a marble marked "2" is 1 · 2/15. Note also that the scales are different on the bottom and on the side. This is allowable since the relations between the probabilities remain clear. As a final example, let us construct a histogram for the two dice experiment:

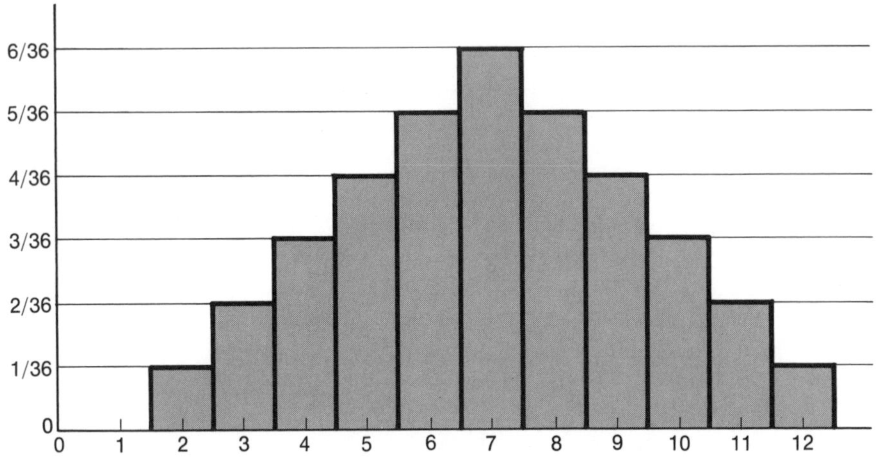

Again, each probability is represented by the area of the rectangle and each discrete value of the variable is represented by an interval; for example, "8" is represented by 7.5 to 8.5, so the rectangle has a base equal to one, height equal to 5/36, and an area equal to 5/36 square units. The sum of the area of all the rectangles representing the distribution adds up to one.

Problems

1. Determine whether or not each of the following is a probability distribution. Make a histogram for those which are.

   (a) $P(x) = x/25$ for $x = 1, 3, 5, 7, 9$
   (b) $P(3) = 5/7$, $P(5) = 1/7$, for $x = 3, 5$
   (c) $P(x) = 1/8$ for $x = 1, 2, 3, 4, 5, 6, 7, 8$
   (d)

x	$P(x)$
1	3/80
3	27/80
7	11/80
9	17/80
13	13/80
14	9/80

(e) $P(x) = x/10$ for $x = 0, 1, 2, 3, 4$
(f) $P(x) = 1/5$ for $x = 0, 1, 2, 3, 4$

2. A box contains four good light bulbs, two defective ones. A bulb is selected at random and put in the socket. If it works, it is left there; if defective, it is discarded and another bulb selected. The process is continued until a good bulb is found. Give the probability distribution for the number of bulbs which will be tried.
3. A box contains four good light bulbs, two defective ones, as in problem 2, but this time there are two sockets to fill. The same process is used as in problem 2. Give the probability distribution for the number of bulbs which will be tried.
4. A probability distribution is defined by $P(x) = (x^2 + 1)/n$ for $x = 1, 2, 3, 4$. Determine n. Make a histogram for the distribution.
5. Give the probability distribution for the number of samples which are analyzed on a particular day in problem 12, section 3.4.
6. A sociologist observes the social behavior of two young children over a period of several hours by checking their positions in a play area every five minutes. The play area is divided into four sections linearly and the children are observed to see which sections they are in. If they are playing in the same section, "0" is recorded; in adjacent sections, a "1." If they are separated by one section, a "2" is recorded, and a "3" if they are at opposite ends of the play yard. In this way information can be gathered concerning the socialization of this pair of children. Determine the probability distribution for the children's location (the variable) if the children behave totally independently and pay no attention to where the other child is located.
7. In a study of oxygen deprivation effects, four guinea pigs were born in a litter which had been subjected to a 0.15 reduction in the amount of oxygen to the placenta. Theoretically, one-third of the group would show effects of this treatment. Let x represent the number of guinea pigs showing effects of this deprivation. Complete the following probability distribution.

x	0	1	2	3	4
$P(x)$		32/81		8/81	

4.2 EXPECTED VALUE

A game of chance is being played at the carnival. A wheel of fortune is spun and if the number you bet on is the winner, you get your choice of any prize worth \$5.00. There are 25 equally likely numbers on the wheel, and it costs 25 cents to take a chance. Is it worth it? Since the probability of your winning with one number is 1/25, you have purchased 1/25 of \$5.00 or 20 cents for your quarter. That is, your expectation is \$5.00 · (1/25) or 20 cents. Another way to look at it is to note that to be *sure* of winning you must select 25 numbers. Since each of the 25 numbers has equal probability of winning, each one is worth the same amount (before the spin) of the \$5.00; that is, each is worth 20 cents. Unfortunately, the cost, 25 cents, is more than the value of the bet, 20 cents, so the investment is not a sound one. A third way to look at it is to note that if you bet a great many times, you would expect to win \$5.00 about 1/25 of the time so that you would win an *average* of about 20 cents per bet. Since you pay 25 cents per bet, on the *average* you would have a net loss of 5 cents per bet. This value per event is called the **expected value** of one event.

A similar analysis applies to a drawing in which you hold one of a hundred tickets. In general, your ticket is worth 1/100 the value of the prizes. If first prize is \$10.00, there are two second prizes of \$4.00, and ten third prizes of \$1.00 each, your probability of winning first prize is 1/100, of winning second prize is 2/100, and of winning a third prize is 10/100. That is, you have a probability of 1/100 of winning \$10.00, 2/100 of winning \$4.00, and 10/100 of winning \$1.00, so that the expected value of your ticket is (1/100)(\$10) + (2/100)(\$4) + (10/100)(\$1) or 28 cents. If the expected value is 28 cents and the ticket costs 50 cents your expected profit is −22 cents. Another way to figure your expected profit is to determine the net gain on each possible outcome. Your net gain on first prize is \$9.50, on second prize is \$3.50, on third prize \$.50, otherwise −50 cents. Thus, the expected value of your profit is 1/100(\$9.50) + 2/100(\$3.50) + 10/100(\$.50) + 87/100(−\$.50) or −22 cents. The *average* ticket, therefore, is worth a net loss of 22 cents. This is consistent with the fact that \$50 was collected and \$28 will be paid in prizes, so the purchasers lose \$22 on 100 tickets, an average of 22 cents per ticket. Note that a negative profit is a loss.

This leads to the following definition:

EXPECTED VALUE

If a variable has values $x_1, x_2, x_3, \ldots, x_n$ and probabilities $P(x_1), P(x_2), P(x_3), \ldots, P(x_n)$ respectively, then the expected value of the variable, x, is given by

$$E(x) = x_1 \cdot P(x_1) + x_2 \cdot P(x_2) + \cdots + x_n \cdot P(x_n)$$

Thus for any random variable, x, $E(x) = \sum x \cdot P(x)$ for each value of the random variable x.

EXAMPLE 1 An ordinary die is rolled once. What is the expected value of the number of dots on the uppermost face?

SOLUTION The random variable, x, has possible values 1, 2, 3, 4, 5, and 6. Each value has probability 1/6. The expected value is thus

$$E(x) = 1 \cdot (1/6) + 2 \cdot (1/6) + 3 \cdot (1/6) + 4 \cdot (1/6) + 5 \cdot (1/6) + 6 \cdot (1/6)$$
$$= 3.5$$

DISCUSSION The number obtained, 3.5, is impossible to obtain on one roll of a die and is not, in fact, a possible value of the random variable. The term "expected value," sometimes called "mathematical expectation" or simply "expectation," is not something we expect, but rather a weighted average—each of the possible outcomes weighted by its own probability.

EXAMPLE 2 A promoter has scheduled an outdoor event which will be cancelled in case of rain. If it rains he will lose $1,000. An insurance company agrees to cover his loss, if it rains, for a premium of $50. If the premium includes $10 administrative costs, and the premium is "fair" to both parties, what probability of rain does the insurance company expect?

SOLUTION The insurance company is wagering $960 against the promoter's $40 that it will not rain. In order for the bet to be "fair," the expected value for each party must be zero; that is, expected profit must equal expected loss. If we denote the probability of rain by p and the probability that it doesn't rain by $(1 - p)$, then the insurance company's expected profit $(40)(1 - p)$ must equal its expected loss $960p$. Then $960p = 40(1 - p)$; that is, $1{,}000p = 40$, so $p = 40/1{,}000 = 0.04$. Thus the probability of rain is 0.04.

EXAMPLE 3 A coin is tossed until a head appears, or three times, whichever comes first. Determine the expected number of tosses and the expected number of tails on one experiment.

SOLUTION Let us set up a sample space for the experiment.

$$S = \{(H), (T, H), (T, T, H), (T, T, T)\}$$

We list each point below, with its probability. The probabilities are obtained by noting that outcomes of successive tosses of the coin are independent. Note that the sample points are *not all* equally likely.

Outcome	H	T, H	T, T, H	T, T, T
Probability	1/2	1/4	1/8	1/8

Now let x represent the number of tosses on one experiment. The values that x can take on are 1, 2, and 3. Only one point corresponds to $x = 1$ and its probability is 1/2, so $P(1) = 1/2$. One point corresponds to $x = 2$ and its probability is 1/4, so $P(2) = 1/4$. Two points correspond to $x = 3$, namely (T, T, H) and (T, T, T), and each has probability 1/8, so $P(3) = 1/4$. These results are summarized in the following table.

x	$P(x)$
1	1/2
2	1/4
3	1/4

Then $E(x) = 1 \cdot (1/2) + 2 \cdot (1/4) + 3 \cdot (1/4) = 7/4$ or 1.75. This would be the mean number of tosses which would be expected to occur if the experiment were repeated a large number of times.

Now suppose we represent the number of tails in one experiment by the random variable y (to distinguish it from x, the number of tosses on the same experiment). Then y can have values 0, 1, 2, or 3. Reference to the table for the sample space gives the following probability distribution for y.

y	$P(y)$
0	1/2
1	1/4
2	1/8
3	1/8

We then have $E(y) = 0 \cdot (1/2) + 1 \cdot (1/4) + 2 \cdot (1/8) + 3 \cdot (1/8) = 7/8$.

Problems

1. In a lottery, a first prize of $10,000 and two second prizes of $5,000 each will be awarded. What is the "value" of each ticket if 30,000 tickets have been sold?

2. The probabilities that 0, 1, 2, 3, or 4 accidents will occur in the Holland Tunnel between 7:30 and 10:30 on a Monday morning are, respectively, 0.92, 0.04, 0.02, 0.01, and 0.01. How many accidents would be expected during this period on a particular Monday morning? During 100 such periods?
3. Mr. Jones is selling his business. A realtor promises that the probability he will make $20,000 is 0.20, that he will make $12,000 is 0.35, that he will make $4,000 is 0.10, that he will break even is 0.15. He concedes, however, that the probability Mr. Jones will lose $6,000 is 0.15, and that there is even a possibility he will lose $12,000. He claims, however, that there are no other possibilities. If he is correct, what is Mr. Jones expected profit on this sale?
4. On a roulette wheel, there are 38 slots, equally spaced, numbered 00, 0, 1, 2, 3, . . . , 36. For a $1.00 bet, if your number wins you receive $35.00 (in addition to your $1.00 being returned); otherwise, you lose your $1.00. Determine the expected value, on one bet, of a gambler who bets on a number.
5. On the roulette wheel described in problem 4, 18 slots are red, 18 slots are black, 00 and 0 are green. A person can, if he desires, place a bet on red or black. For a $1.00 bet, if he wins, he receives an additional $1.00; if he loses, he is out $1.00. Calculate the expected value, on one bet, of a gambler who bets one color? (The same applies to odd and even, 00 and 0 are considered neither.)
6. If one team is approximately twice as strong as the other in a World Series, the probabilities that the series will end in 4, 5, 6, or 7 games are, respectively, 0.21, 0.30, 0.27, and 0.22. What is the expected number of games in such a series? If Mr. Black has tickets only to the fourth and fifth games, what is the expected number of games he will see?
7. An urn contains 6 white and 3 red marbles. If 4 marbles are drawn at random without replacement, determine the expected number of red marbles by finding the probability distribution and applying the rule for expected value. Do you note a simpler way?
8. A game is played in which we may draw two cards, at random, without replacement, from an ordinary deck. For each black card that we draw, we receive $5.00. How much should we pay for the privilege of playing the game, if the game is to be fair?
9. A psychologist wants to send out attitude questionnaires to a random sample of working adults. She sends questionnaires to a factory which employs 115 men and 100 women and to a bank which employs 65 women and 50 men. If she receives 71 questionnaires from the factory and 87 from the bank, what would be the expected number of men and women respondees, if men and women are equally likely to respond?
10. A biologist collects 20 specimens from each of 5 colonies of *Anopheles* mosquitoes. The proportions of those bearing a recessive gene in each colony under study are 0.1, 0.3, 0.2, 0.1, 0.3. What is the expected number of mosquitoes in the sample bearing the gene, if the collection procedures assure random selection?

11. A population containing only high school graduates has 28% of the population who completed grade 12, 23% who completed grade 13, 36% who completed grade 14, and the remainder who completed grade 15. Fifty persons are selected at random from the population. If we add together the number of grades each has finished, what is the expected total?

4.3 MEAN AND VARIANCE OF A PROBABILITY DISTRIBUTION

Most of us are familiar with the concept of **average**. The arithmetic average of a set of values is equal to their sum divided by the number of values. It is, in fact, the expected value of the set since each of n values has a probability of $1/n$. A more precise term for this value is **mean**. In terms of a probability distribution, the concept of the mean still holds. For instance, if we toss a coin four times, the probabilities of getting 0, 1, 2, 3, or 4 heads, respectively, are 1/16, 4/16, 6/16, 4/16, and 1/16. From section 4.2, the expected number of heads is $(1/16) \cdot 0 + (4/16) \cdot 1 + (6/16) \cdot 2 + (4/16) \cdot 3 + (1/16) \cdot 4$ or exactly 2. This value is also the mean of the distribution. The mean of a probability distribution is usually symbolized by the Greek letter $\boldsymbol{\mu}$ (mu).

MEAN OF A PROBABILITY DISTRIBUTION	For any discrete probability distribution, $\mu = E(x)$; that is, $\mu = \sum x \cdot P(x)$ for each value of the random variable x.

EXAMPLE 1 Find the mean value of one toss of a single die.

SOLUTION Each face has an equal probability (1/6) so

$$\mu = 1 \cdot (1/6) + 2 \cdot (1/6) + 3 \cdot (1/6) + 4 \cdot (1/6) + 5 \cdot (1/6) + 6 \cdot (1/6)$$

$$= \frac{1 + 2 + 3 + 4 + 5 + 6}{6} = \frac{21}{6} = 3.5.$$

EXAMPLE 2 A die is rolled three times, or until a 6 appears, whichever occurs first. Determine the mean of the probability distribution for the number of times rolled.

SOLUTION Let x represent the random variable representing the number of times the die is rolled. We must first determine the probability distribution for x. The possible outcomes are as follows. A 6 can be obtained on the first roll, in which case the experiment ceases. The probability of this occurring is 1/6. If a 6 is not obtained (probability 5/6), the die is rolled again. If a 6 is obtained (probability 1/6 on the second roll), the experiment ceases. The probability that this will occur is $(5/6) \cdot (1/6)$ or 5/36. The remaining two possibilities both assume that neither the first nor second roll was a 6. The probability of that happening is $(5/6) \cdot (5/6)$ or 25/36. If desired, you may determine the separate probabilities, but each outcome yields three tosses. If the first two are not 6 and the third is, the probability of that occurrence is $(5/6) \cdot (5/6) \cdot (1/6)$ or 25/216. The probability that none of the three is a 6 is $(5/6) \cdot (5/6) \cdot (5/6)$ or 125/216. Both of these points correspond to three rolls, so $P(3) = 25/216 + 125/216 = 150/216 = 25/36$. Thus we have the following probability distribution for x.

x	$P(x)$
1	1/6
2	5/36
3	25/36

Then $\mu = 1 \cdot (1/6) + 2 \cdot (5/36) + 3 \cdot (25/36) = 91/36$ or $2\ 19/36$.

EXAMPLE 3 Find the mean of the following probability distribution:

x	$P(x)$
1	1/9
2	1/18
3	1/3
4	5/18
5	1/6
6	1/18

SOLUTION

$$\begin{aligned}\mu &= 1 \cdot (1/9) + 2 \cdot (1/18) + 3 \cdot (1/3) + 4 \cdot (5/18) + 5 \cdot (1/6) + 6 \cdot (1/18)\\ &= 1/9 + 1/9 + 1 + 10/9 + 5/6 + 1/3\\ &= 3.5.\end{aligned}$$

In both example 1 and example 3, the distribution had integral values from 1 through 6 and a mean of 3.5. Yet these distributions are quite different. Histograms for the two distributions show this:

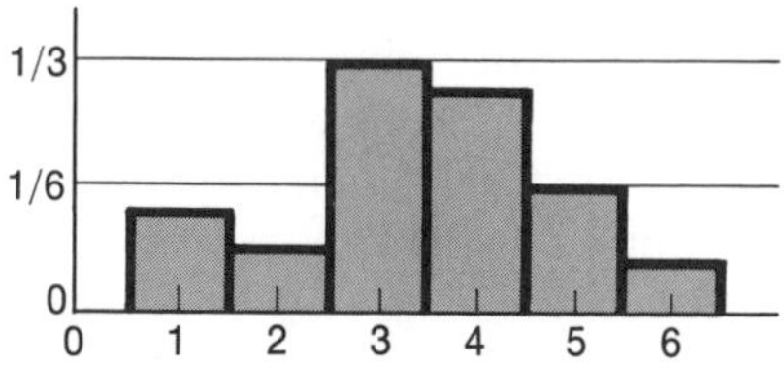

The second of these distributions is "heavier" in the vicinity of the mean. In fact, 11/18 of the distribution lies within one unit of the mean in contrast to only 1/3 (or 6/18) of the first distribution. This cannot be seen from the mean, however, or from the set of values. A number of different measures have been used to record the actual dispersion of a distribution, but the most effective are two related measures, the variance and the standard deviation. What is needed is a measure that will differentiate between distributions clustered closely about the mean and those more widely scattered. The mean squared deviation, that is, the mean of the squares of the distance of each score from the mean, has the merit that scores widely different from the mean carry proportionately much more weight than those near it. This measure is called the **variance** and is symbolized by $\boldsymbol{\sigma}^2$, where σ is the Greek letter sigma (lowercase).

The definition of the variance which will apply to any probability distribution is the **expected value** of the squared deviations from the mean, $E(x - \mu)^2$. For a discrete probability distribution, the special rule is given here.

VARIANCE OF A PROBABILITY DISTRIBUTION	For any discrete probability distribution, $\sigma^2 = E(x - \mu)^2$; that is, $\sigma^2 = \sum (x - \mu)^2 P(x)$ for each value of the random variable x.

EXAMPLE 4 Find the variance of the probability distributions given in examples 1 and 3 above.

SOLUTION A table with appropriate headings is often useful. First, of course, we must find the mean, then find $(x - \mu)$ for each x and square it, multiply by the probability, and then sum it.

x	$P(x)$	$x \cdot P(x)$	$x - \mu$	$(x - \mu)^2$	$(x - \mu)^2 \cdot P(x)$
1	1/6	1/6	−5/2	25/4	25/24
2	1/6	2/6	−3/2	9/4	9/24
3	1/6	3/6	−1/2	1/4	1/24
4	1/6	4/6	1/2	1/4	1/24
5	1/6	5/6	3/2	9/4	9/24
6	1/6	6/6	5/2	25/4	25/24
		$\mu = 3.5$ (or 7/2)			$\sigma^2 = 70/24 \doteq 2.92$

Thus, $\sigma^2 \doteq 2.92$ for the distribution of example 1.

In general, the above procedure is not the most efficient one to use in finding the variance or standard deviation. Instead it can be shown that $E(x - \mu)^2 = E(x^2) - \mu^2$. For discrete probability distributions, the formula becomes

$$\sigma^2 = \left(\sum x^2 P(x)\right) - \mu^2$$

This says to square each value, multiply by its probability, then sum it. This sum must be corrected by subtracting the square of the mean.

We use this method to find the variance of the probability distribution of example 3.

x	$P(x)$	$x \cdot P(x)$	x^2	$x^2 \cdot P(x)$
1	1/9	1/9	1	1/9
2	1/18	1/9	4	2/9
3	1/3	1	9	3
4	5/18	10/9	16	40/9
5	1/6	5/6	25	25/6
6	1/18	1/3	36	2
		$\mu = 3.5$ (or 7/2)		251/18

Then $\sigma^2 = \frac{251}{18} - \left(\frac{7}{2}\right)^2 = \frac{502}{36} - \frac{441}{36} = \frac{61}{36} = 1\frac{25}{36} \doteq 1.69$. Thus, $\sigma^2 \doteq 1.69$. Note that the variances are consistent with the fact that *more* of the area in the first histogram is away from the mean than in the second. Thus, if two distributions have different variances, a smaller variance indicates that more of the distribution tends toward the center (or mean) of the distribution.

Useful as the variance is, it has one major drawback—it is not in the same units as the values of the distribution. Its units are the *square* of the value units. Thus, it cannot be plotted on the graph (or histogram) of the distribution. The simple procedure of taking the square root of the variance gives a number which can be plotted. This measure is called the standard deviation and is symbolized by $\boldsymbol{\sigma}$. In terms of the distribution itself, it is the square root of the mean squared deviation.

STANDARD DEVIATION OF A PROBABILITY DISTRIBUTION

$\sigma = \sqrt{\sum (x - \mu)^2 P(x)}$ for each x in the distribution.

The variance and standard deviation both play roles in the development of statistics, and will be encountered often hereafter.

EXAMPLE 5 Find the standard deviation of the distributions given in example 4.

SOLUTION Since the variances are about 2.92 and 1.69, respectively, the standard deviations are $\sqrt{2.92}$ and $\sqrt{1.69}$. Referring to Table 1, we see that $\sqrt{2.92} \doteq 1.71$ and $\sqrt{1.69} = 1.30$. These values, then, are the required standard deviations.

Problems

1. A probability distribution has $P(-1) = 0.3$, $P(0) = 0.2$, $P(1) = 0.3$, and $P(2) = 0.2$. Determine its mean, variance, and standard deviation.
2. A probability distribution has $P(9) = 0.3$, $P(10) = 0.2$, $P(11) = 0.3$, and $P(12) = 0.2$. Determine its mean, variance, and standard deviation. Compare with problem 1.
3. *If $P_1(x) = P_2(x + c)$ with c some constant where P_1 and P_2 define two different probability distributions, with means μ_1 and μ_2 and standard deviations σ_1 and σ_2, respectively, show that $\mu_2 = \mu_1 + c$, and $\sigma_1 = \sigma_2$.

* Fairly difficult.

4. Suppose that the probabilities that a world series will last 4, 5, 6, or 7 games, are respectively, 0.1, 0.2, 0.4, and 0.3. Determine the mean, variance, and standard deviation of the distribution.

5. A probability distribution has $P(-3) = 0.3$, $P(0) = 0.2$, $P(3) = 0.3$, and $P(6) = 0.2$. Determine its mean, variance, and standard deviation. Compare with problem 1.

6. * If $P_1(x) = P_2(cx)$ with c some constant, where P_1 and P_2 define two different probability distributions, with means μ_1 and μ_2 and standard deviations σ_1, σ_2 respectively, show that $\mu_2 = c\mu_1$ and $\sigma_2 = |c|\sigma_1$. NOTE: $|c|$ means "absolute value" of c and can be defined as the positive square root of c^2 if c^2 is not negative.

7. Two dice are tossed and the sum of the top faces recorded (see section 4.1). Calculate the mean and variance of this probability distribution.

8. Calculate the mean and standard deviation of each distribution:

(a)

x	13	17	21	24	25
$P(x)$	0.2	0.4	0.2	0.1	0.1

(b)

x	107	114	121	128	135
$P(x)$	0.22	0.27	0.16	0.21	0.14

(c)

x	5	11	13	18	22	27	33	34	39
$P(x)$	0.20	0.14	0.23	0.11	0.13	0.09	0.05	0.04	0.01

(d)

x	1,000	1,100	1,200	1,300
$P(x)$	0.1	0.2	0.3	0.4

9. Determine the mean and variance of the probability distribution for the number of samples analyzed on a particular day given by problem 4, section 4.1.

10. Determine the mean and standard deviation of the probability distribution for the behavior of the children in problem 6, section 4.1.

11. Find the mean and variance of the probability distribution for the number of guinea pigs showing the effects of deprivation in problem 7, section 4.1.

4.4 SUMMARY

In this chapter we have studied characteristics of probability densities and particularly of discrete probability distributions. For a discrete probability distribution the probability, $P(x)$, is positive for each value of x, and the sum of the probabilities is one.

The most important characteristics of a probability distribution are the mean and standard deviation. These special formulas are given here again.

MEAN AND STANDARD DEVIATION OF A PROBABILITY DISTRIBUTION

$$\mu = E(x) = \sum x \cdot P(x)$$

$$\sigma = \sqrt{E(x-\mu)^2} = \sqrt{\left(\sum x^2 P(x)\right) - \mu^2}$$

Problems

1. Find the mean and standard deviation of the probability distribution defined by the following table:

x	$P(x)$
1	3/80
3	27/80
7	11/80
9	17/80
14	13/80
16	9/80

2. A probability distribution is defined by

$$P(x) = \frac{x^2 - 1}{n} \quad \text{for} \quad x = 2, 3, 4, 5$$

   (a) Determine n.
   (b) Make a histogram for the distribution.
   (c) Calculate the mean of the distribution.
   (d) Calculate the standard deviation of the distribution.

3. Eight pints of blood are stored in the hospital laboratory. It is known that exactly three of these are type O, but it is not known which three. Two pints of Type O blood are needed. One pint at a time is removed from storage and typed. If it is Type O, it is used; if it is not, it is set off to one side and the next pint tested.

(a) Make a probability distribution for the number of pints it will be necessary to type in order to obtain two of Type O.
(b) Determine the mean and standard deviation for this distribution.

4. A probability distribution is defined by the following table:

Class	*Probability*
3.5–6.5	0.08
6.5–9.5	0.17
9.5–12.5	0.31
12.5–15.5	0.19
15.5–18.5	0.12
18.5–21.5	0.06
21.5–24.5	0.04
24.5–27.5	0.02
27.5–30.5	0.01

(a) Draw a histogram for the distribution.
(b) Draw a smooth curve approximating the histogram.
(c) Using class marks, calculate the mean and standard deviation for the distribution.

5. A man wishes to buy a one-year term insurance policy on his life for $50,000. The insurance company calculates the probability that he will die during this period to be 0.0042. If the premium includes $30 administrative costs, what will be the premium for the policy?

6. A coin is tossed four times or until a head appears, whichever comes first. Calculate the expected number of tosses on one experiment; calculate the expected number of tails.

7. A radiologist is interested in saving her company money. She needs a sample of radioactive materials and goes to a cut-rate supply house to buy them. The proprieter tells her that he has four samples on hand, but two of them have been around long enough to be useless for the radiologist's purposes. He agrees to let her have two of them at only 80% of the price for one. If she selects two of them at random, determine the probability distribution for the number of samples which would be satisfactory. Give the mean and variance for the distribution. Do you think she has made a good bargain?

8. In a certain part of a bay, the concentration of plankton varies according to time of day. At 6 A.M. we would expect 600 per cc. By noon the concentration has increased to 1,000 per cc. At 6 P.M. the concentration is 1,200 per cc. At midnight the concentration has dropped to 800 per cc. A total of 100 cc of bay water contains 25 cc taken at 6 A.M., 35 cc taken at noon, 30 cc taken at 6 P.M., and 10 cc taken at midnight. What would be the expected concentration of plankton in the 100 cc?

PART III

THE BINOMIAL AND NORMAL DISTRIBUTIONS

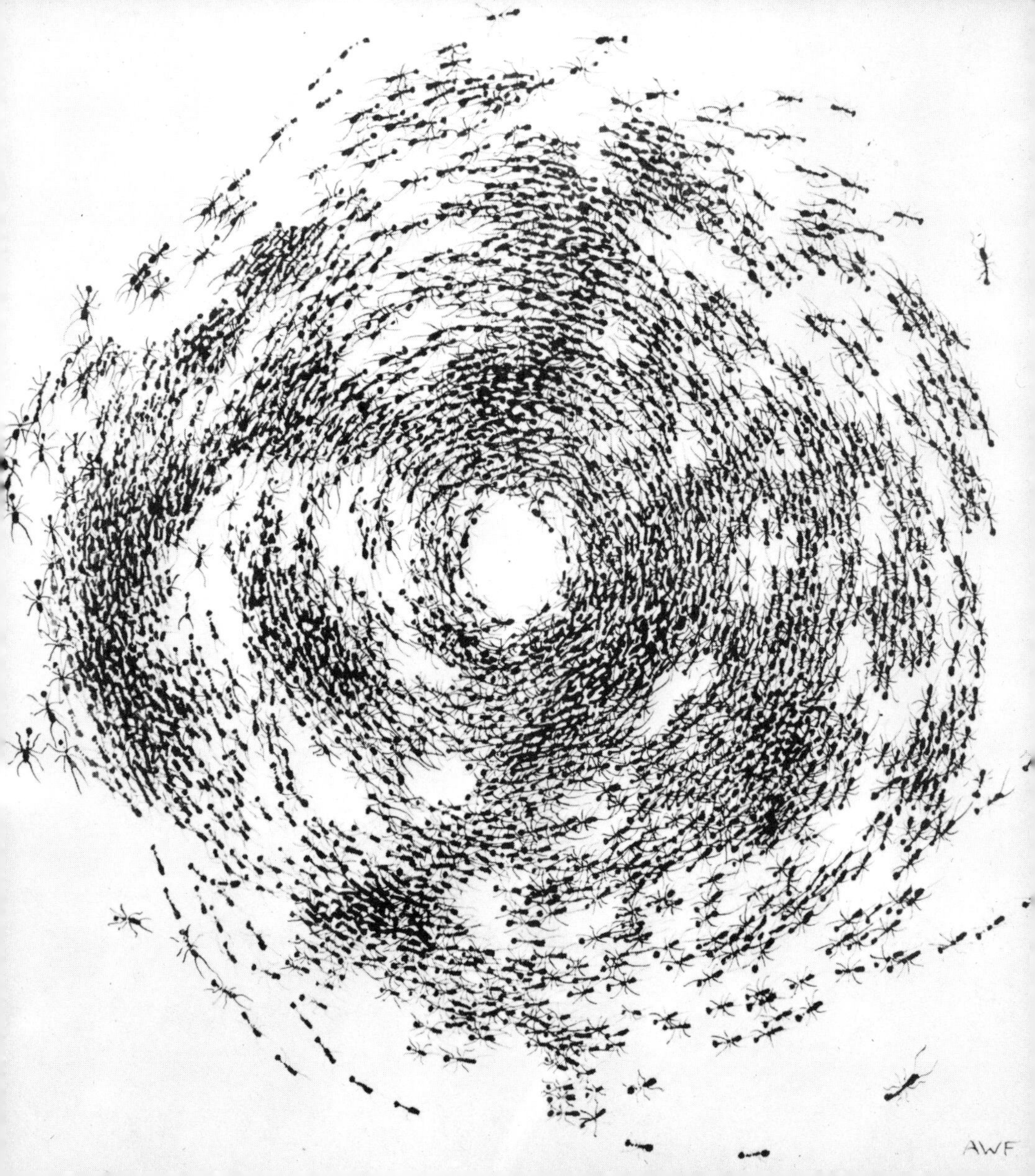

Part III Photo: Courtesy of the American Museum of Natural History

Chapter V

The Binomial Distribution

5.1 THE BINOMIAL DISTRIBUTION

On a certain television game show, a contestant has to choose one of twenty-five boxes. One box contains a check for a large sum of money. If the contestant selects the box at random, he has exactly one chance in twenty-five of selecting the prize box. In an entire week of games, what is the probability that no one will win the prize or that all the contestants will win the prize or that two prizes will be won?

This is an example of a **binomial experiment**.

BINOMIAL EXPERIMENT	A **binomial experiment** is an experiment in which 1. there are exactly two possible outcomes; 2. probabilities remain constant from trial to trial; 3. the trials are independent; 4. the number of trials is fixed.

The **probability of a success** is usually symbolized by p in a binomial experiment, and the **probability of failure** is symbolized by $1 - p$ or by q, with the understanding that $p + q = 1$. In the above example, if success is choosing the prize box, then $p = 1/25$. The probability of failure is $1 - p$ or 24/25. The probability of there being five winners in a week of games is (1/25)(1/25)(1/25)(1/25)(1/25) or $(1/25)^5$. The probability that no one will win is $(24/25)^5$. How do we calculate the probability that there will be two winners in one week? If there were two winners during one week, there would be two successes and three failures. You might conclude that the probability of this happening would be (1/25)(1/25)(24/25)(24/25)(24/25), but this would be wrong since it assumes that the outcomes happen in a particular order. The two successes could come on Monday and Tuesday, Wednesday and Friday, or on any two days of the week. The number of different combinations of two days selected from the five days of the week is $\binom{5}{2}$, or ten. Each particular combination has the same probability as that mentioned before: $(1/25)^2(24/25)^3$. Since there are ten of these combinations, the probability of two successes is $10(1/25)^2(24/25)^3$ or about 0.014.

In general, if a binomial experiment is repeated n times, the probability that x successes will occur in a particular order is $p^x(1 - p)^{n-x}$, while there are $\binom{n}{x}$ possible orders. This leads to the following rule.

BINOMIAL PROBABILITY	If a binomial experiment is repeated n times and the probability of "success" on one trial is p, the probability of exactly x successes is given by $$P(x) = \binom{n}{x} p^x(1 - p)^{n-x}$$

If a binomial experiment is repeated a number of times and the probabilities that one outcome will occur for each possible number of times are calculated, the resulting distribution of probabilities is called a **binomial distribution**. The values of x for a binomial distribution are 0, 1, 2, 3, ..., n where n is the number of times it is repeated and the probabilities are given by the above formula.

In order for the binomial probability distribution to apply, it is necessary that the probability p be constant from trial to trial and that the trials be independent. If, for instance, one white marble were mixed with nine black ones and a drawing was made, the probability of obtaining a white marble on one trial would be 0.1. On the second trial, however, it would make a difference whether the marble drawn were replaced. If the first marble drawn were black and not replaced, the probability of drawing a white marble on the second draw would be 1/9—the conditional probability of white on the second draw, given black on the first. If the first marble drawn were white, the probability of drawing a white marble on the second trial, if it were not replaced, would be zero. In neither case is the probability 0.1 as it was on the first trial. Thus the binomial probability distribution can be used in cases where successive trials are truly independent, as in tossing coins or dice or where conditions are restored exactly at the end of each trial to their state at the beginning. This can be accomplished by replacing everything drawn.

In many experiments this is impossible. In such cases the hypergeometric distribution may be used (see section 5.5).

EXAMPLE 1 In a certain experiment five animals are given a drug. One-fifth of all animals given the drug develop certain symptoms. Determine the probability distribution for this experiment.

SOLUTION The number of animals that can develop symptoms is 0, 1, 2, 3, 4, or 5. The outcome is binomial since an animal either develops symptoms or does not. Thus, the experiment is binomial and the results may be tabulated as follows:

$$n = 5, \quad p = 0.2 \quad 1 - p = 0.8$$

x	$P(x)$
0	$\binom{5}{0} \cdot (0.2)^0 \cdot (0.8)^5 = 0.32768$
1	$\binom{5}{1} \cdot (0.2)^1 \cdot (0.8)^4 = 0.40960$
2	$\binom{5}{2} \cdot (0.2)^2 \cdot (0.8)^3 = 0.20480$
3	$\binom{5}{3} \cdot (0.2)^3 \cdot (0.8)^2 = 0.05120$
4	$\binom{5}{4} \cdot (0.2)^4 \cdot (0.8)^1 = 0.00640$
5	$\binom{5}{5} \cdot (0.2)^5 \cdot (0.8)^0 = 0.00032$

DISCUSSION Note that since five animals are given the drug and one-fifth of all animals given the drug develop certain symptoms, it is tempting to conclude that exactly one animal $[(5) \cdot (1/5)]$ will develop symptoms. Reasoning such as this is similar to misapplications of the law of large numbers which were discussed in Chapter 3. This reasoning has a certain kind of merit, but a glance at the table also shows that although $P(1)$ is the greatest value, it is still less than 0.5 and far from certain.

In each case, the binomial distribution is a probability distribution, so the mean and standard deviation can be calculated. In the case of the die example, the calculations look like this:

x	$P(x)$	$x \cdot P(x)$	x^2	$x^2 \cdot P(x)$
0	16/81	0	0	0
1	32/81	32/81	1	32/81
2	24/81	48/81	4	96/81
3	8/81	24/81	9	72/81
4	1/81	4/81	16	16/81
		$\mu = 108/81 = 4/3$		$\sum x^2P(x) = 216/81 = 8/3$

Then $\sigma^2 = 8/3 - (4/3)^2 = 24/9 - 16/9 = 8/9$, so $\sigma = (1/3)\sqrt{8}$ or about 0.943. Now note a very interesting phenomenon. The number of trials was 4, the probability of success was 1/3, and $4(1/3) = 4/3 = \mu$. Further, $4(1/3)(2/3) = 8/9 = \sigma^2$. If this were a valid formula, the work for a binomial distribution would certainly be reduced. In example 1, for instance, we would have $\mu = 5(0.2) = 1$ and $\sigma^2 = 5(0.2)(0.8) = 0.8$. Using the usual calculations we have:

x	$P(x)$	$x \cdot P(x)$	x^2	$x^2 \cdot P(x)$	
0	0.32768	0	0	0	
1	0.40960	0.40960	1	0.40960	
2	0.20480	0.40960	4	0.81920	
3	0.05120	0.15360	9	0.46080	$\sigma^2 = 1.8 - (1)^2 = 0.8000$
4	0.00640	0.02560	16	0.10240	
5	0.00032	0.00160	25	0.00800	
		$\mu = 1.00000$		1.80000	

Surprisingly enough, the results check out. These formulas can be proved, but they will be stated here without proof.

MEAN AND VARIANCE OF A BINOMIAL DISTRIBUTION

If a binomial experiment is repeated n times and the probability of "success" on one trial is p, the mean of the distribution is given by $\mu = np$, and the variance is given by $\sigma^2 = np(1 - p)$.

Thus, the standard deviation is given by $\sigma = \sqrt{np(1-p)}$, and these values are quite simple to compute.

EXAMPLE 2 A coin is tossed 6 times. What is the mean and standard deviation of the distribution of heads?

SOLUTION Here $n = 6$, $p = 1/2$, $1 - p = 1/2$, so $\mu = 6(1/2) = 3$, and $\sigma^2 = 6(1/2)(1/2) = 3/2 = 1.5$, so $\sigma = \sqrt{1.5} \doteq 1.22$. We check this by calculating the distribution.

x	$P(x)$
0	$\binom{6}{0}\left(\frac{1}{2}\right)^0\left(\frac{1}{2}\right)^6 = \left(\frac{1}{2}\right)^6 = 1/64$
1	$\binom{6}{1}\left(\frac{1}{2}\right)^1\left(\frac{1}{2}\right)^5 = 6(1/2)^6 = 6/64$
2	$\binom{6}{2}\left(\frac{1}{2}\right)^2\left(\frac{1}{2}\right)^4 = 15(1/2)^6 = 15/64$
3	$\binom{6}{3}\left(\frac{1}{2}\right)^3\left(\frac{1}{2}\right)^3 = 20(1/2)^6 = 20/64$
4	$\binom{6}{4}\left(\frac{1}{2}\right)^4\left(\frac{1}{2}\right)^2 = 15(1/2)^6 = 15/64$
5	$\binom{6}{5}\left(\frac{1}{2}\right)^5\left(\frac{1}{2}\right)^1 = 6(1/2)^6 = 6/64$
6	$\binom{6}{6}\left(\frac{1}{2}\right)^6\left(\frac{1}{2}\right)^0 = \left(\frac{1}{2}\right)^6 = 1/64$

x	$P(x)$	$x \cdot P(x)$	x^2	$x^2 \cdot P(x)$	
0	1/64	0	0	0	
1	6/64	6/64	1	6/64	
2	15/64	30/64	4	60/64	
3	20/64	60/64	9	180/64	$\sigma^2 = 10.5 - (3)^2$
4	15/64	60/64	16	240/64	$= 1.5$
5	6/64	30/64	25	150/64	
6	1/64	6/64	36	36/64	
		$\mu = 192/64 = 3$		$672/64 = 10.5$	

EXAMPLE 3 On a multiple choice test with 5 answers (1 correct, 4 incorrect) to each question, a student randomly selects an answer to 225 questions. Determine the mean and standard deviation of this distribution.

SOLUTION Here $n = 225$, $p = 1/5$, $1 - p = 4/5$, so $\mu = 225(1/5) = 45$, $\sigma = \sqrt{225(1/5)(4/5)} = \sqrt{36} = 6$. Checking this by calculating the distribution is not a rewarding task, but the reader may do so if he desires. Remember, x can take on all values from 0 to 225 and each probability is given by

$$\binom{225}{x}\left(\frac{1}{5}\right)^{x}\left(\frac{4}{5}\right)^{225-x}$$

Methods for approximating individual probabilities will be given in sections 5.3 and 6.4.

NOTE If a sample is drawn without replacement from a finite population, the theoretical model to be used is the **hypergeometric distribution**, which is discussed in section 5.5. If the sample is relatively small with respect to the population, however, it is possible to use the binomial distribution as a good approximation to the hypergeometric distribution. If one person is selected at random from a group of which 1/5 have visual problems, the probability that person will have visual problems is 1/5, or 0.200. If we replace the person selected and select a second person, the probability that that person will have visual problems is also 0.200. If, as is usually the case, the first person is not replaced before the second is selected, the acts are not independent, so the binomial model is not appropriate. In a group of ten people, for example, if the first person has the visual defects, then the probability that the second one does also is 1/9; if the first person does not have the visual defect, the probability that the second one does is 2/9. Obviously the probabilities are quite different. Examine the effect of group size, however, on the probability of the second person having the visual defect. We let $S \mid F$ stand for the event "second has visual defect if the first one selected does" and $S \mid F'$ denote the event "second has visual defect if the first one selected does not."

N = *group size*	$P(S\|F)$	$P(S\|F')$
10	0.1111	0.2222
50	0.1837	0.2041
100	0.1919	0.2020
500	0.1984	0.2004
1,000	0.1992	0.2002
5,000	0.1998	0.20004
10,000	0.1999	0.20002

We can see that for a sufficiently large population size, the probabilities do not change much. For a group of 10,000, for example, if a sample of 100 were taken, the probability that a person has the visual defect if the first 99 did not is raised to 0.2020, which is not much of a change, but may be important. If the sample is relatively small, however, and the population is large, minor discrepancies are generally ignored and the binomial model can be used.

Problems

In problems 1–4 determine the probability distribution and the mean and variance of each distribution:

1. Two dice are thrown. A success is 7 or 11. The experiment is repeated 4 times.
2. An urn contains 4 red and 6 blue balls. A ball is drawn at random and replaced before the next drawing. Five drawings are made. Examine the probability distribution for the number of blue balls drawn in the 5 drawings.
3. Two baseball teams play a 3 game series. The probability that the home team will win a particular game is 0.55. Assume that the outcomes are independent.
4. A coin is tossed 10 times and a head is regarded as a success.
5. Calculate the mean and standard deviation of the probability distribution of the number of successes for an experiment in which the cards are shuffled and the top and bottom cards examined to see if at least one of them is an ace; the experiment is repeated 200 times.
6. The probability that a particular seed of red fescue grass will germinate is 0.75. A man buys a bag which contains 10,000 seeds. There is a probability distribution for the number of seeds (from 0 to 10,000) which will germinate. Calculate the mean and standard deviation of the probability distribution.

c7. If the sex ratio of humans at birth is 100 females to 105 males, what are the expected proportions of 3, 2, and 1 male in samples of 5 newborns? What is the mean number of males for the distribution?

8. One-third of the subjects studied have taken a hallucinogen at least once. If the population studied is large enough to overlook the fact that sampling is done without replacement so that the binomial model can be used, what is the probability that exactly 1/3 of a group of 6 subjects chosen at random from the population studied will have taken a hallucinogen?
9. In a medical school, standard X rays are kept for use by students. It is hypothesized that second-year students have a probability of 4/5 of diagnosing the X ray correctly if there is a pathological condition, but only 3/5 of diagnosing correctly if the X ray shows normal conditions. Three students examine an X ray and draw their conclusions. What is the probability that exactly one student will make the correct diagnosis if (a) a pathological condition is present; (b) if the X ray shows normal conditions?
10. Four students on campus are sampled at random and asked if they qualify for the college work-study program. If 1/10 of all students on the campus qualify for work-study, give the probability distribution for x, the number of students qualifying for work-study. Determine the mean and standard deviation for x.

5.2 APPLICATIONS OF THE BINOMIAL DISTRIBUTION

In practice, it is not unusual to encounter problems involving the binomial distribution, although often the questions concern ranges of values, such as the probability of 3 to 7 successes or of 4 or more or of less than 6, and so on. For extensive use, tables of values of individual terms and of cumulative terms are available for $n \leq 100$ and various values of p. Some of these tables will be discussed in the next section.

Applications of the binomial distribution rest on the fact that events represented by different values of the random variable are mutually exclusive so that the probabilities can be added. A few examples will be given to illustrate these applications.

EXAMPLE 1 A die is tossed three times and the occurrence of a "1" or "2" is called a "success." Determine the probability of (a) at least one success; (b) no more than two successes; and (c) at least one, but less than three successes.

SOLUTION The following probabilities are given by the binomial formula

x	$P(x)$
0	$\binom{3}{0}\left(\frac{1}{3}\right)^0\left(\frac{2}{3}\right)^3 = 8/27$
1	$\binom{3}{1}\left(\frac{1}{3}\right)^1\left(\frac{2}{3}\right)^2 = 12/27$
2	$\binom{3}{2}\left(\frac{1}{3}\right)^2\left(\frac{2}{3}\right)^1 = 6/27$
3	$\binom{3}{3}\left(\frac{1}{3}\right)^3\left(\frac{2}{3}\right)^0 = 1/27$

Then we have

(a) $$P(x \geq 1) = P(1) + P(2) + P(3) = 12/27 + 6/27 + 1/27 = 19/27$$

or

$$P(x \geq 1) = 1 - P(x < 1) = 1 - P(0) = 1 - 8/27 = 19/27$$

(b) $P(x \leq 2) = P(0) + P(1) + P(2)$
$= 8/27 + 12/27 + 6/27 = 26/27$

or

$$P(x \leq 2) = 1 - P(x > 2)$$
$$= 1 - P(3)$$
$$= 1 - 1/27 = 26/27$$

(c) $P(1 \leq x < 3) = P(1) + P(2)$
$= 12/27 + 6/27 = 18/27 = 2/3$

EXAMPLE 2 Referring to example 1 in the preceding section, in a certain experiment 5 animals are given a drug. One-fifth of all animals given the drug develop certain symptoms. Determine the probability that (a) 3 or 4 of the animals will develop symptoms; (b) at least 3 of the animals will develop symptoms; (c) less than 3 of the animals will develop symptoms; (d) some, but not all, of the animals will develop symptoms.

SOLUTION For convenience, the distribution is reproduced:

x	0	1	2	3	4	5
$P(x)$	0.32768	0.40960	0.20480	0.05120	0.00640	0.00032

(a) $P(x = 3 \text{ or } 4) = P(3) + P(4)$
$= 0.05120 + 0.00640$
$= 0.05760$

(b) $P(x \geq 3) = P(3) + P(4) + P(5)$
$= 0.05120 + 0.00640 + 0.00032$
$= 0.05792$

(c) $P(x < 3) = P(0) + P(1) + P(2)$
$= 0.32768 + 0.40960 + 0.20480$
$= 0.94208$

Note that $P(x < 3) = 1 - P(x \geq 3)$; since $P(x \geq 3) = 0.05792$ (from part b), $P(x < 3) = 1 - 0.05792$ or $P(x < 3) = 0.94208$.

(d) In mathematical terms, "some" means "at least one." Thus, some but not all of 5 possibilities would be 1, 2, 3, or 4. It would also be all except 0 and 5. So we have two procedures

$$\begin{aligned} P(1 \leq x \leq 4) &= P(1) + P(2) + P(3) + P(4) \\ &= 0.40960 + 0.20480 + 0.05120 + 0.00640 \\ &= 0.67200 \end{aligned}$$

or

$$\begin{aligned} P(1 \leq x \leq 4) &= 1 - [P(0) + P(5)] \\ &= 1 - (0.32768 + 0.00032) \\ &= 1 - (0.32800) \\ &= 0.67200 \end{aligned}$$

DISCUSSION Note that there are often several different ways to obtain these probabilities. It is up to the reader to determine the most efficient method of obtaining a solution. By the nature of these distributions, even the least efficient correct method will yield the correct solution. It can, however, be quite tiring to always do things by the least efficient methods.

Problems

1. The probability that a bowler will make a score of 200 or more in a given game is 0.4. What is the probability that he will score 200 or more in at least 3 of 5 games?
2. A gas station has been grossing over $1,000 a day on the average of 8 days in 10 over the past several months. Assuming this to be an accurate measure, what is the probability that the station will gross over $1,000 at least 5 times in the next 6 days?
3. One-third of the calls a certain door-to-door salesman made have led to his being let in the door. Of these calls he made sales one-fourth of the time. A certain block contains 6 houses. What is the probability that he will be allowed into no more than 2 houses in the block? What is the probability that he will make at least one sale?
4. A coin is tossed 6 times. What is the probability that there will be more than 1 head, but less than 5 heads?
5. A soft drink company claims that their cola tastes "unique." Four different brands of cola are set before a taster who is to choose the one which is "different" (supposedly, of course, the company's brand). Now suppose that there really is no difference in the way that they taste, but that each of three tasters picks one of the drinks anyway, not knowing which is which since they are in

identical containers. What is the probability that at least two of the tasters will select the "correct" drink by chance alone?

6. Of every 1,000 parts produced by a machine, on the average 10 are defective. What is the probability that some, but not all, of a sample of 3 of these parts are defective?

c7. The sex ratio of humans at birth is 100 females to 105 males. What is the probability that, in 6 single births, at least half the babies born are females? Compare this with the result which would be obtained if we use a sex ratio of 1 to 1.

8. Referring to problem 9, section 5.1, what is the probability that at least one of the three students diagnoses the X ray correctly (a) if there is a pathological condition; (b) if there is no such condition?

9. Suppose that 5 different students examine the X ray as in the previous problem. What is the probability that at least 2 will correctly diagnose a pathological condition?

c10. A random sample is drawn from a group of Spanish-American people in a large city. If the group is large enough to use the binomial model and if one-fifth of the group are unable to speak English, what is the probability that at least half of seven people selected from the group will be able to speak English?

5.3 USING BINOMIAL TABLES (Optional)

Even the relatively small numbers in the experiments in the preceding section make the task of computing the probabilities of a range of values quite tedious. Extensive tables for the probabilities in the binomial distribution are available. These tables generally come in two parts. One part gives individual probabilities: the probability of obtaining exactly x successes in n trials of a binomial experiment. The other part gives cumulative probabilities: the probability of obtaining *r or more* successes in n trials of a binomial experiment. Since the entries for individual probabilities can be found easily from the cumulative tables, but the converse is not true, we will restrict our attention solely to the latter part of the table.

A short table of cumulative probabilities is given in Table 9. Values are given for $n = 2, 3, 4, \ldots, 25$, and for $p = 0.01, 0.05, 0.10, 0.20, 0.30, 0.40, 0.50, 0.60, 0.70, 0.80, 0.90, 0.95$, and 0.99. Each three-digit entry in the table should be read as a decimal; that is, an entry of 983 is a probability of 0.983. The symbol $1-$ means that the probability is less than 1 but larger than 0.9995, while the symbol $0+$ means a positive probability less than 0.0005. There are actually twenty-four such tables which are presented consecutively since the column headings do not change.

Now suppose we wish to compute the probability of 5 or more successes in 8 trials when $p = 0.4$. We have $n = 8, r = 5, p = 0.4$, so we go to the corresponding entry in the table and find the entry 174. This tells us that $P(x \geq r) = 0.174$. We can check this by the method of the preceding section as follows.

$$\begin{aligned} P(x \geq 5) &= P(5) + P(6) + P(7) + P(8) \\ &= \binom{8}{5}(0.4)^5(0.6)^3 + \binom{8}{6}(0.4)^6(0.6)^2 \\ &\quad + \binom{8}{7}(0.4)^7(0.6) + \binom{8}{8}(0.4)^8 \\ &= 56(0.01024)(0.216) + 28(0.004096)(0.36) \\ &\quad + 8(0.0016384)(0.6) + (0.00065536) \\ &= 0.12386304 + 0.04128768 + 0.00786432 + 0.00065536 \\ &= 0.1736704 \\ &\doteq 0.174 \end{aligned}$$

Notice how much simpler it is to look in the tables.

To find the probability of fewer than r successes, note that $P(x < r) = 1 - P(x \geq r)$. Thus the probability of fewer than 5 successes in 8 trials with $p = 0.4$ is about $1 - 0.174$ or 0.826.

We could also look at the probability of failure. Fewer than 5 successes is equivalent to 4 or more failures. The probability of failure is 0.6, so we can look in the table for $n = 8$, $r = 4$, $p = 0.6$ and directly read 0.826, so that $P(x < 5) \doteq 0.826$.

Finally, to find the probability of exactly 5 successes in 8 trials for $p = 0.4$, we can find $P(x \geq 5)$ and $P(x \geq 6)$ and subtract to get $P(5)$. We already have $P(x \geq 5) \doteq 0.174$; from Table 9, with $n = 8, r = 6, p = 0.4$, we find $P(x \geq 6) \doteq 0.050$. Thus $P(5) \doteq 0.174 - 0.050 \doteq 0.124$. Direct calculation gives $P(5) = 0.12386304 \doteq 0.124$.

EXAMPLE 1 Find $P(x \geq 10)$, $P(x \leq 7)$, and $P(8)$ for $n = 12$ with (a) $p = 0.20$, (b) $p = 0.60$, (c) $p = 0.95$.

SOLUTION From Table 9, with $n = 12$, $r = 10$, $p = 0.20$, we find $P(x \geq 10) \doteq 0+$, which is less than 0.0005. Since $P(x \leq 7) = 1 - P(x \geq 8)$, we have $P(x \leq 7) \doteq 1 - 0.001 \doteq 0.999$. Alternatively, using $p = 0.80$ and finding $P(12 - x \geq 5)$, where $12 - x$ represents failures, we have

$P(12 - x \geq 5) \doteq 0.999$. This is equal to $P(x \leq 7)$, since if $x \leq 7$, $12 - x \geq 5$. To determine $P(8)$, we find $P(x \geq 8)$, which is 0.001, and $P(x \geq 9)$, which is $0+$. Since we cannot find the difference exactly, we simply conclude that $P(8) \doteq 0.001$.

Using $n = 12$, $p = 0.60$, we find that, for $r = 10$, we get $P(x \geq 10) \doteq 0.083$; using $r = 8$, we have $P(x \geq 8) \doteq 0.438$, so $P(x \leq 7) \doteq 0.562$. Now $P(x \geq 9) \doteq 0.225$ and since $P(x \geq 8) \doteq 0.438$, we have $P(8) \doteq 0.438 - 0.225 \doteq 0.213$.

Finally, using $p = 0.95$, for $n = 12$, $r = 10$, we have $P(x \geq 10) \doteq 0.980$; for $r = 8$, $P(x \geq 8) \doteq 1-$, so $P(x \leq 7) \doteq 0+$, less than 0.0005, and thus we cannot directly determine $P(8)$. $P(x \geq 9)$, however, is 0.998, so $P(8)$ is less than 0.002.

EXAMPLE 2 A medical study showed that, of a survey of 15,000 men who had a heart attack and recovered and then subsequently died, 60% died of a second heart attack, while 40% died of other causes. The case histories of 20 men who have had a heart attack and recovered are under study. Making the assumption that the binomial distribution may be used here, even though the sampling is done without replacement, determine the probability that 10 or fewer of these men will die of a second heart attack. (Note that the assumption means that the results will be approximate, but the fact that the sample is quite small compared with the population makes the approximation a very good one.)

SOLUTION We apply the empirically determined probability, $p = 0.60$, to the sample with $n = 20$, $r = 11$, to obtain $P(x \geq 11) \doteq 0.755$. Then $P(x \leq 10) \doteq 0.245$.

EXAMPLE 3 From past experience a golfer knows that the probability that he will hit a drive into a sand trap from the tee is about 0.25 on any given hole. Using this assumption, what is the probability that he will hit the ball into a sand trap on exactly 4 of the first 9 holes?

SOLUTION With the proper table, one would look for $n = 9$, $r = 4$, and $r = 5$, for $p = 0.25$. Our limited tables do not have $p = 0.25$, however, so we must *interpolate*. This means that we can find the values if $p = 0.20$ and if $p = 0.30$ and then find the value halfway between them for $p = 0.25$. Thus, for $n = 9$, $p = 0.20$, $P(x \geq 4) \doteq 0.086$ and $P(x \geq 5) \doteq 0.020$, so $P(5) \doteq 0.066$.

For $n = 9$, $p = 0.30$, $P(x \geq 4) \doteq 0.270$ and $P(x \geq 5) \doteq 0.099$, so $P(5) \doteq 0.171$. The mean of 0.066 and 0.171 is 0.1185. Rounding off to the even digit, we have $P(4) = 0.118$ for $n = 9$, $p = 0.25$, so the probability that he will get into exactly 4 sand traps in the first 9 holes is 0.118. Using the binomial formula we have $P(4) = \binom{9}{4}(0.25)^4(0.75)^4 \doteq 0.117$, which indicates that interpolation from the tables gives a very good approximation.

Problems

In problems 1–10, x is the number of successes in n trials of a binomial experiment with probability of success equal to p. Determine the probabilities in problems 1–6.

1. $P(x \geq 11)$ and $P(11)$ for $n = 20$, $p = 0.7$.
2. $P(x \leq 5)$ and $P(5)$ for $n = 12$, $p = 0.2$.
3. $P(x > 18)$ and $P(20)$ for $n = 22$, $p = 0.8$.
4. $P(x > 14)$ and $P(x < 14)$ for $n = 25$, $p = 0.4$.
5. $P(x \geq 7)$ and $P(7)$ for $n = 15$, $p = 0.45$.
6. $P(x < 12)$ and $P(12)$ for $n = 25$, $p = 0.75$.
7. For $n = 20$, find the value of p for which $P(x \geq 14) = 0.250$.
8. For $n = 15$, $p = 0.6$, find the value of r for which $P(x \geq r) < 0.05$.
9. For $n = 25$, $p = 0.4$, find the value of r for which $P(x \geq r) < 0.10$.
10. For $n = 18$, $p = 0.35$, find the value of r for which $P(x \geq r) < 0.05$.
11. The probability of a bowler getting a strike on the first ball of each frame is 1/5. In a game of 10 frames, what is the probability that he will get at least 2 strikes? Exactly 3 strikes?
12. If 30% of the viewers in a town watch a particular television show, what is the probability that a majority of 25 persons sampled will not watch the show?
13. A drug has deleterious side effects on 15% of the patients on which it is used. What is the probability that exactly 3 out of 20 patients who receive the drug will have these side effects?
14. Although the sex ratio of humans to birth is 100 females to 105 males, it is common to use a 1 to 1 ratio for computational purposes. In a group of 24 births, the probability of exactly 12 females and 12 males can be computed to be 0.160 if we use the actual ratio. What is the probability if we use $p = 0.5$?
15. The probability that a doctor will be able to distinguish nervous tension from a brain tumor in a patient who has nervous tension is about 0.95. If 10 different doctors examine a patient suffering from nervous tension, what is the probability that all 10 correctly distinguish it from a brain tumor? What is the probability that 8 or 9 will do so, but not all 10?

16. Thirty percent of children who suffer feelings of racial deprivation react with aggressive behavior. A random sample consisting of 15 students suffering feelings of racial deprivation are given a testing instrument to measure aggressiveness of behavior. Assuming the instrument accurately measures the sought response, what is the probability that at least half the sample will be classified by the instrument as aggressive?

5.4 THE BINOMIAL MODEL IN HYPOTHESIS TESTING

In a great many problems in everyday life we are dealing with a binomial population. Many questions concerning everyday situations have exactly two answers. Should we buy a new television set or not? Is a polio vaccine effective or not? Such questions can often be handled effectively by using the binomial distribution. Generally our samples are sufficiently small so that ignoring the lack of replacement does not materially affect the validity of using the binomial model.

Suppose that we wish to test the effectiveness of a new treatment for eczema. Research in the past has shown that if white mice are given a case of exczema, 2/5 of them will be symptom-free naturally within four weeks. We interpret this to mean that if a mouse is given a case of eczema and left untreated, the probability is 2/5 (or 0.4) of recovering within four weeks. To test the new treatment, we use it on a sample of mice who have been given eczema, observe the number of symptom-free mice after four weeks, and draw conclusions about the effectiveness of the treatment based on the probability that this number of symptom-free mice would appear if the treatment had no effect.

Suppose, for example, that the sample consisted of fifteen mice which were infected with eczema and then given the treatment. Without any treatment, the expected number of symptom-free mice is $(2/5)(15) = 6$, so if more than six are symptom-free after four weeks, this may be considered to be some evidence that the treatment is effective. The more mice that are symptom-free, the stronger is the evidence that the treatment is effective. At some point we must be able to say, "Yes, the evidence is sufficient to indicate that the treatment is effective." But even if all fifteen mice were free of symptoms after four weeks, there is still a possibility that this outcome is due to chance. The probability of this is $(2/5)^{15}$ or about 0.00000107. This probability is quite small—about one chance in a million—but it could happen. Thus we can *never* be *certain* about the effectiveness of the treatment. We must always take some chances. *Determining when to take these chances and controlling the risk involved is the essence of statistical hypothesis testing.*

In the eczema treatment experiment, we wanted to test the effectiveness of the treatment. Our **research hypothesis** was that the treatment is effective in combating eczema. This means that we would expect more than 2/5 of the mice to show no symptoms after four weeks. But how many more? We have no way of guessing what to look for, so we approach the problem another way. We suppose that the treatment has no effect; that is, we assume that the probability of a particular mouse being symptom-free in four weeks is 2/5. This is called the **null hypothesis**. We then base our conclusions about the outcome of the experiment on the assumption that the null hypothesis is true. After performing our experiment, we use the null hypothesis to determine the probability of the actual outcome; that is, we determine the probability that *at least* this many mice would be symptom-free in four weeks. This probability, designated α (alpha), *is the probability that we would make an error by concluding that the treatment is effective.*

If we find that r mice are symptom-free after four weeks, we can calculate α for $n = 15$, $p = 0.4$, either directly or from Table 9. We have the following table of probabilities for $r \geq 6$.

r	α
6	0.597
7	0.390
8	0.213
9	0.095
10	0.034
11	0.009
12	0.002
13	0+
14	0+
15	0+

If as many as eight mice are symptom-free after four weeks, there is about one chance in five that the results are due to chance. This may be, and usually is, considered to be too great a chance and the results are not considered conclusive. If ten or more mice, however, are symptom-free after four weeks, the probabilities that the results are due to chance are much smaller; only one chance in thirty exists that the outcome would occur if the treatment had no effect. At this level of α, we would probably conclude that the treatment was effective.

In most cases an acceptable α is selected beforehand and the possible outcomes are divided into two groups. One group makes up the **rejection region**, while the other forms the **acceptance region**. Suppose that we decide that a suitable α is 0.05. This means that we are willing to take one chance in twenty of incorrectly concluding that the treatment is effective. We would then say that the rejection region is composed of all r values of 10 or more since $P(x \geq 10) < 0.05$, while any value of r less than 10 is in the acceptance region.

The names of these regions are derived from the fact that if we get a number which falls in the rejection region we are justifying in rejecting the hypothesis that the treatment is ineffective and concluding that it is effective (with a probability of 0.05 of being wrong). If the obtained value falls in the acceptance region, we are justified in accepting the hypothesis that the treatment is ineffective. If we do conclude this, however, we run a different risk. We may *incorrectly* conclude that the treatment is ineffective. This is possible because the treatment may be effective, even though our results failed to show this. The probability of this happening is designated β (beta) and is dependent upon the actual probability of a mouse being sympton-free after four weeks because of the effect of the treatment. Since this probability is not known, β cannot be calculated directly, but it is known that β is inversely dependent on α for a given sample size. Thus, decreasing α, to be more certain that we do not conclude that the treatment is effective when it is not, will increase β, the likelihood of making the opposite error. If α is fixed by preselection, the only way to decrease β is to increase the sample size. This is a reason for using samples which are as large as possible. Further discussions of hypothesis testing can be found in section 6.5 and in Chapter 9.

EXAMPLE 1 A businessman carries two products, *A* and *B*, which are identical except for brand name. His inventory has grown to the point where he wishes to discontinue one of the brands. Brand *A* does a great deal of advertising, so he is tempted to carry this brand only. Past records show that about equal amounts of the two brands are sold, but he feels that his customers may be showing a preference for brand *B* in recent weeks. He decides to set up an experiment based on the next 25 sales of the product.

SOLUTION He will order brand *A* unless he is convinced brand *B* is preferred by more customers. Suppose we let p represent the probability that brand *B* is preferred by his customers. This is numerically equal to the proportion of customers preferring brand *B*. If more than half his customers prefer brand *B*, he will order brand *B*. If fewer than half prefer brand *B*, he will order brand *A*. If exactly half prefer each brand, he will stock brand *A* to take advantage of the advertising. Thus the crucial value of p is 0.5. He wishes to order brand *A* unless he feels sure that $p > 0.5$. He tests the hypothesis that $p = 0.5$ by examining the consequences of this hypothesis with respect to his sample of 25. He is willing to accept a probability of 0.10 of being wrong in rejecting brand *B*. (This is a bit higher than usual, but acceptable.) He uses the binomial probability tables and finds that if $n = 25$, $p = 0.5$, then $P(x \geq 17) \doteq 0.054$ and $P(x \geq 16) \doteq 0.115$. He then concludes that if 17 or more customers buy brand *B*, he will stock brand *B*, but if 16 or fewer buy brand B, he will stock brand A. Since the random variable can take on values 0, 1, 2, ..., 25, this means that the rejection region is 17, 18, 19, ..., 25, while the acceptance region is 0, 1, 2, ..., 16.

DISCUSSION Suppose that the true probability is 0.6. If this is the case, $P(x \geq 17) \doteq 0.274$, so the probability that he will make a mistake by stocking brand A is $\beta = 0.726$, the probability that fewer than 17 customers will buy brand B, even though brand B is preferred by about 60% of the customers. If $p = 0.8$, however, $\beta \doteq 0.047$ since then $P(x \geq 17) \doteq 0.953$. A danger of small samples, then, is that the tests may not be sensitive enough to detect relatively small deviations from the tested hypothesis. It would be better to use a larger number of customers spread out over a period of several days.

Problems

1. A psychologist claims that he has trained rats to be able to select the proper path that leads to food at least 80% of the time by "intuition." A colleague disputes this and challenges him to prove it. Twenty-three rats are exposed to a previously unknown maze. How many must select the correct path in order for it to be considered that he has proved his point, allowing a chance of 0.20 of being in error?
2. The mortality rate for a certain disease is 0.30 within two years after diagnosis. A medical laboratory has developed a drug which lowers the mortality rate in laboratory animals. They are now ready to test the drug on human beings, and have selected 25 patients from a group of volunteers. The number of deaths is given by X.
   (a) If $\alpha = 0.05$, what values of X will lead to the conclusion that the drug is effective?
   (b) If $\alpha = 0.10$, what values of X will lead to the conclusion that the drug is effective?
   (c) Suppose that three patients drop out of the program for various reasons. Now, for what values of X, with $\alpha = 0.05$, can it be concluded that the drug is effective?
   (d) Suppose that 4 of the 25 patients die. What conclusion will be reached with $\alpha = 0.05$? What is the value of β if actually the drug is effective with a mortality rate of 0.20?
3. A manufacturer samples his output daily for acceptability. He is willing to accept 5% defectives for the total output. He tests a sample of 20 units daily and records the number of defectives. Letting $\alpha = 0.10$, he will accept the daily output unless his sample leads him to believe it contains more than 5% defectives. What is his rejection region?

5.5 OTHER IMPORTANT DISTRIBUTIONS (Optional)

Although the binomial distribution is the most important distribution encountered in an elementary discussion, there are a number of other discrete distributions which have practical applications. Four of them will be discussed in this section.

A binomial experiment is one in which each trial has exactly two outcomes, the trials are independent, and the probabilities remain constant from trial to trial. An experiment with several outcomes in which the probabilities remain constant from trial to trial is called a **multinomial experiment**. Independence of trials is still required. A **multinomial distribution** assigns probabilities to all the possible outcomes. Suppose these outcomes are $A_1, A_2, A_3, \ldots, A_k$ with probabilities $p_1, p_2, p_3, \ldots, p_k$ respectively. If the experiment is performed n times, we will have a certain number (zero or more) of each possible outcome. For instance, there may be an experiment with four outcomes, A_1, A_2, A_3, A_4, whose probabilities are p_1, p_2, p_3, p_4, respectively, and the experiment is performed, say, nine times. One possible event is to obtain A_1 three times and each of the other outcomes two times. One such outcome is $A_1 A_2 A_1 A_3 A_2 A_4 A_4 A_1 A_3$. The probabilities of exactly this outcome in exactly this order is $p_1^3 p_2^2 p_3^2 p_4^2$. The number of different orders to obtain the same outcome is $9!/3!2!2!2!$ (see Permutation Rule B, section 3.2), so the probability of obtaining exactly 3 A_1's, 2 A_2's, 2 A_3's, and 2 A_4's in 9 trials is exactly $9!/3!2!2!2!\ p_1^3 p_2^2 p_3^2 p_4^2$. This can be generalized as follows:

MULTINOMIAL DISTRIBUTION

If an experiment has possible outcomes $A_1, A_2, A_3, \ldots, A_k$ with probabilities $p_1, p_2, p_3, \ldots, p_k$, respectively, and is repeated n times, the probability of obtaining A_i exactly x_i times is given by

$$P(x_1, x_2, \ldots, x_k) = \frac{n!}{x_1!x_2!x_3!\cdots x_k!} p_1^{x_1} p_2^{x_2} p_3^{x_3} \cdots p_k^{x_k}$$

where $\sum p_i = 1$ and $\sum x_i = n$

EXAMPLE 1 An urn contains 1 red marble, 2 white marbles, and 3 blue marbles. One marble is drawn at random and replaced. The experiment is repeated 4 times. Figure out the probability distribution of the outcomes.

SOLUTION Here $p_1 = P(\text{red}) = 1/6$; $p_2 = P(\text{white}) = 1/3$; $p_3 = P(\text{blue}) = 1/2$, and we have the following table:

Red	*White*	*Blue*	*Probability*	
4	0	0	$1 \cdot (1/6)^4$	$= 1/1296$
0	4	0	$1 \cdot (1/3)^4$	$= 1/81$
0	0	4	$1 \cdot (1/2)^4$	$= 1/16$
3	1	0	$4 \cdot (1/6)^3 \cdot (1/3)$	$= 1/162$
3	0	1	$4 \cdot (1/6)^3 \cdot (1/2)$	$= 1/108$
1	3	0	$4 \cdot (1/6) \cdot (1/3)^3$	$= 2/81$
0	3	1	$4 \cdot (1/3)^3 \cdot (1/2)$	$= 2/27$
1	0	3	$4 \cdot (1/6) \cdot (1/2)^3$	$= 1/12$
0	1	3	$4 \cdot (1/3) \cdot (1/2)^3$	$= 1/6$
2	2	0	$6 \cdot (1/6)^2 \cdot (1/3)^2$	$= 1/54$
2	0	2	$6 \cdot (1/6)^2 \cdot (1/2)^2$	$= 1/24$
0	2	2	$6 \cdot (1/3)^2 \cdot (1/2)^2$	$= 1/6$
2	1	1	$12 \cdot (1/6)^2 \cdot (1/3) \cdot (1/2)$	$= 1/18$
1	2	1	$12 \cdot (1/6) \cdot (1/3)^2 \cdot (1/2)$	$= 1/9$
1	1	2	$12 \cdot (1/6) \cdot (1/3) \cdot (1/2)^2$	$= 1/6$

DISCUSSION It is interesting to note that the sum of the multinomial coefficients always equals k^n. This can be a useful check in determining if all the cases have been discovered.

So far we have discussed experiments with replacement. Many cases, such as choosing 3 people from a group of people, involve sampling at random *without* replacement. The probability distribution which arises in a binomial population in which sampling is done without replacement is called the **hypergeometric distribution**. Suppose these 3 people are to be selected from a group of 5 men and 3 women. What is the probability that exactly 2 of these will be men and 1 of them a woman? Note that 2 men can be selected from 5 in $\binom{5}{2}$ ways, and 1 woman can be selected from 3 in $\binom{3}{1}$ ways. Thus, the total number of ways we can select 2 men and 1 woman is $\binom{5}{2} \cdot \binom{3}{1}$. The total number of ways that 3 people can be selected from 8 is $\binom{8}{3}$. Thus, the probability of selecting exactly 2 men and 1 woman from this group is

$$\frac{\binom{5}{2}\binom{3}{1}}{\binom{8}{3}} = \frac{\frac{5 \cdot 4}{2 \cdot 1} \cdot \frac{3}{1}}{\frac{8 \cdot 7 \cdot 6}{3 \cdot 2 \cdot 1}} = \frac{10 \cdot 3}{56} = \frac{30}{56} = \frac{15}{28}$$

Generalizing this we obtain the following:

HYPERGEOMETRIC DISTRIBUTION

If a population consists of A objects of one kind and B objects of a second kind, the probability of selecting (at random, without replacement) x objects of type A and $n - x$ objects of type B is given by

$$P(x) = \frac{\binom{A}{x}\binom{B}{n-x}}{\binom{A+B}{n}}$$

EXAMPLE 2 An urn contains 12 red, 15 white marbles. Five marbles are drawn, at random, without replacement. Give the probability distribution.

SOLUTION Here $A = 12$, $B = 15$, $x = 5, 4, 3, 2, 1, 0$, and $n = 5$. So we have the following table:

Red	*White*	*Probability*
5	0	$\frac{\binom{12}{5}\binom{15}{0}}{\binom{27}{5}} = \frac{\frac{12 \cdot 11 \cdot 10 \cdot 9 \cdot 8}{5 \cdot 4 \cdot 3 \cdot 2 \cdot 1} \cdot 1}{\frac{27 \cdot 26 \cdot 25 \cdot 24 \cdot 23}{5 \cdot 4 \cdot 3 \cdot 2 \cdot 1}} = \frac{792}{80{,}730} = \frac{44}{4{,}485}$
4	1	$\frac{\binom{12}{4}\binom{15}{1}}{80{,}730} = \frac{495 \cdot 15}{80{,}730} = \frac{7{,}425}{80{,}730} = \frac{55}{598}$
3	2	$\frac{\binom{12}{3}\binom{15}{2}}{80{,}730} = \frac{220 \cdot 105}{80{,}730} = \frac{23{,}100}{80{,}730} = \frac{770}{2{,}691}$
2	3	$\frac{\binom{12}{2}\binom{15}{3}}{80{,}730} = \frac{66 \cdot 455}{80{,}730} = \frac{30{,}030}{80{,}730} = \frac{77}{207}$
1	4	$\frac{\binom{12}{1}\binom{15}{4}}{80{,}730} = \frac{12 \cdot 1{,}365}{80{,}730} = \frac{16{,}380}{80{,}730} = \frac{14}{69}$
0	5	$\frac{\binom{12}{0}\binom{15}{5}}{80{,}730} = \frac{1 \cdot 3{,}003}{80{,}730} = \frac{3{,}003}{80{,}730} = \frac{77}{2{,}070}$

EXAMPLE 3 A bag contains 5 red and 4 yellow apples. Five children take an apple, at random, from the bag. What is the probability that the children took 3 of one color, and 2 of the other?

SOLUTION The distribution is hypergeometric, but we are interested only in $P(3) + P(2)$. If we let $A = 5$, $B = 4$, $x = 3$ or 2, and $n = 5$, we have

$$P(3) = \frac{\binom{5}{3}\binom{4}{2}}{\binom{9}{5}} = \frac{\frac{5 \cdot 4 \cdot 3}{3 \cdot 2 \cdot 1} \cdot \frac{4 \cdot 3}{2 \cdot 1}}{\frac{9 \cdot 8 \cdot 7 \cdot 6 \cdot 5}{5 \cdot 4 \cdot 3 \cdot 2 \cdot 1}} = \frac{10 \cdot 6}{126} = \frac{60}{126}$$

$$P(2) = \frac{\binom{5}{2}\binom{4}{3}}{\binom{9}{5}} = \frac{\frac{5 \cdot 4}{2 \cdot 1} \cdot \frac{4 \cdot 3 \cdot 2}{3 \cdot 2 \cdot 1}}{126} = \frac{10 \cdot 4}{126} = \frac{40}{126}$$

$P(3) + P(2) = \frac{100}{126} = \frac{50}{63} \doteq 0.79$, so the probability that the children took 3 of one color, 2 of a second, is about 0.79.

So far we have discussed distributions which have a finite number of points. There are a number of distributions which, at least theoretically, have an infinite number of points. One such distribution is the **geometric distribution**. It arises from the geometric sequence usually studied in elementary algebra. An example is the following situation—a die is tossed until a "six" appears. Theoretically a six may never appear (though in all practicality one may assume that it will or else we stop tossing it). The probability that a six will appear on the first toss is 1/6; that it will appear for the first time on the second toss is $(5/6) \cdot (1/6)$; that it will appear for the first time on the tenth toss is $(5/6)^9 \cdot (1/6)$. Generalizing we have the following:

GEOMETRIC DISTRIBUTION

If a binomial experiment is repeated until a "success" occurs, and the probability of "success" on one trial is p, then the probability that the first success will occur on trial x is

$$P(x) = (1 - p)^{x-1} \cdot p$$

EXAMPLE 4 On the average, about 1 of 10 persons passing a shop enters the shop. Suppose we wish to determine the probabilities for the number of people who will pass the shop after it opens in the morning, before someone enters the shop. Let x represent the number of the first such person to enter the shop after $x - 1$ persons have passed without entering. Compute probabilities for each value of x to 11, then determine $P(x > 11)$.

SOLUTION This is a geometric distribution with $p = 1/10$ demonstrating that a particular person passing the shop enters it. Then $P(x)$ is the probability that $x - 1$ persons have passed the shop and the xth person enters. That is, $P(1) = 1/10$, $P(2) = (9/10)(1/10)$ and in general, $P(x) = (9/10)^{x-1}(1/10)$. Thus, we have

x	$P(x)$
1	$1/10 = 0.1$
2	$(9/10)\ (1/10) = 0.09$
3	$(9/10)^2\ (1/10) = 0.081$
4	$(9/10)^3\ (1/10) = 0.0729$
5	$(9/10)^4\ (1/10) = 0.06561$
6	$(9/10)^5\ (1/10) = 0.059049$
7	$(9/10)^6\ (1/10) = 0.0531441$
8	$(9/10)^7\ (1/10) = 0.04782969$
9	$(9/10)^8\ (1/10) = 0.043046721$
10	$(9/10)^9\ (1/10) = 0.0387420489$
11	$(9/10)^{10}(1/10) = 0.03486784401$
>11	$= 0.31381059609$

The mean and standard deviation of the geometric distribution are given by $\mu = 1/p$, $\sigma = (1/p)\sqrt{1 - p}$. In the geometric distribution of example 4, then $\mu = 10$, $\sigma \doteq 9.49$, and the mean number of persons passing before someone enters is 9.

Another important distribution is known as the **Poisson distribution**. This distribution is used to determine the expected number of occurrences per interval of time or unit of space when the average number is small. Examples include the number of telephone calls passing through a switchboard, the number of electron tubes requiring replacement in an assembly per period of time, number of fish caught per man in a unit of time. It can also be used as an approximation to the binomial when the mean is small. One formula for the distribution is this one:

POISSON DISTRIBUTION

$$P(x) = \frac{e^{-\mu}\mu^x}{x!} \quad \text{for } x = 0, 1, 2, 3, \ldots$$

where μ is the mean or expected value of x, the random variable.

One of the important reasons for using the Poisson distribution is that its value depends on μ alone. Most statistical calculators include the exponential function on the keyboard. A short table of Poisson probability distributions has been calculated and is reproduced in Table 10. For example, if $\mu = 3$, we have the following distribution:

x	$P(x)$
0	0.051
1	0.149
2	0.224
3	0.224
4	0.168
5	0.101
6	0.050
7	0.022
8	0.008
9	0.003
10	0.001

For $x > 10$, the probabilities are less than 0.0005, but may be computed if needed. For example, $P(12) \doteq 0.000055$ and $P(13) \doteq 0.000013$. In the table all probabilities less than 0.0005 are omitted.

EXAMPLE 5 The number of tropical storms in a given area is approximated by a Poisson distribution in which μ is the mean or expected value. If the mean number of tropical storms in an area of Florida has been 5 per year, determine the probability that the area will have, this year, (a) from 5 to 8 such storms; (b) fewer than 3 such storms.

SOLUTION Since $P(5 \leq x \leq 8) = P(5) + P(6) + P(7) + P(8)$, from Table 10 with $\mu = 5$ we have

$$P(5 \leq x \leq 8) \doteq 0.175 + 0.146 + 0.104 + 0.065 \doteq 0.490.$$

Now $P(x < 3) = P(0) + P(1) + P(2) = 0.007 + 0.034 + 0.084 \doteq 0.125$.

Problems

1. Three pool balls are randomly selected from among 15 numbered 1 through 15, and placed in pockets of the table. Classifying the numbers on the balls as (a) odd, single digit; (b) even, single digit; and (c) two digit, compute the distribution of the 3 balls by classification.

2. In a box are the names of 8 students with grade point averages above 3.5; 5 with averages between 3.2 and 3.5; 4 with averages between 3.0 and 3.2, and 3 with averages between 2.8 and 3.0. Four names will be selected at random and scholarships awarded. What is the probability that at least one student in each of the first three categories will be selected? Assume replacement.
3. What is the probability that a 5 card poker hand contains exactly 2 aces? One ace? No aces?
4. Twenty radio tubes are sent to a store. Five are sold to a customer. It turns out that 4 of the 20 are defective. Compute the probability distribution for the number of defective tubes sold to the customer.
5. Five red and 3 blue marbles are placed in an urn and 3 marbles are drawn without replacement. Figure out the probability distribution for the number of red marbles drawn, and calculate its mean and standard deviation. Compare with the hypergeometric formulas

$$\mu = \frac{nA}{A + B}, \quad \sigma^2 = \frac{nAB(A + B - n)}{(A + B)^2(A + B - 1)}$$

where n is the total number drawn.
6. A coin is tossed until a head appears. Calculate the probability distribution for the number of tosses (up to 10) and make a histogram. Compute the mean and standard deviation of the distribution.
7. The mean number of microorganisms on a slide of untreated material is found to be 6 per square centimeter. If the material is treated, what is the probability that a particular slide will have 4 microorganisms per square centimeter if the treatment is ineffective? Assume the Poisson distribution. What is the probability that there will be more than 6 microorganisms per square centimeter if the treatment is ineffective? If the treatment is effective and reduces the number of microorganisms to 3 per square centimeter, what is the probability that there will be more than 6 microorganisms per square centimeter on the treated slide?
8. The Poisson distribution can be used as an approximation to the binomial when the mean is small. Use the Poisson distribution to calculate the probability of 7 or more successes in 10 trials with $p = 0.6$. Compare with the normal approximation to the binomial distribution and the exact value as calculated or taken from Table 9. Repeat for the probability of having 4 or fewer successes in 25 trials for $p = 0.2$.

c9. A random sample of 5 persons is selected from a group of 300 persons, in which 100 favor capital punishment and 200 are opposed. Determine the probability that exactly 2 of those selected favor capital punishment. Use the hypergeometric distribution. If we use the binomial distribution, assuming replacement, what would the probability be? Do you think that the extra effort involved in using the correct model, the hypergeometric, yields extra value?

10. One-third of the bulbs sold by a seed company under the heading "Tulips—Mixed Colors" are red, one-ninth are purple, the rest are white. Assuming that the population is large enough to ignore the lack of replacement, what is the probability that a set of 9 bulbs bought by a customer will have exactly 3 red, 1 purple, 5 white.

11. Automobiles passing through an intersection generally follow a Poisson distribution. If the mean number of automobiles proceeding north through a particular intersection between the hours of 2 A.M. and 6 A.M. on a Monday morning is 9, what is the probability that there will be anywhere from 6 to 10 cars (inclusive) proceeding north through the intersection between those hours on a particular Monday morning?

5.6 SUMMARY

This chapter has been devoted primarily to the study of the binomial distribution. A binomial distribution is the result of sampling from a binomial (two-valued) population. If the probability of success on one trial is p, p is constant from trial to trial, and the trials are independent, the binomial formula given below generates the binomial distribution.

BINOMIAL FORMULA

$$P(x) = \binom{n}{x} p^x (1 - p)^{n-x}$$

The binomial distribution also has special formulas for the mean and standard deviation, although the formulas for the general probability distribution given in Chapter 4 are still valid.

MEAN AND STANDARD DEVIATION OF A BINOMIAL DISTRIBUTION

$$\mu = np$$

$$\sigma = \sqrt{np(1 - p)}$$

Other distributions discussed were the multinomial, hypergeometric, geometric, and Poisson.

In addition, we introduced the concept of hypothesis testing, using the binomial distribution. For a sample of size n, we wished to show, with some reasonable assurance, that a probability of some outcome is greater than a certain value, p. To do this we selected α, the acceptable probability of concluding incorrectly that the desired outcome had occurred. Then we divided the possible outcomes into acceptance and rejection regions. We did this by finding a number r such that, for a given n and p, $P(x \geq r) > \alpha$ and

$P(x \geq r - 1) < \alpha$, if the probability is actually p. Then if our sample contained at least r favorable outcomes, we rejected the hypothesis that the probability of these outcomes occurring by chance is P and conclude that it is greater than p, with the acceptable probability α of being wrong. If our sample had fewer than r favorable outcomes, we concluded that the true probability is not greater than p (or is equal to p), subject to a finite probability that we are making an error in concluding this.

Problems

1. In an experiment with three dice, tossing "10" or "11" is considered a success. If the experiment is repeated ten times,
   (a) compute the mean and standard deviation of the resulting probability distribution;
   (b) calculate the probability of obtaining at least one success;
   (c) calculate the probability of obtaining fewer than three successes.
2. Four aces are tossed into a hat—one ace from each suit. A card is drawn, observed, and returned to the hat. Assuming consecutive trials are independent, how many drawings must be made so that the probability of drawing at least one ace of spaces is greater than four-fifths?
3. It is known that the mortality rate for a certain disease is 60%. At a particular hospital 9 of the last 10 patients admitted with the disease have been cured. Does this indicate that the results are consistent with the known mortality rate, or should an investigation be made to see if some new treatment is in use at this hospital?
4. A public opinion poll is being taken on the issue of free housing for indigent laborers. If actually 70% of the population favors the issue, what is the probability that, of 15 persons interviewed, 9, 10, 11, or 12 favor the free housing
   c (a) calculating the probabilities directly;
   (b) using the probability tables.
5. Repeat problem 4 if 65% of the population favors the issue.
6. A biologist wished to test if the presence of a birth defect can be associated with a certain drug. On the average, it is present in one-tenth of the litter. He injected 25 pregnant guinea pigs with the drug and observed the number of litters in which the defect was present. If he is willing to accept 1 chance in 20 of incorrectly concluding that the association between the drug and the birth defect is present, give his null and alternate hypotheses and acceptance and rejection regions. Repeat the problem if he is willing only to accept 1 chance in 100 of making such an error.
7. A large number of applicants for a position included one-tenth minority race applicants. Seven persons were asked for an interview. Assuming that the

number of applicants was sufficiently large to justify use of the binomial distribution and that the qualifications of all individuals were the same, what is the probability that at least one minority applicant would be called for an interview? Suppose that actually two minority members were called for an interview. What is the probability that more than one minority member would be called for an interview? If three or more minority members were called, would the employer be open to charges of reverse discrimination, say, with a significance level of 0.01?

8. Among persons contracting leukemia, spontaneous remission occurs in about 1% of the cases. Among 25 persons given a new treatment at a local hospital, 3 experience complete remission. If we do not believe the remissions are caused by the new treatment and claim that the result is due to chance, what is the probability that more than 2 would have spontaneous remission if it is indeed due to chance? Would the new treatment merit further study involving the allocation of a million dollars for research? Suppose the allocating agency wants a 0.01 level of significance in the pilot study.

9. Psychological testing devices are frequently used to test the presence of "extrasensory perception." In one test, for mind-reading ability, subject A and subject B are in different rooms. Subject A has 4 cards, a triangle, a square, a cross, and a circle. Subject A randomly selects 1 of the cards and concentrates on it while subject B writes down what she thinks subject A has selected. After 20 trials, the lists are compared. A test for clairvoyance involves keeping the cards face down. Subject B writes down what she thinks the card is after it is selected but before subject A looks at it. The cards are scrambled after each selection. Binomial probability tables are used to compute the degree of subject B's ESP ability by calculating the probability of guessing *at least* as many correctly by chance alone. The level of significance determined in this way would be used to classify the person's ESP ability. Classifications vary, but one way would be to call 0.25 or less "some degree of ESP," 0.10 or less "a marked degree of ESP," 0.05 or less "a high degree of ESP," and 0.01 or less "a very high degree of ESP." Significance levels of 0.001 or less would be rare enough to be called "a remarkable degree." Suppose that a subject called 12 of 20 cards correctly on the mind-reading test and 7 of 20 on the clairvoyance test. Classify the subject according to ESP ability.

10. A die is rolled 8 times. What is the probability of obtaining a six 3 times, and each other number exactly once? What is the probability that each number will occur at least once?

11. A certain mathematics department has 12 members, 5 of senior rank, 7 of junior rank. If a committee of 4 is chosen at random, determine the probability it will not have more senior rank members than junior rank.

12. The hypergeometric distribution can easily be extended to a multinomial population. A mother goes to the grocery store and takes 6 cans of baby food off a shelf containing 6 cans of applesauce, 4 cans of peaches, and 3 cans of pears. If she

takes the 6 cans at random, what is the probability she will have taken 2 cans of each? What is the probability that she will have taken at least one of each kind?

13. Two dice are tossed until a total of 7 on the two dice is obtained. What is the probability that more than 4 tosses will be required?
14. On the average, a man will catch 2 fish per hour in Perdido Bay. If the distribution of catches is Poisson, what is the probability that a man will catch 3 or more fish in an hour in Perdido Bay?

Chapter VI

The Normal Distribution

6.1 CONTINUOUS AND DISCRETE VARIABLES

The random variables studied in the previous chapter were discrete; that is, they could take on only a unique and countable number of values. Countable in this sense means that there are the same number of elements in the distribution as there are in the set of positive integers. Another important type of random variable is the **continuous random variable**. This is not a good place to discuss the mathematical notions of the varieties of infinities or the precise definitions of "continuous." Intuitively, if a variable is continuous, it can be subdivided infinitely. Thus there is no "smallest" measure such as a toss, a cent, or another discrete quantity. Examples of continuous variables are time, weight, and temperature. These variables are continuous, although they can be made discrete artificially by measuring them to the nearest unit—say to the nearest pound. A probability distribution based on a continuous variable is called a **probability density**. One of the major drawbacks to studying probability densities at an elementary level is that this study requires more than a nodding acquaintance with calculus. In view of this fact, we shall study only the basic properties of a few probability densities.

In the case of probability distributions, we could construct histograms to portray the distribution. For instance, the distribution of bowling scores among 10,000 randomly selected games might be given by the following table.

Score (pins)	*Number of Games*	*Probability*
Under 126	812	0.0812
126–150	1,764	0.1764
151–175	2,433	0.2433
176–200	1,911	0.1911
201–225	1,646	0.1646
226–250	1,037	0.1037
251–275	294	0.0294
276–300	103	0.0103

A histogram representing these data is given here. One unit on the horizontal scale represents 25 pins. Each interval extends 0.5 in each direction. The second bar from the left, for example, extends over the interval from 125.5 to 150.5 pins.

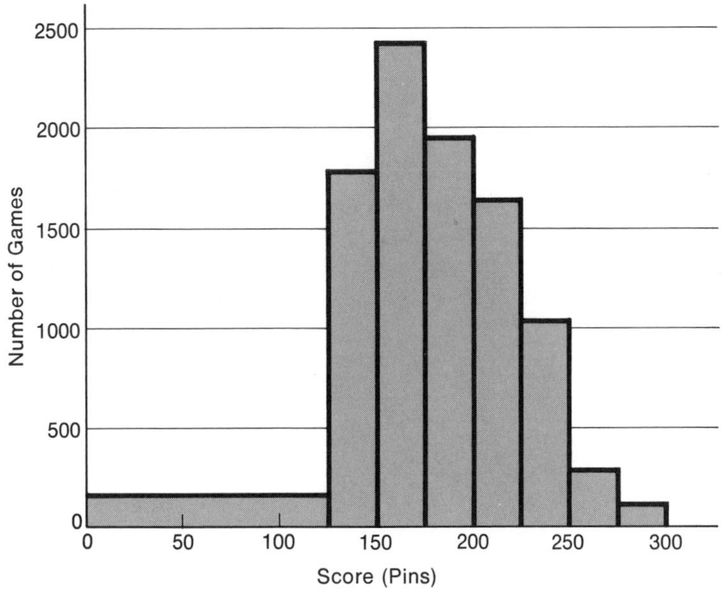

Since the first interval (0 to 125) is five times the size of the other intervals, its height is 162.4, or one-fifth of 812. This is because the proportion of scores is represented by the *area* of the rectangle. Now since the unit of difference in scores (one pin) is relatively small in relation to the range of the distribution (300 pins), the histogram may be approximated by a smooth curve. A curve derived from this histogram is given here.

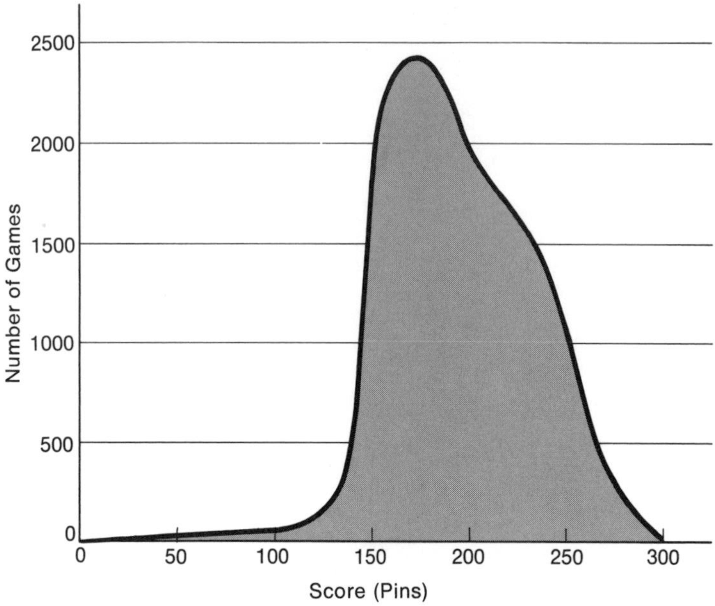

This curve will be a good approximation of the probability distribution if it gives an accurate impression of the data. The curve must have approximately the same area in the intervals as the histogram; that is, 0 to 125 should contain 0.0812 of the area, 126 to 150 must contain 0.1764 of the area, and so on. Note one essential difference—in the raw data, there is nothing between 125 and 126, nothing between 200 and 201, and so on. With the continuous approximation, the following **correction for continuity** must be made: each number in the original (discrete) distribution is represented in the continuous approximation by an interval from one-half unit below the number to one-half unit above the number. Thus, 50 would be represented by the interval 49.5 to 50.5 and the area under the curve from 49.5 to 50.5 would be equal to the proportion of cases (see following figure).

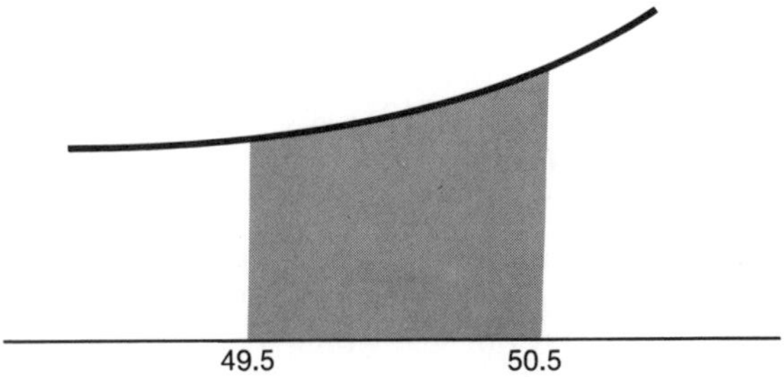

Thus, if 3 of the original 10,000 lines bowled were scores of 50, the area under the curve between 49.5 and 50.5 would be 0.0003 of the area under the entire curve. It would be too much, of course, to expect an approximation to agree in such fine detail. In the bowling data given, 125 would be represented by the interval 124.5 to 125.5. Since 125 belongs in the first interval given, but 126 in the second, all scores under 125 would belong to the interval 0 to 125.5. Then the areas would be given by the following continuous approximation:

Interval	*Area (probability)*
0–125.5	0.0812
125.5–150.5	0.1764
150.5–175.5	0.2433
175.5–200.5	0.1911
200.5–225.5	0.1646
225.5–250.5	0.1037
250.5–275.5	0.0294
275.5–300.5	0.0103

The most striking feature of a probability density (i.e., a probability function over a continuous variable) is that the area under the curve is always equal to one, and the probability that the variable will have a value between x_1 and x_2 is the area under the curve between x_1 and x_2. That is, $P(x_1 < x < x_2) =$ shaded area.

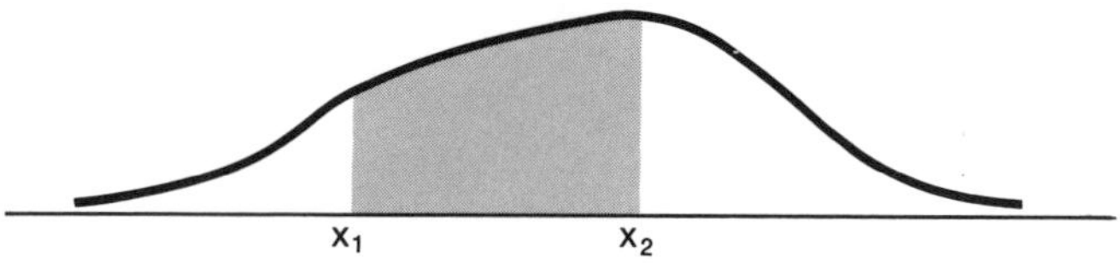

There is no difference between $P(x_1 < x < x_2)$ and $P(x_1 \leq x \leq x_2)$, because the area under the curve above x_1 is technically zero; this is the area of a line segment.

6.2 NORMAL DISTRIBUTIONS

One of the most important and useful set of continuous distributions in statistics is the set of **normal distributions**. Reference is often made to *the* normal distribution. This means the *standard* normal distribution, which will be discussed later in this section.

This curve was first developed to deal with errors in experimental work and was then found to have many other applications, as we shall see. A graph of a normal distribution is given.

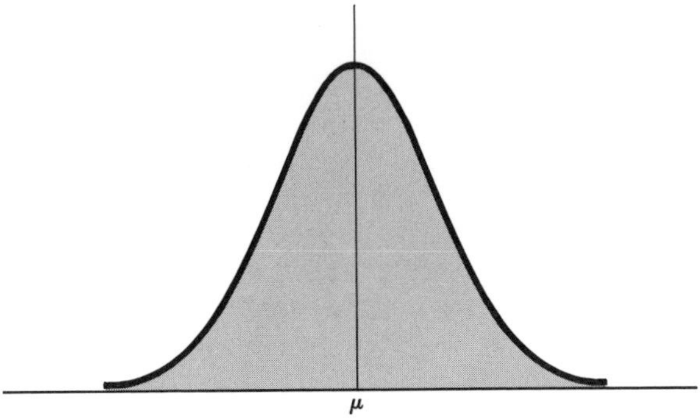

The normal distribution curve is determined entirely by the mean and the standard deviation. Because of this, graphs of normal distributions with the same mean, but different standard deviations, differ only in amount of dispersions, as shown in the following figure.

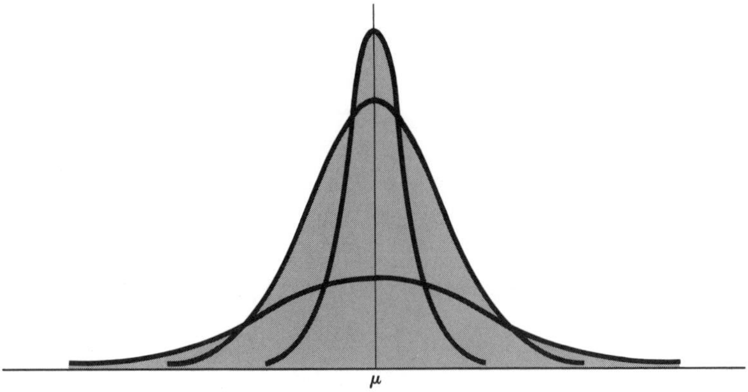

These curves are similar because they reach their highest point about the same mean, and taper off equally in both directions.

Normal distributions with the same standard deviation, but different means, look identical in shape and differ only in their placement on the x-axis as shown here:

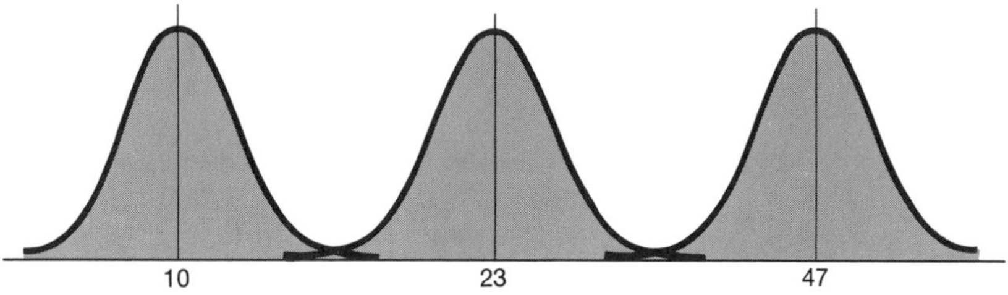

Theoretically, although it may not be apparent from these small drawings, the curve never touches the x-axis. However, it approaches it so closely when the value of x is about four standard deviations from the mean that any area lying further from the mean than that is, for practical purposes, negligible.

One of the properties of a probability density function is that the total area under its curve is equal to one. Thus the normal distribution, which is a probability density function, has a total area of one. A normal distribution with $\mu = 0$, $\sigma = 1$ is called the **standard normal distribution**. Any normal distribution can be scaled as follows:

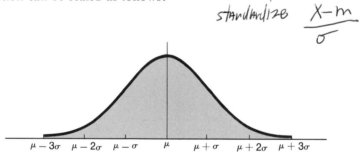

The relationship of any value of x to μ can be expressed in terms of the number of standard deviations distant from the mean. For example, if $\mu = 32$, $\sigma = 9$, a score of 45.5 is exactly 13.5 units above the mean, or (13.5)/9, that is, 1.5 standard deviations above the mean. A score of 18.5 is 13.5 units below the mean, or 1.5 standard deviations below the mean. This can be symbolized as -1.5 as opposed to $+1.5$ for 45.5. As mentioned in Chapter 2, these scores are called **standard scores**, and referred to as ***z*-values** or ***z*-scores.** The name comes from the fact that such scores are standard deviations of the standard normal distribution; that is, a score of 1.37 for the standard normal distribution would be 1.37 standard deviations above the mean. Thus, values of x for any normal distribution can be related to standard scores by the following formula:

STANDARD SCORE FORMULA $$z = \frac{x - \mu}{\sigma}$$

EXAMPLE 1 Determine standard scores for $x = 18.3, 27.9, 34.4, 39.3$ in the normal distribution for which $\mu = 30.1$, and $\sigma = 2.4$.

SOLUTION If

$$x = 18.3,\ z = \frac{18.3 - 30.1}{2.4} = \frac{-11.8}{2.4} \doteq -4.92$$

if

$$x = 27.9,\ z = \frac{27.9 - 30.1}{2.4} = \frac{-2.2}{2.4} \doteq -0.92$$

if

$$x = 34.4,\ z = \frac{34.4 - 30.1}{2.4} = \frac{4.3}{2.4} \doteq 1.79$$

if

$$x = 39.3,\ z = \frac{39.3 - 30.1}{2.4} = \frac{9.2}{2.4} \doteq 3.83$$

EXAMPLE 2 Determine values of x in the distribution of example 1 for which the standard scores are $z = -3.07, -1.04, 0.73, 2.44$.

SOLUTION If $z = (x - \mu)/\sigma$, then $z\sigma = x - \mu$ or $z\sigma + \mu = x$; if you prefer $x = z\sigma + \mu$. Thus, if

$$z = -3.07,\ x = (-3.07)(2.4) + 30.1 \doteq -7.4 + 30.1 = 22.7$$

if

$$z = -1.04,\ x = (-1.04)(2.4) + 30.1 \doteq -2.5 + 30.1 = 27.6$$

if

$$z = 0.73,\ x = (0.73)(2.4) + 30.1 \doteq 1.8 + 30.1 = 31.9$$

if

$$z = 2.44,\ x = (2.44)(2.4) + 30.1 \doteq 5.9 + 30.1 = 36.0$$

EXAMPLE 3 In a normal distribution, a value of 42.1 is 1.3 standard deviations above the mean of 31.7. What is the standard deviation of the distribution?

SOLUTION In this case we know x, z, μ, but not σ. Since $z = (x - \mu)/\sigma$, $\sigma = (x - \mu)/z$, so if $x = 42.1$, $z = 1.3$ and $\mu = 31.7$,

$$\sigma = \frac{42.1 - 31.7}{1.3} = \frac{10.4}{1.3} = 8.0$$

EXAMPLE 4 A normal distribution has a standard deviation of 1.7. If a value of 11.3 lies 2.1 standard deviations below the mean, determine the mean of the distribution.

SOLUTION Here we know x, z, σ, but not μ. Since $z = (x - \mu)/\sigma$, $z\sigma = x - \mu$. Therefore $z\sigma + \mu = x$, or $\mu = x - z\sigma$. Then if $x = 11.3$, $\sigma = 1.7$, $z = -2.1$, $\mu = 11.3 - (-2.1)(1.7) \doteq 11.3 + 3.6 = 14.9$.

The fact that any normal distribution can be related to the standard normal distribution is very important. Because of this, the standard normal distribution has been studied in detail and the results can be transferred to any normal distribution. The table of areas found in Table 2 is used to work with the standard normal distribution. Most tables of this sort are arranged like this, so if the student becomes proficient with this table, he will also be able to use similar tables.

In Table 2, the entries on the left and top correspond to values of z given to two decimal places. The integer value and the first decimal value are given in the column at the left, and the second decimal value in the top row. The entries in the body of the table are the areas under the normal curve between the mean, 0, and the given value of z correct to four decimal places. Remember that the standard normal curve is also a probability density, so that the area under the curve is also the probability that the random variable has a value between μ and the given z. Of course most of the applications of the normal curve involve approximations so that results obtained by using Table 2 should also be viewed as approximations.

For instance, if $z = 1.62$, to find the corresponding area look down the left column to find 1.6; look along the top row to find 0.02. The entry which is in both the row of 1.6 and the column of 0.02 is 0.4474. This means that the area under the standard normal curve between 0 and 1.62 is 0.4474 as illustrated:

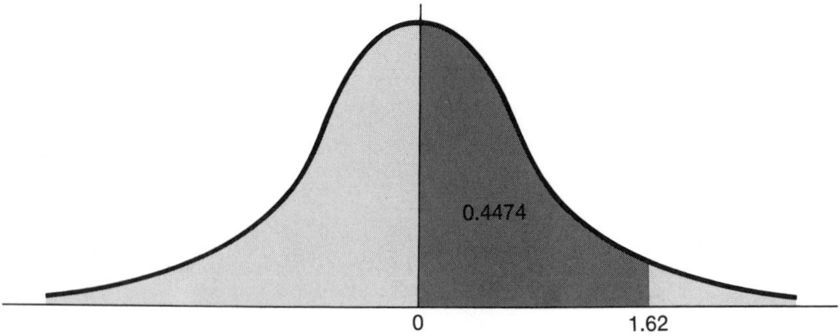

Several other pieces of information can be derived from this observation. Since the normal curve is symmetric, each side contains exactly 1/2 of the area of the curve. Thus, the area under the curve to the right of 1.62 is 0.5000 − 0.4474 or 0.0526. This is also illustrated here:

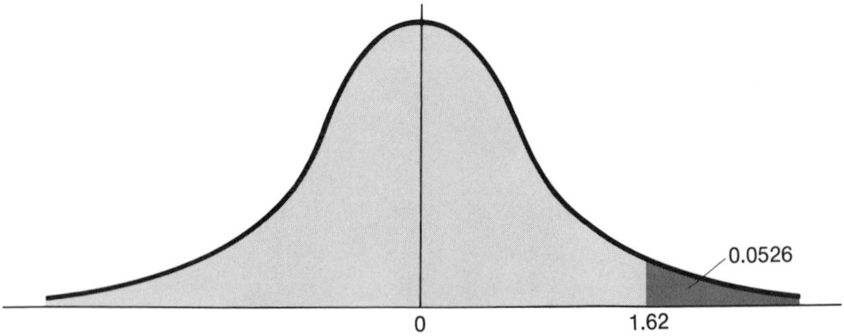

In addition, since the normal curve is symmetric, the area between $-z$ and 0 is equal to the area between 0 and z. Therefore, in this case the area between -1.62 and 0 is also 0.4474 and the area to the left of -1.62 is 0.0526 as shown:

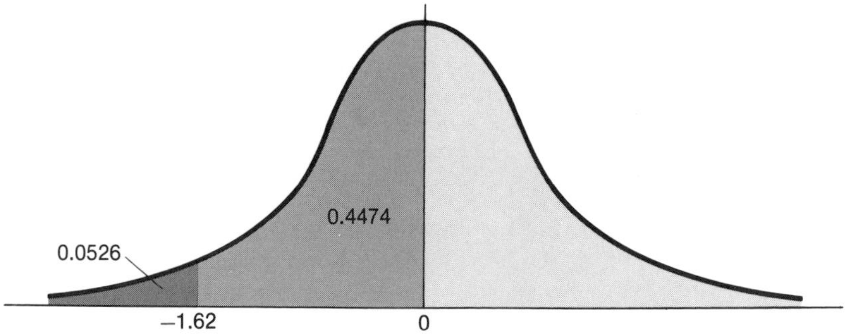

Finally, the area between -1.62 and 1.62 is therefore twice the area between 0 and 1.62 or 0.8948. The probability that the variable in a standard normal distribution has a value between -1.62 and 1.62 is equal to 0.8948. This is illustrated by the following:

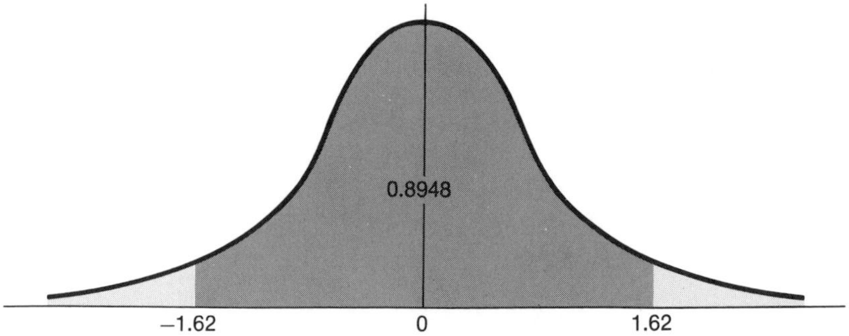

Various combinations can be obtained with these mathematical tools which apply to the standard normal curve.

EXAMPLE 5 Find the area under the standard normal curve between 0 and z if $z = 0.07, 0.83, 1.70, 2.56, -0.24, -1.12, -3.01$.

SOLUTION From the table we read the following values:

z	*Area between 0 and z*
0.07	0.0279
0.83	0.2967
1.70	0.4554
2.56	0.4948
−0.24	0.0948
−1.12	0.3686
−3.01	0.4987

DISCUSSION Remember that the *area* under the curve must always be positive. Remember also that the entries on the edge of the table (left and top) represent standard deviations distant from the mean (standard scores) while the entries in the body of the table represent areas under the standard normal curve between the mean and the given standard score (z-value). The student should verify example 5 and the remainder of the examples in this section, and keep at it until he is proficient in the use of Table 2.

EXAMPLE 6 Find the standard score for which the area under the standard normal curve between it and the mean is 0.2019, 0.3621, 0.4345, 0.4599, 0.4908.

SOLUTION From the table we have

Area	z
0.2019	0.53
0.3621	1.09
0.4345	1.51
0.4599	1.75
0.4908	2.36

DISCUSSION For 0.4908, note that no entry in the table is precisely 0.4908. The two entries closest to it are 0.4906 and 0.4909. Since 0.4908 is closer to 0.4909 than 0.4906 we use the value for 0.4909. An approximation method known as interpolation can be used if desired, but this is not necessary since two decimal accuracy for standard scores is sufficient for our purposes. In cases where the given area is midway between two table entries, such as 0.3953, in the absence of other information (such as more complete tables) we shall follow the rule of using the closest value of z for which the second decimal is even, that is, 0, 2, 4, 6, or 8. For 0.3953, then, $z = 1.26$ to two decimal places.

EXAMPLE 7 Determine the value of z for which 0.3300 of the area under the standard normal curve is to its left.

SOLUTION In solving these problems, sketching the curve is always useful. Here we know that *less* than half the area under the curve is to the left of the desired z-score so that z must be negative, since exactly half the area under the curve is to the left of zero. Furthermore, the table gives areas between z and the mean.

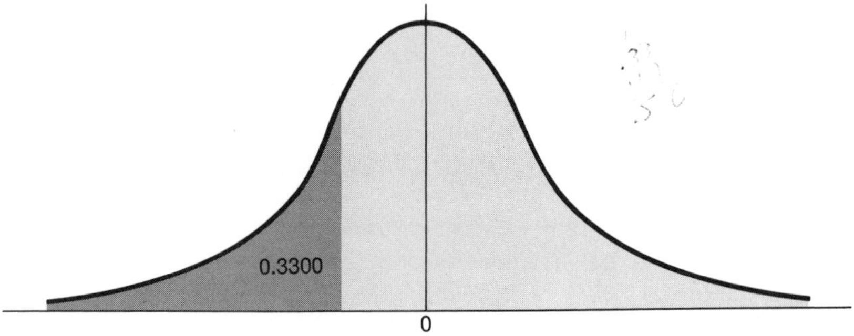

The area we will look up is the complement of the area between z and the mean; that is, $0.5000 - 0.3300$ or 0.1700 of the area under the curve lies between z and the mean. The corresponding z-score in the table is 0.44. In the light of the other information, then, we have $z = -0.44$ as illustrated by the following figure.

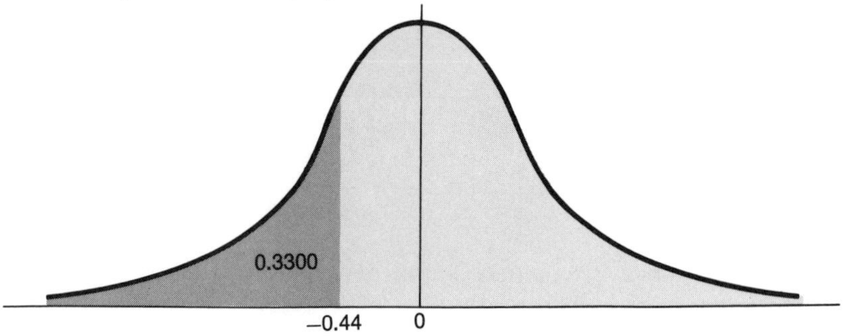

EXAMPLE 8 Find the area under the normal curve between $z = -1.34$ and $z = 0.57$; between $z = 0.59$ and $z = 1.27$.

SOLUTION For $z = -1.34$ and $z = 0.57$, the values of the areas from the table are, respectively, 0.4099 and 0.2157. Since these are on *opposite* sides of the mean, they must be added together.

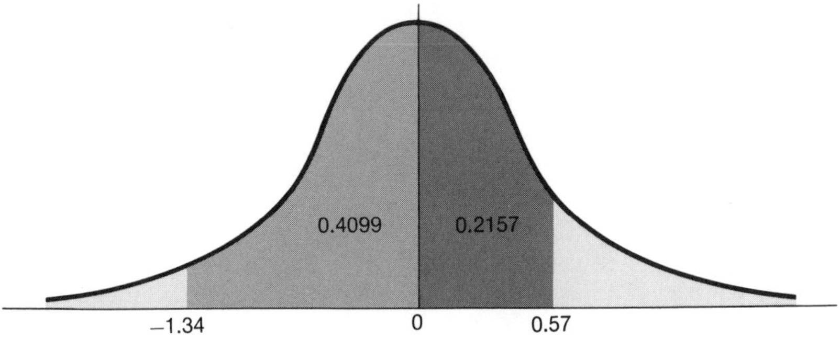

Thus, the area under the normal curve between $z = -1.34$ and $z = 0.57$ is 0.6256.

For $z = 0.59$ and $z = 1.27$ the corresponding areas are 0.2224 and 0.3980. Since they are on the *same* side of the mean, however, their difference is the desired area.

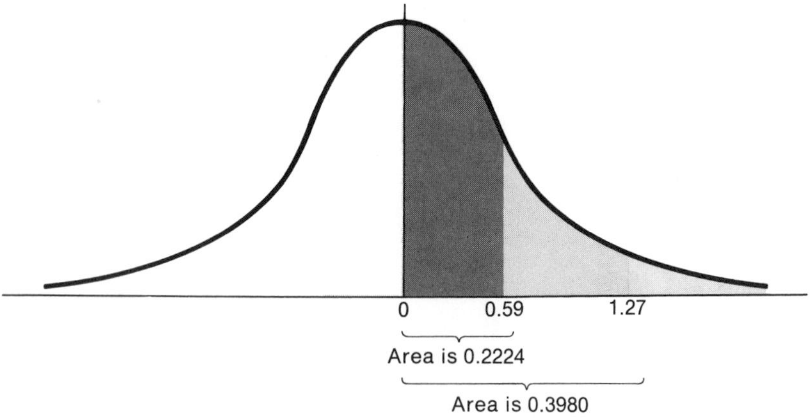

Thus, the area under the normal curve between $z = 0.59$ and $z = 1.27$ is 0.1756.

EXAMPLE 9 For a normal distribution with mean 38.7, standard deviation 10.2, estimate the probability that a value will fall between 29.6 and 44.8.

SOLUTION Each score must be converted to a standard score. In the first place,

$$z = \frac{29.6 - 38.7}{10.2} = \frac{-9.1}{10.2} \doteq -0.89;$$

in the second,

$$z = \frac{44.8 - 38.7}{10.2} = \frac{6.1}{10.2} \doteq 0.60.$$

The corresponding areas under the standard normal curve are 0.3133 and 0.2257. Since the z-scores are on opposite sides of the mean (as are the original values) the areas must be added to determine the area between the scores. Adding we obtain $P(29.6 < x < 44.8) = 0.5390$.

EXAMPLE 10 A normal distribution has a mean of 133 with a standard deviation of 21. Determine a number such that 80% of all scores fall within that number of the mean.

SOLUTION If 80% of all scores fall within the desired number of the mean, then 40% fall between the mean, 133, and $133 + n$, where n is the number. This is because the normal curve is symmetric. The z-score corresponding to 0.4000 is 1.28, to two decimals. Then we have $1.28 = (x - 133)/21$, so that $x = 1.28(21) + 133$ or $x \doteq 160$. Thus 40% of the scores fall between 133 and 160, so that $n = 27$, and 80% of all scores fall within 27 of the mean; that is, 80% of all scores lie between 106 and 160.

Problems

1. Estimate standard scores for the following values of x in a normal distribution with mean 284.7 and standard deviation 14.6.

(a) $x = 261.4$
(b) $x = 303.7$
(c) $x = 259.3$
(d) $x = 280.4$
(e) $x = 293.9$
(f) $x = 321.2$

2. Give values of x, for the following standard scores, in a normal distribution with $\mu = 10.4$, $\sigma = 11.8$.

   (a) $z = 1.64$
   (b) $z = 2.07$
   (c) $z = -2.16$
   (d) $z = 0.50$
   (e) $z = -0.13$
   (f) $z = 1.14$

3. Find the area under the standard normal curve between

   (a) $z = 0$ and $z = 2.18$
   (b) $z = -1.04$ and $z = 1.54$
   (c) $z = 1.56$ and $z = 2.93$
   (d) $z = -0.49$ and $z = -0.12$
   (e) $z = -3.04$ and $z = 1.63$
   (f) $z = -0.43$ and $z = 2.09$

4. Find the area under the standard normal curve

   (a) to the right of $z = 1.43$
   (b) to the left of $z = -1.03$
   (c) to the right of $z = -0.77$
   (d) to the left of $z = 2.01$

5. Find the area under the standard normal curve between z and $-z$ if

   (a) $z = 1$
   (b) $z = 2$
   (c) $z = 3$
   (d) $z = 1.28$
   (e) $z = 1.64$
   (f) $z = 1.96$
   (g) $z = 2.33$
   (h) $z = 2.58$

6. Find the value of z for which 0.1230 of the area under the standard normal curve lies to the right of z.

7. If a normal distribution has mean 193.4 and standard deviation 13.7, calculate the probability that a particular value of the variable lies

   (a) above 210.4
   (b) below 186.5
   (c) above 179.9
   (d) below 200.0
   (e) between 173.4 and 186.8
   (f) between 166.1 and 207.9
   (g) between 200.0 and 210.0
   (h) above 200.0 or below 180.0

8. The mean of a normal distribution is 100.0. If the probability that the variable assumes a value greater than 121.0 is 0.1446, what is the standard deviation of the distribution?

9. A normal distribution has a standard deviation of 134. The probability that the variable takes a value less than 1,072 is 0.7734. What is the mean of the distribution?

10. Find to three decimal places, if possible, the value of z such that exactly 50% of the area under the standard normal curve lies between $-z$ and z. This value, once in use but now out of favor, is called the **probable error** of z for the standard normal distribution. Using the obtained value, estimate the probable error of x in a normal distribution with $\mu = 2{,}750$, $\sigma = 40$.

6.3 APPLICATIONS OF THE NORMAL DISTRIBUTIONS

Many sets of data have distributions which are approximately normal. For these sets, the methods of section 6.2 are applicable.

EXAMPLE 1 A certain brand of spaghetti sauce is packed in cans which are supposed to have a net weight of 15 1/2 ounces. Since the packing is done by machine and weight will vary from can to can, the machine will usually be set to average a little more than the stated net weight. Suppose the machine is set to fill the cans with a mean of 16 ounces and the distribution of fills has a standard deviation of 0.3 ounces. How many of a lot of 10,000 cans would you expect to find with less than the stated net weight of 15 1/2 ounces?

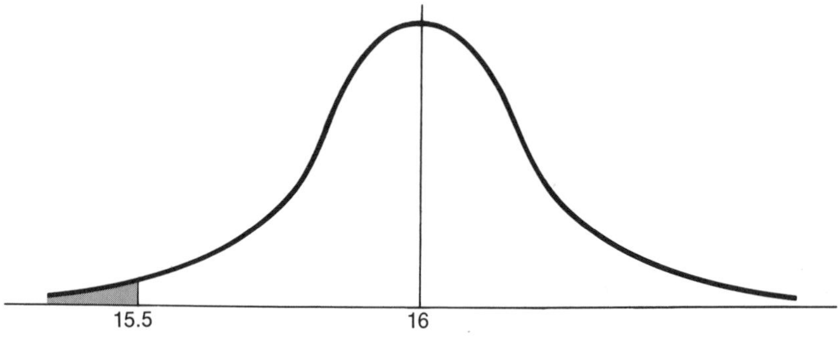

SOLUTION The standard value is obtained by the formula

$$z = \frac{15.5 - 16}{0.3} = \frac{-0.5}{0.3} \doteq -1.67$$

so that this is about 1.67 standard deviations below the mean. According to Table 2, 0.4525 of the area under the curve lies between 15.5 and 16, so 0.0475 of the area lies to the left of 15.5. Since the area under the curve represents (in this example) 10,000 cans, you would expect about 475 of the 10,000 cans to weigh less than 15.5 ounces. On the other hand, about 475 cans would weigh more than 16.5 ounces. This is a practical problem and, since the standard deviation is relatively difficult to adjust on such a machine, any changes to be made would be made on the mean.

EXAMPLE 2 Among workers in a certain industrial plant, the mean age is 45 years with a standard deviation of 4 years. One worker is stopped at random and asked to fill out a questionnaire. What is the probability that he is between 48 and 50 years of age?

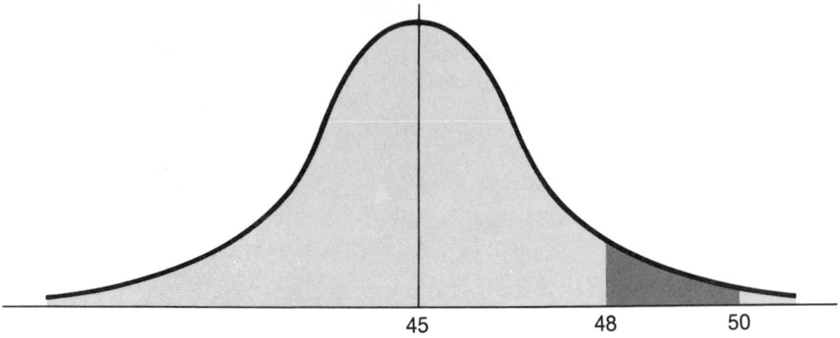

SOLUTION If z_1 denotes the standard score for 48 and z_2 the standard score for 50, we have

$$z_1 = \frac{48 - 45}{4} = 0.75, \; z_2 = \frac{50 - 45}{4} = 1.25$$

The areas under the normal curve associated with each (from Table 2) are, respectively, 0.2734 and 0.3944. The area between 48 and 50, then, is 0.3944 − 0.2734 or 0.1210. Thus, $P(48 < x < 50) = 0.1210$ where x is the age of the worker.

DISCUSSION Although age is measured by time and is thus measured continuously, age is often given to the last birthday. The way the previous example was answered gave the probability that the worker was between his forty-eighth and fiftieth birthdays. If the form had only age in years rather than date of birth, then our data could be considered discrete and we might have used a continuity correction. When age is measured only to the nearest year, we might represent it by the midpoint of an interval and *here* "between 48 and 50" would be represented by 48.5 to 49.5 and would change the result considerably. The question of when to and when not to use the continuity correction has not been satisfactorily resolved. The recommendation in this text will be to use it whenever the data are truly discrete (without an underlying continuous variable as in example 2) and are measured in integers. Thus, data measured in dollars should be corrected for, but data measured in dollars and cents should not. This interpretation will not satisfy all statisticians or even authors of elementary statistics texts, but it is sufficient for our purposes and does at least standardize our responses.

EXAMPLE 3 A machine produces light bulbs. They are examined in lots of 1,000. On the average, a lot will have 10 defective bulbs; the distribution of defective bulbs is approximately normal with a standard deviation of 3.16. What is the probability that a certain lot will have at least 3 but not more than 6 defective bulbs? What is the probability that it will have more than 15 defective bulbs?

SOLUTION Since the data (number of defective bulbs) are discrete and measured in integers, a continuity correction must be applied. Since the interval representing 3 is 2.5 to 3.5 and the interval representing 6 is 5.5 to 6.5, the interval representing at least 3, but not more than 6 is 2.5 to 6.5 since both 3 and 6 are included.

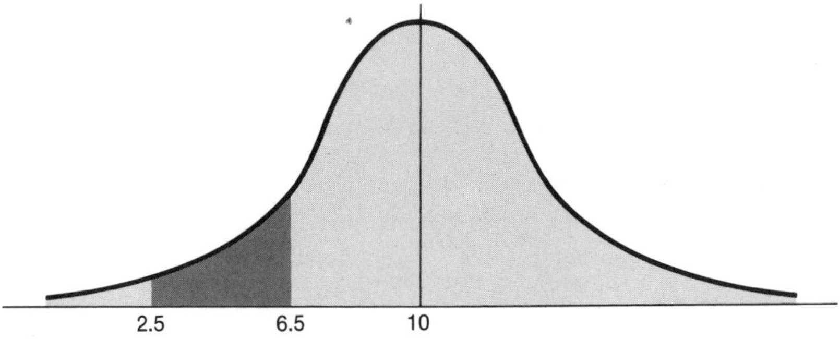

Then, if z_1 and z_2 represent the standard scores for 2.5 and 6.5, respectively, we have

$$z_1 = \frac{2.5 - 10}{3.16} = \frac{-7.5}{3.16} \doteq -2.37 \quad \text{and} \quad z_2 = \frac{6.5 - 10}{3.16} = \frac{-3.5}{3.16} \doteq -1.11$$

The associated areas (from Table 2) are 0.4911 and 0.3665. The area between 2.5 and 6.5, then, is 0.4911 − 0.3665 or 0.1246. Thus, $P(2.5 < x < 6.5) = 0.1246$. In terms of the original, discrete data, $P(3 \le x \le 6) = 0.1246$.

To determine $P(x > 15)$, note that 15 is represented by the interval 14.5 to 15.5. To be greater than 15 means, after application of the continuity correction, greater than 15.5, since 15 is not to be included.

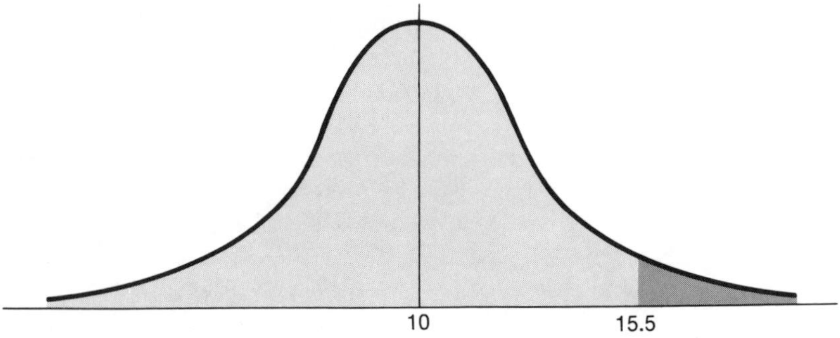

Here

$$z = \frac{15.5 - 10}{3.16} = \frac{5.5}{3.16} = 1.74.$$

The associated area is 0.4591, so

$$P(x > 15) = 0.0409$$

EXAMPLE 4 In a certain high-rent district, the monthly rental for apartments is approximately normally distributed with a mean of $384.22 and a standard deviation of $126.40. Above what value is the highest 30% of the monthly rentals in this district?

SOLUTION Although the variable is discrete, its values are not given in integers so we do not apply the continuity correction. According to Table 2, 20% of the values are between the mean and z for $z \doteq 0.52$. At this point, 30% of the values are above it.

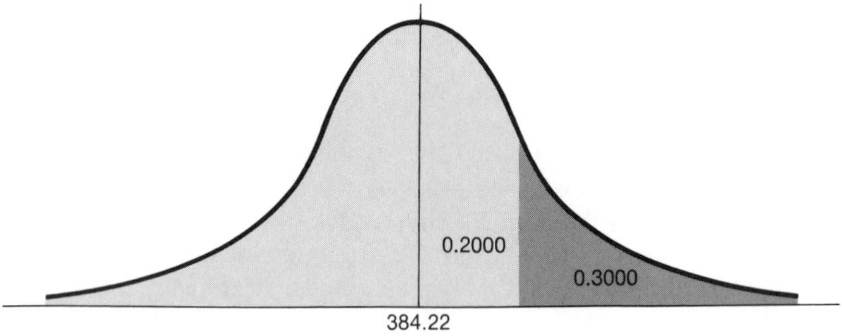

Thus, we have $0.52 = (x - 384.22)/126.40$, so $(0.52)(126.40) = x - 384.22$ or $x = (0.52)(126.40) + 384.22$ or about 449.95. Thus, about 30% of the rentals are above \$449.95. A slightly more accurate figure could be obtained with more detailed tables.

Problems

1. Number 10 cans of peaches are supposed to hold 72 ounces. The canner sets the machine to fill the can with, on the average, 72.6 ounces. The distribution of fill weights is approximately normal with a standard deviation of 0.4 ounces. (a) How many of 100,000 cans will have less than 72 ounces? (b) The manufacturer feels that this is too many. Assuming the standard deviation remains unchanged, what mean should the manufacturer set if he wants no more than 300 of the cans to contain less than 72 ounces?
2. Light bulbs have lives that are approximately normally distributed except for those that don't work at all. If a shipment of good light bulbs is known to have a mean life of 916 hours with a standard deviation of 32 hours, what is the probability that a light bulb selected at random will last more than 1,000 hours?
3. The efficiency rating of certain machines is calculated each day. Over the years, machine A's ratings have had a mean of 0.873 with a standard deviation of 0.032, and machine B's ratings have had a mean of 0.846 with a standard deviation of 0.038. On a certain day, what is the probability that machine A will have a rating less than the mean for machine B? What is the probability that machine B will have a rating greater than the mean for machine A?
4. A basket of peaches in a certain fruit stand will have about 84 peaches. This varies a bit, however, and is normally distributed with the mean number of peaches per basket at 84 and a standard deviation of 2 peaches. What is the probability that the basket will contain between 80 and 90 peaches? (NOTE: "Between" does not include the end points.) What is the probability that it will contain exactly 86 peaches? How many of 10,000 such baskets would you expect to contain exactly 86 peaches?
5. Dowel rods for a certain product must have a diameter of 4 mm with a tolerance of 0.020 mm. Rods with diameters greater than 4.020 mm or less than 3.980 mm are not usable. Company A guarantees that the standard deviation on a lot with mean 4 mm will be 0.015 mm and will charge \$400 per lot of 10,000. Company B guarantees that the standard deviation on a lot with mean 4 mm will be 0.012 mm and will charge \$460 per lot of 10,000. Which company should be selected for the order if the major criterion is cost per usable dowel rod?
6. To test whether a process is in control, a reading is taken on a machine. If the reading is greater than 860.0, the machine is retooled. If the process is in control, the daily readings will have a mean of 832.4 with a standard deviation of 10.2. What is the probability of retooling a machine by mistake? That is, what is the probability of getting a reading above 860.0 when the process is in control?

7. A survey organization regularly sends out questionnaires. The number of replies on a mailing of 1,000 is approximately normally distributed with a mean of 785 and a standard deviation of 41. If a mailing of 1,000 questionnaires is made, what is the probability, on the basis of past experience, of receiving more than 850 replies?
8. Weights of male students in a large university are approximately normally distributed. Estimate the mean and standard deviation of the distribution if 6.68% of the students weigh less than 125 pounds and 15.87% weigh more than 170 pounds.

6.4 THE NORMAL APPROXIMATION TO THE BINOMIAL

One of the most important distributions in statistics is the binomial. Yet as the number of trials increases, it becomes more and more cumbersome to use. To determine the probability, say, of 30 or more heads in 50 tosses of a coin might require 21 separate calculations of individual probabilities. Fortunately, as n (the number of trials) increases, if the probability of a success is about 0.50, the normal distribution is satisfactory as a continuous approximation to the binomial. Below is a graph of a binomial distribution for $n = 10$, $p = 0.5$, with a continuous distribution superimposed upon it.

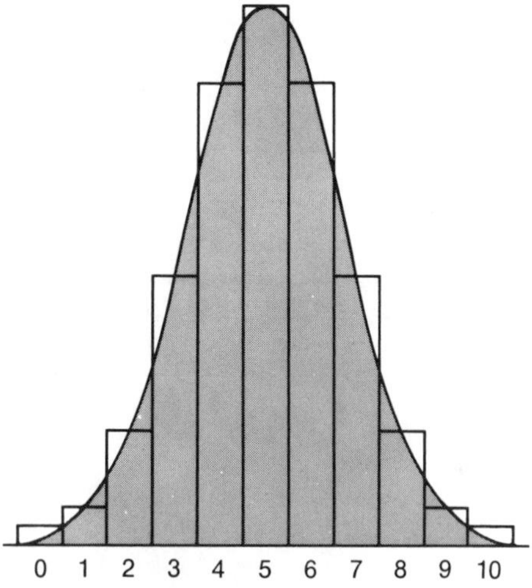

As the number of trials becomes greater and greater, the binomial distribution becomes closer and closer to the normal distribution and for very large numbers of trials, the probability of success can differ substantially from 0.50. The accuracy of the approximation depends, of course, on the number of trials and the actual probabilities involved. As a rule of thumb, the continuous approximation should not be used unless both np and $n(1 - p)$ are greater than 5 where n is the number of trials and p is the probability of success.

In the event these conditions are satisfied, a binomial distribution can be approximated by a normal distribution with the same mean and standard deviation. Since the mean and standard deviation of a binomial distribution are given by np and $\sqrt{np(1-p)}$ respectively, this information is always readily available, and since the data are discrete and measured in integers, a continuity correction must always be applied.

Example 3 of the previous section is actually a binomial distribution with $p = 0.01$. Thus the histogram for the data from 3 to 6 would appear as the following graph.

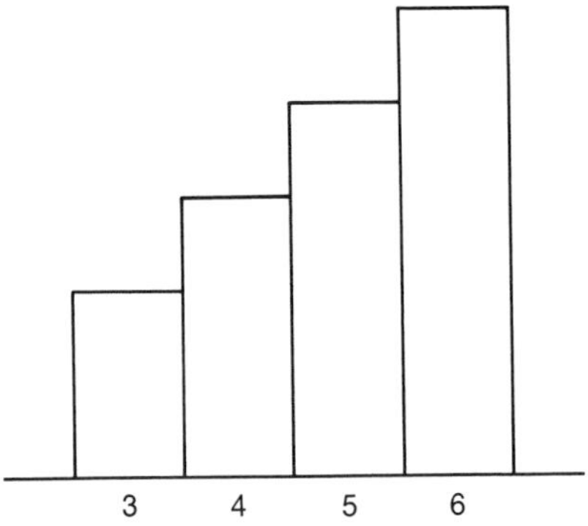

The normal curve, superimposed on the binomial distribution would then appear this way, with the class limits indicated.

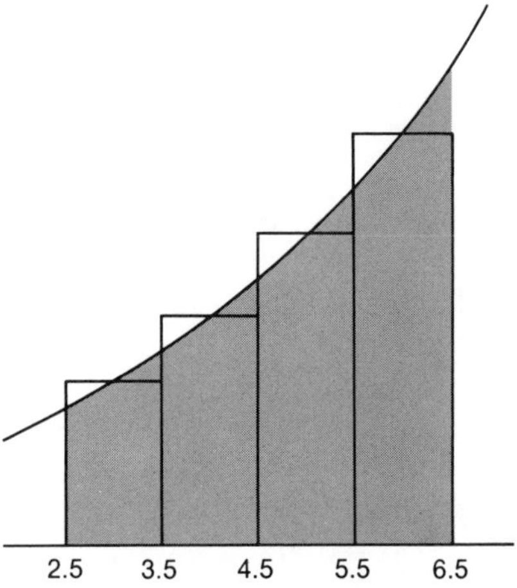

Thus the shaded area on the graph is as close an approximation as possible to the area in the rectangles—the areas associated with 3, 4, 5, and 6 defective bulbs. Hence the proportion of the area under the curve which is shaded is approximately $P(3 \leq x \leq 6)$ which was calculated to be about 0.1246. Direct calculation using the binomial formula yields a value of 0.1262 which agrees relatively satisfactorily, particularly since the probability is far removed from 0.5.

EXAMPLE 1 What is the probability of obtaining 30 or more heads in 50 tosses of a coin?

SOLUTION This is a binomial experiment with $n = 50$ and $p = 1/2$. Its distribution can be approximated by a normal distribution with $\mu = 50(1/2) = 25$ and $\sigma = \sqrt{50(1/2)(1/2)} = \sqrt{12.5} \doteq 3.54$. Now 30 is represented on the continuous distribution by the interval 29.5 to 30.5; since 30 is included, we must determine the probability that x is greater than 29.5. The appropriate standard score is

$$z = \frac{29.5 - 25}{3.54} = \frac{4.5}{3.54} \doteq 1.27$$

This corresponds to an area of 0.3980 between 29.5 and the mean (see Table 2). Since we are interested in the probability of obtaining 30 or more, this probability is 0.5000 − 0.3980 or 0.1020. Thus, $P(x \geq 30) \doteq 0.1020$.

EXAMPLE 2 Find the probability of obtaining 5 or more successes in 25 trials if the probability of success on a given trial is 0.20. Compare with the true probability. Repeat for $p = 0.40$.

SOLUTION Using the normal approximation with $n = 25$, $p = 0.2$, we have $\mu = np = 25(0.2) = 5$, $\sigma = \sqrt{25(0.2)(0.8)}$, or $\sigma = \sqrt{4} = 2$. Since 5 is represented by the interval 4.5 to 5.5, we want the probability that x is greater than 4.5. The standard score is

$$z = \frac{4.5 - 5}{2} = \frac{-0.5}{2} = -0.25$$

which corresponds to an area of 0.0987. Since the entire area above the mean is included, we have $P(x \geq 5) = 0.5987$. From tables of binomial probabilities, the true probability is $P(x \geq 5) = 0.579$. In this case, the normal approximation is not particularly good except for rough guessing. Note that $np = 5$ and that our rule of thumb stated that the normal approximation should not be used unless np is greater than 5.

For $n = 25$, $p = 0.4$, we have $\mu = 25(0.4) = 10$, $\sigma = \sqrt{25(0.4)(0.6)} = \sqrt{6} \doteq 2.45$. The standard score in this case is

$$z = \frac{4.5 - 10}{2.45} = \frac{-5.5}{2.45} \doteq -2.24.$$

The area under the normal curve in this case is 0.4875, so $P(x \geq 5) = 0.9875$. The true binomial probability is 0.991, so that the approximation here is better than for $p = 0.20$.

EXAMPLE 3 Determine the probability of getting from 6 to 10 "sevens," inclusive, in 45 rolls of a pair of dice. Estimate the probability of obtaining exactly 3 "sevens."

SOLUTION The probability of obtaining a seven on one roll is 1/6, so $\mu = 45(1/6) = 7.5$ and $\sigma = \sqrt{45(1/6)(5/6)} = 2.5$. Since both 6 and 10 are to be included, we wish to determine the probability of the interval from 5.5 to 10.5. The standard scores are

$$z_1 = \frac{5.5 - 7.5}{2.5} = \frac{-2}{2.5} = -0.80 \quad \text{and} \quad z_2 = \frac{10.5 - 7.5}{2.5} = \frac{3}{2.5} = 1.20$$

The corresponding areas are 0.2881 and 0.3849. Since they are on opposite sides of the mean, we have $P(6 \leq x \leq 10) = 0.6730$.

To determine the probability of exactly 3 "sevens," note that 3 is represented by the interval 2.5 to 3.5. Thus, we have

$$z_1 = \frac{2.5 - 7.5}{2.5} = \frac{-5}{2.5} = -2.00 \quad \text{and} \quad z_2 = \frac{3.5 - 7.5}{2.5} = \frac{-4}{2.5} = -1.60$$

The corresponding areas are 0.4772 and 0.4452. Since these values are on the same side of the mean, we must subtract them, so $P(3) = 0.0320$.

Problems

1. Find the probability of obtaining 3 heads on 12 tosses of a coin using the normal approximation and check by using the binomial formula.

2. A coin is tossed 25 times. If 15 or more heads appear, a player receives \$20. If 15 or more tails appear, the player must pay \$10. What is the player's expected profit (or loss) for one game of 25 tosses.

3. A test has 225 multiple choice questions, each with 1 correct and 4 incorrect answers. Each question is answered randomly, without regard to the true state of things. What is the probability of obtaining more than 60 correct answers?

4. If it is known that 25% of statistics students study for a test, what is the probability that less than one-third of a random sample of 48 students will study for tomorrow's test?

5. A poll is taken on the subject of new antipollution laws. If actually 60% of the population support the laws, what is the probability that the majority of a random sample of 96 persons will nevertheless be opposed to the law?

6. A thumbtack is tossed 84 times. It falls point up, on the average, about 30% of the time. What is the probability that it will fall point up at least 10 times, but no more than 25 times?

7. Rats were tested to see if they preferred a square or a triangle. In an experiment, the rat could step on a square or a triangle and would receive food in either case. The figures were the same in area, color, and composition, so that the only difference was shape. If there were no preference, it can be assumed that the rat would choose each shape with a probability 0.5. What is the probability that, if there really is no preference, a rat would choose the triangle 40 or more times in 60 trials?

8. Approximately 6% of the male population is color-blind. If a random sample of 250 male adults were sampled, what is the probability that at least 10, but no more than 20 of the subjects would be color-blind?

9. A test for right and left preference of salmon was conducted prior to construction of a "ladder" to aid salmon swimming upstream to get around a dam. Artificial channels were constructed to observe whether salmon swimming upstream would prefer left or right channel. A total of 732 salmon was observed. Of them, 348 swam up the left channel and the remainder used the right channel. Determine the probability that 348 or fewer would select the left channel by chance if there really were no preference.

10. It has been found that sex ratios in families are conditional. That is, if a couple has a daughter, it is more likely that the second child will be a daughter. If a run of children of the same sex occurs, the probability of the same sex increases with each additional child. (You may recall a famous entertainer who had 9 sons in a row.) Suppose that the probability of having a son if the first 2 children are girls is 0.44. Of 100 families having 2 girls then having a third child, what is the probability that at least one-half of the children born (assuming single births) will be boys?

11. In a certain area of a large city, incomes classified as below minimum subsistence levels are said to be possessed by 45% of the families. A random sample of 300 families is interviewed. If we use the binomial model, what is the probability that more than half the families will have incomes so classified if the claim is true?

6.5 MORE ON HYPOTHESIS TESTING

In section 5.4 we investigated an experiment concerning eczema in mice. The use of the normal approximation to the binomial allows us to extend the number in the sample without extensive binomial tables.

Remember that in this experiment we had a treatment for eczema whose effectiveness we wished to assess. Without the treatment we expected about 2/5 of the mice to be free of eczema after four weeks. Thus our *null hypothesis* was that $p = 2/5$, where p is the probability of a particular mouse being eczema-free after four weeks. We can use the normal approximation to the binomial for sample sizes which meet the criteria; that is, both np and $n(1 - p)$ must be greater than 5. (Note that, for cases for which this is not true, such as $n = 100$, $p = 0.01$, it is often possible to use the Poisson distribution as an approximation to the binomial. See section 5.5 for details.)

Now suppose we wish to decide whether the treatment is effective, but we wish to take a chance of only 0.05 of being mistaken in concluding that it is effective. Since 0.05 is the proportion in the upper tail for $z = 1.64$, we let

$$1.64 = \frac{x - \mu}{\sigma} = \frac{x - np}{\sqrt{np(1 - p)}}$$

for $p = 2/5$ and our sample size. If the sample is, say, 84, then $np = 33.6$, $\sqrt{np(1 - p)} = \sqrt{20.16} \doteq 4.49$, so

$$1.64 = \frac{x - 33.6}{4.49}, \quad \text{or} \quad x = (1.64)(4.49) + 33.6 \doteq 40.96.$$

Thus our *rejection region* is for $x > 40.96$. If 41 or more mice in the sample are eczema-free after four weeks, we can say that the treatment is effective with a probability of only 0.05 of being mistaken.

EXAMPLE 1 In the example of section 5.4, a businessman carries two products identical except for brand names, brands A and B. His inventory requires that he carry only one of the brands. Brand A does a great deal of advertising, so he is tempted to carry this brand only. Past records show about equal amounts of the two brands sold, but he feels that his customers may be showing a preference for brand B in more recent weeks. He sets up an experiment based on the next 25 sales of the product. His null hypothesis is $p = 0.5$, where p is the probability that brand B is preferred. He is willing to accept a probability of 0.10 of being wrong in rejecting brand B and finds that he will stock brand B only if 17 or more of the next 25 customers buy brand B; otherwise, he will stock brand A. One problem is that, if the true probability is 0.6, he will have a probability of 0.726 of stocking brand A

mistakenly. One way to improve this situation is to increase the sample size. Suppose he decides to take into account all sales of the product for the next two weeks. The formal method of testing this hypothesis is set forth in Chapter 9, and the particular technique is given in section 10.3. In the formal methods the rejection region is given in terms of z values rather than x values. Nonetheless, if the sample size is known, the rejection region can be determined. In this case, if the sample is 163, then, for $p = 0.5$, $np = 81.5$, $\sqrt{np(1-p)} \doteq 6.38$, and 0.10 of the area is in the tail for $z = 1.28$, so we have

$$1.28 = \frac{x - 81.5}{6.38}$$

or $x = (1.28)(6.38) + 81.5 \doteq 89.67$. Thus if 90 or more customers buy brand B, he would stock it; otherwise, he would stock brand A. Now, if the true probability is 0.6, β, the probability of landing in the acceptance region, can be determined. That is, for $p = 0.6$, $np = 97.8$, $\sqrt{np(1-p)} \doteq 6.25$ and the appropriate z value for $x = 89.5$ (since we must use a continuity correction) is

$$z = \frac{89.5 - 97.8}{6.25} \doteq -1.33$$

Only 0.0918 of the area under the normal curve is to the left of this value for z, so the probability of fewer than 90 customers buying brand B if the probability is 0.6 that brand B is actually preferred is $\beta = 0.0918$.

Problems

1. Determine β for example 1 in this section if $p = 0.7$.
2. A biologist working with a new strain of virus finds that rats injected with the virus develop symptoms similar to those of the common cold. He wishes to compare the duration of these symptoms with that of the cold, so he injects 100 rats with the virus and observes the duration of the symptoms. The symptoms of the common cold last for 14 days, on the average. If he hypothesizes that the symptoms last longer than the cold, he is testing the null hypothesis that the symptoms last 14 days. The alternate hypothesis is that the symptoms last longer than 14 days, or $p = 0.5$ that the symptoms last 14 days. What is his rejection region for $\alpha = 0.05$? That is, how many rats would have to have symptoms lasting more than 14 days for him to conclude, with the probability of being wrong equal to 0.05, that the symptoms do last more than 14 days? What is his rejection region for $\alpha = 0.02$?
3. A television network is offering a new show and expects at least 40% of the viewers to watch the premier performance. A poll is taken by telephone of 400

viewers and it is found that 189 are watching the new show. Set up null and alternate hypotheses for this experiment, determine the rejection region for $\alpha = 0.05$, and draw conclusions based on the results.

4. Patients who contract a certain rare disease have only 0.01 probability of surviving with old methods of treatment. A new technique is being tried throughout the country. A total of 400 patients have been treated with this technique and 10 have survived. Using the Poisson approximation to the binomial with $\alpha = 0.05$, test whether or not we can say that the new treatment is more successful than previous treatments.

6.6 SUMMARY

In this chapter we have discussed continuous probability distributions with special emphasis on the normal distributions.

A normal distribution is related to the standard normal distribution ($\mu = 0$, $\sigma = 1$) by the formula

STANDARD SCORE FORMULA $$z = \frac{x - \mu}{\sigma}$$

The values of z are used to enter Table 2 to obtain the area under the standard normal curve between $|z|$ and 0.

The binomial distribution may be approximated by a normal distribution if np and $n(1 - p)$ are both greater than 5. In these cases, the normal distribution approximating the binomial will have $\mu = np$, $\sigma = \sqrt{np(1 - p)}$.

In the normal approximation to the binomial and all other continuous approximations to discrete data distributions, a continuity correction must be applied if the variable is truly discrete and measured in integers. In these cases, each integer, x, is represented by the interval from $x - 0.5$ to $x + 0.5$.

It is also possible to extend the hypothesis testing technique discussed in Chapter 5 by using the normal approximation to the binomial.

Problems

1. On a certain IQ test, scores are approximately normally distributed with a mean of 100 and a standard deviation of 16. Assuming an underlying continuous distribution determine the following:

(a) the proportion of scores falling between 80 and 110
(b) the proportion of scores falling above 130
(c) the proportion of scores falling below 85
(d) the score above which lie 25% of the scores
(e) the score corresponding to a standard score of 1.12
(f) the score corresponding to a standard score of −0.87
(g) the probability that a score selected at random will lie between 100 and 120
(h) the probability that a score selected at random will be less than 110

2. A test for manual dexterity is found to have approximately a normal distribution. The mean is found to be 37.6 and the probability that a score is below 30.0 is 0.0918. Calculate the standard deviation.

3. A memorization test has a standard deviation of 11.62. If 69.5% of the scores exceed 70.34, what is the mean score for the test?

4. A machine which fills fruit juice cans can be set to fill the cans with a mean of anywhere from 44.80 ounces to 48.20 ounces. The standard deviation of the fills will be 0.32 ounces regardless of the mean. If the manufacturer, in order to comply with federal regulations, wishes no more than 100 cans of every 10,000 to contain less than 46 ounces, what should he set as the mean fill? If he wants no more than 100 of every 100,000 cans to fall below 46 ounces, at what value should the mean fill be set?

5. A true-false test is answered by tossing a coin for each question. The test consists of 100 questions. What is the probability of getting 60 or more correct answers?

6. In a large community only 45% of the registered voters are in favor of a bond issue. If only 600 voters go to the polls and vote, what is the probability that the issue will pass? (A tie vote counts as a defeat.)

7. A die is rolled 180 times:

   (a) What is the probability of obtaining exactly 20 sixes?
   (b) What is the probability of getting more than 40 threes?
   (c) What is the probability of getting at least 50 numbers divisible by 3?

8. On the average 60% of the graduates of a certain high school go on to college. This year's graduating class has 150 students in it. Using past records as a criterion, what is the probability that fewer than 75 of the graduates will attend college? One hundred or more? Exactly 100?

9. To test the hypothesis that fewer than 30% of all consumers will choose a higher priced but more attractively packaged version of a product, both versions are displayed prominently in a store and a careful record is kept of the sales of each type. The experiment will be based on the results of the first 200 sales. What rejection region leads to the conclusion that actually fewer than 30% of all consumers will buy the higher priced version, for $\alpha = 0.05$? What conclusion will be drawn if actually 53 persons buy the higher priced type? If the actual probability is 0.25, what is β for this experiment?

10. An amateur gardener decides to put in a border of petunias along his driveway. He buys a package of mixed petunia seeds. One-fifth of the seeds should grow into pink petunias and four-fifths into red and white variegated petunias. Assume that the seeds were selected at random from a population mixed in this proportion and that the binomial model applies. If there are 200 seeds in the package, what is the probability that there will be at least 30 but not more than 50 pink petunias if all the seeds germinate?

11. A biologist needs a minimum of 10 specimens of a species of annelid with an abnormal alimentary dysfunction. On the average, about 20% of all members of this species possess this characteristic. If she goes on an expedition to collect specimens, it is fairly costly to collect more than needed but impractical to test each for the dysfunction in the field. She brings back a sufficient number of specimens to give her a high probability of having at least ten of the needed specimens. Since 10 is 20% of 50, the inexperienced researcher might feel that it is sufficient to bring back a sample of 50. Using the normal approximation, what is the probability that a sample of 50 would contain at least 10 with the specified dysfunction? If she wants to be 99% confident of having enough, how many should she bring back? (This part is difficult.)

12. In an experiment concerning sex bias, 50 personnel managers were given the same 5 files each and asked to select a person for a job from these 5. Although the same files were given, the names which indicated sex were varied among the files. The qualifications were approximately equal so that the applicants should have had equal chances of being selected and the randomization of names was done to further equalize the chances. Three of the names were male and 2 were female. Seventeen of the respondents chose a female name, 33 chose a male name. Using a 0.05 level of significance, do you feel that this shows sex bias on the part of the personnel managers?

PART IV

SAMPLES

Part IV Photo: Verna R. Johnston from National Audubon Society

Chapter VII

Sampling and Sampling Distributions

7.1 SAMPLES

For the past several chapters we have been dealing with the characteristics of a complete statistical population. In Chapter 2 we defined a **population** as a set of data which consists of all possible or hypothetically possible values which can be assigned to a certain set of observations.

Probability, in a broad sense, deals with the application of population characteristics to obtain information about a **sample**, or subset of the population. More often, perhaps most often, our data are incomplete. We know the characteristics of a sample but not the population from which it was drawn. The extent to which we can apply our knowledge of this sample to estimate the characteristics of the whole population constitutes, in a broad sense, the field of **statistical inference**.

To learn about a population—workers of a certain type, students in a certain grade, customers of a certain store, fertilizers, motion picture films—it is quite often expedient to select a sample and generalize the results to the population from which the sample was taken. A ready-made sample, such as the set of students in a class, is called an **incidental sample**. Since the value of a sample depends on the degree to which it represents the original population, an incidental sample should not be used if better samples are available.

Statisticians have differing opinions on what constitutes a good sample, but much current research tends to indicate that samples obtained *at random* from a population may be considered representative of the population. A sample of a population is a **random sample** if *every sample of a given size is equally likely*, or, if the population contains N measures, there are $\binom{N}{n}$ samples of size n and each sample then has probability $1/\binom{N}{n}$ of being selected.

If, for instance, we wish to poll 10% of the student body, we may not be able to obtain a random sample by asking our questions of every tenth student. Consider the students standing in a line to register. If we chose students 10, 20, 30, 40, and so on, we would have 10% of the students. If the students were in the line at random, the sample would be random. On the other hand, if the students were grouped together in some way—for example, if all the mathematics majors were together—some *bias* would be introduced into the method of selection. A random sample could be obtained by using a spinner with numbers one through ten on it, which are equally likely to show up selecting a "lucky" number, say "one," and spinning the spinner every time a student appears. If a "one" shows up, we would poll the student; otherwise, we would not. Other methods involve the use of tables of random numbers which may be found in many statistics texts or in specially prepared books (see Table 11). In reality, of course, although one of these methods may be used beforehand, we would probably preselect which students should be picked. If we wanted to ask, on the average, every twelfth student in the cafeteria line if he was satisfied with the service, we could use a pair of dice and call "10" a success, since $P(10) = 1/12$. Then we could toss the dice, counting the tosses, and record the number of the toss on which "10" occurred. Such a list might look like this:

$$11, 17, 24, 42, 44, 50, 57, 78, 103, 104, 119, 143, \ldots$$

We would continue until we had enough. Then, when the students were in the line, we would simply count and ask students 11, 17, 24, 42, 44, and so on, and be assured that our sample was a random sample (as nearly as possible) of the students in the line that day. If we could be sure, however, that the students were randomly arranged, it would be all right to choose every twelfth one. As a general rule, a random sample can be obtained from any population by a random selection procedure, and from a randomly arranged population by a systematic selection procedure. If a sample is obtained from a biased population in a systematic way, however, it cannot be considered random, and will probably not represent the population very well. If one were to poll the faculty of a university by stopping at every tenth office, the result would probably not be random. If the mathematics faculty were arranged in nine consecutive offices, it might be that they had *no* chance of inclusion.

A good procedure is to use a **table of random numbers**, a sample of which is given in Table 11. If we do we are assured of the randomness of our procedure. In such a table the digits 0, 1, 2, 3, 4, 5, 6, 7, 8, 9 are arranged in sequence completely at random. The sequence is often generated by a computer. In this way the probability of a particular digit being in a particular place is the same for each digit. The numbers can be used singly (0, 1, . . . , 9), as two-digit numbers (00, 01, . . . , 99), or however needed. To use the table to select a sample, we determine the size of the population from which the sample is to be selected. We then use the number of digits necessary to assign every member of the population a number. For example, if we are to select a sample of 50 from a population of 5,054 students, we must use four-digit numbers. The numbers 0,001 to 5,054 will represent the students. There are often specific procedures for entering the table, but these are usually given in the book of tables itself. We might open the book at random and put our finger anywhere in the table. The first digit (if the table has between one and ten pages) will specify the page on which we are to begin. The next two digits identify the line and the final two digits give us the column of the number on which we should begin. Now suppose we do this and obtain the digits 10,927. We would turn to page one, line nine, twenty-seventh digit from the left. Then suppose the entries from this point are 70866005086192670496 . . . . We use these digits four at a time to obtain 7086, 6005, 0861, 9267, 0496. Numbers for which there are no students are ignored. Thus, if we have numbered the students from 1 to 5,054, we would select students numbered 0861 and 0496 as our first selections. When we run out of numbers in the line at which we entered the table we go to the next line. We would stop when 50 different students have been selected.

A table of random numbers can also be used to simulate experiments. A toss of twenty-five coins can be simulated by entering the table and calling odd digits "heads" and even digits "tails." These are equally likely. If, in another case, we want to make the probability of success 5/12, for example, we can let 01, 02, 03, 04, 05 represent success; 06, 07, 08, 09, 10, 11, 12 represent failure, and all other two-digit representations are ignored. This would be time consuming, however, so it might be better to use any multiple of 12 under 100, such as 96. Then we could let any 40 two-digit numbers represent success, 56 represent failure, and ignore 4. It would be simplest, of course, to let 01 through 40 represent success, 41 through 96 failure, and ignore 97, 98, 99, and 00. The table may be used for any such assignment or experiment. All that is necessary is to be absolutely certain that such use preserves the probabilities.

Many other methods of sample selection exist, such as selecting matched pairs or stratified samples, but for our purposes we shall assume our samples are obtained by random sampling.

After obtaining a sample, performing whatever experiments we desire, and applying appropriate statistical analyses, we must interpret the results. Whatever conclusions we draw about our sample must be generalized to some population—the population (or populations) of which the sample is representative. This population could be the obvious one from which the sample was physically drawn, or a more general one. A sample of three white mice taken from a cage are obviously representative of all the mice in the cage. They also represent all laboratory white mice in the world, as well as the hypothetically infinite population of all mice, past, present, and future. The population to which the results are to be generalized should be determined beforehand and samples drawn, so that this population is represented as accurately as possible.

Problems

1. A newspaper in a metropolitan area of about 250,000 people asks its readers to clip a coupon and express their opinion on a certain issue. On the basis of 506 returned coupons, of which 317 oppose the issue and 189 favor it, the newspaper concludes that citizens of the area overwhelmingly oppose the issue. Discuss the validity of this conclusion.
2. In order to obtain information, it is desired to interview a random sample of families in a certain city. Discuss the validity of each of the following methods:
   (a) selecting every tenth house, or every tenth apartment in an apartment building
   (b) standing on a street corner and interviewing a random sample of people passing by
   (c) sending questionnaires to every person on each postal route, with return postage included
   (d) telephoning a random sample of persons selected from the city telephone directory
   (e) visiting a random sample of persons selected from the city postal directory

   Does each method of sampling yield a truly random sample?
3. Suppose that a sample had been obtained using each of the methods of problem 3. Discuss the population to which the results can be generalized.

7.2 PARAMETERS AND STATISTICS

Characteristics or measures such as the mean and standard deviation, which are descriptive of a total *population*, are called population **parameters**. Characteristics or measures which describe a *sample* are called **statistics**. One of the important uses of statistical inference is the determination of how accurately sample statistics estimate the population parameters.

In Chapter 2, several formulas for the mean, variance, and standard deviation of a sample were given. These formulas are related to the formulas for the corresponding population parameters.

The mean of a probability distribution and of a population is given by $\mu = E(x) = \sum x \cdot P(x)$. If a sample has a total of n pieces of data (which need not be different) the mean is given by

$$\sum x \cdot \frac{1}{n} = \frac{1}{n} \sum x = \frac{\sum x}{n}$$

since each piece of data has a probability $1/n$. We then have the familiar formula for the sample mean.

MEAN OF A SAMPLE

The mean of a sample containing n pieces of data is given by

$$\bar{x} = \frac{\sum x}{n}$$

Similarly the variance of a population is given by $\sigma^2 = E(x - \mu)^2$ which is equal to $\sum (x - \mu)^2 \cdot P(x)$ for a probability distribution. Since, again, each piece of data has probability $1/n$ we obtain a similar formula for standard deviation of a sample. However for sample data we use $n - 1$ rather than n to obtain the sample variance which is an unbiased estimate of the population variance. The formula for the sample variance is given here.

VARIANCE OF A SAMPLE

The variance of a sample with mean $\bar{x}$ and containing n pieces of data is given by

$$s^2 = \frac{\sum (x - \bar{x})^2}{n - 1}$$

From a practical viewpoint, the alternate formulas, presented previously in Chapter 2, may be more useful. These formulas are for use with raw data and are particularly handy if a calculator is available. They also yield less round-off error than using the approximate mean sometimes required by the defining formula. The formulas given here are for the standard deviation, the square root of the variance.

RAW DATA FORMULAS FOR THE STANDARD DEVIATION

$$s = \sqrt{\frac{\sum x^2 - \frac{(\sum x)^2}{n}}{n-1}} = \sqrt{\frac{n\sum x^2 - (\sum x)^2}{n(n-1)}}$$

Further discussion of these formulas, as well as special formulas for use with an assumed mean, were given in Chapter 2. For grouped data, the class marks are used for x.

EXAMPLE 1 The ages of 25 people in a certain income bracket are distributed as follows:

Age	29	33	37	38	39	40	42	43	45	47	50	59	66
Frequency	1	1	3	4	2	3	2	2	3	1	1	1	1

Determine the mean and standard deviation of this sample.

SOLUTION If there are several data points which have the same value, it is obvious that we simply multiply each value by its frequency rather than add it in the required number of times. Thus, the mean is given by:

$$\bar{x} = \frac{29 + 33 + 3 \cdot 37 + 4 \cdot 38 + 2 \cdot 39 + 3 \cdot 40 + 2 \cdot 42 + 2 \cdot 43 + 3 \cdot 45 + 47 + 50 + 59 + 66}{25}$$

$$\bar{x} = \frac{29 + 33 + 111 + 152 + 78 + 120 + 84 + 86 + 135 + 47 + 50 + 59 + 66}{25}$$

$$\bar{x} = \frac{1{,}050}{25} = 42$$

For $\sum x^2$ we have

$$\begin{aligned}\sum x^2 &= (29)^2 + (33)^2 + 3(37)^2 + 4(38)^2 + 2(39)^2 + 3(40)^2 + 2(42)^2 \\ &\quad + 2(43)^2 + 3(45)^2 + (47)^2 + (50)^2 + (59)^2 + (66)^2 \\ &= 841 + 1{,}089 + 4{,}107 + 5{,}776 + 3{,}042 + 4{,}800 + 3{,}528 + 3{,}698 \\ &\quad + 6{,}075 + 2{,}209 + 2{,}500 + 3{,}481 + 4{,}356 \\ &= 45{,}502\end{aligned}$$

Then

$$s^2 = \frac{45{,}502 - \dfrac{(1{,}050)^2}{25}}{24} = \frac{45{,}502 - 44{,}100}{24} = \frac{1{,}402}{24} \doteq 58.4167,$$

so $s \doteq 7.64$.

Problems

Using the best technique determine the mean and standard deviation of each of the samples in exercises 1–4.

1. 20, 22, 23, 26, 29, 30
2. 28, 25, 20, 33, 27, 29, 23, 21, 24, 18, 30, 25
3. 1.34, 1.69, 1.78, 1.89, 2.03, 2.27, 2.39, 2.88, 3.16, 3.34, 4.92, 5.57, 6.83, 7.44, 9.63, 11.82

c4.

x	*Frequency*	x	*Frequency*
134	1	171	12
137	1	173	10
143	1	174	13
150	2	178	8
152	3	181	6
153	1	186	3
161	3	193	2
164	2	201	1
166	8	217	1
169	7	234	1
170	13	246	1

5. Use the data of problem 4 to construct a frequency table with 8 classes, class interval 15; make the smallest class with apparent limits 131–145. Use the class marks to determine the mean and standard deviation and compare with the results of problem 4.
6. The wing lengths of 10 click beetles are given below. Determine the mean and standard deviation for this sample. The lengths are in millimeters: 22.6, 23.7, 19.6, 24.5, 26.2, 25.4, 27.2, 23.3, 24.3, 21.9.

7.3 SAMPLING DISTRIBUTIONS

A basic problem in statistical inference is to infer population parameters from sample statistics with a stated degree of accuracy. For instance, suppose we want to determine the mean number of automobiles which pass a certain corner between 7 A.M. and 11 P.M. for the purpose of deciding whether or not to build a service station on the corner. With the permission of the highway department, we install traffic counters and take daily readings for ten weekdays with the following results.

Day	*1*	*2*	*3*	*4*	*5*	*6*	*7*	*8*	*9*	*10*
Number of Cars	284	386	273	308	317	281	309	290	278	271

These numbers have a mean of 299.7. We may then use this figure as an estimate of the number of cars which pass the corner each weekday. However, there are a number of hazards attached to such a use. Although we do not have information to the contrary, it is obvious that if we used another sample (if the experiment were repeated), we would probably obtain some other mean. Statistical inference provides us with a method of estimating the true mean to a desired degree of accuracy. In order to employ this method, we must study the theoretical sampling distribution.

The sampling distribution of a statistic is the probability distribution of that statistic for all samples of a given size. For instance, suppose that there are 6 litters of guinea pigs and the numbers of young in the litters are 2, 3, 4, 5, 6, and 7. If we took a sample of 4 litters, there are $\binom{6}{4}$ or 15 different possible samples. We determine the mean of each sample of four and list them as a distribution with the sample means as the random variable.

Sample	*Litter sizes*	$\bar{x}$
1	2,3,4,5	3.5
2	2,3,4,6	3.75
3	2,3,4,7	4.0
4	2,3,5,6	4.0
5	2,3,5,7	4.25
6	2,3,6,7	4.5
7	2,4,5,6	4.25
8	2,4,5,7	4.5
9	2,4,6,7	4.75
10	2,5,6,7	5.0
11	3,4,5,6	4.5
12	3,4,5,7	4.75
13	3,4,6,7	5.0
14	3,5,6,7	5.25
15	4,5,6,7	5.50

This distribution of sample means can be treated as a probability distribution with the results listed below.

$\bar{x}$	$P(\bar{x})$
3.50	1/15
3.75	1/15
4.00	2/15
4.25	2/15
4.50	3/15
4.75	2/15
5.00	2/15
5.25	1/15
5.50	1/15

Since this is the distribution of all possible (theoretical) means of samples of size 4, this is an example of a **theoretical sampling distribution of the mean**, the general term for any probability distribution in which the random variable takes on the values of all means of a sample of a given size.

The mean of this distribution, denoted by $\mu_{\bar{x}}$, is 4.50, and the standard deviation, $\sigma_{\bar{x}}$, is about 0.54. The mean of the sampling distribution is equal to the population mean since we have simply counted each value ten times and divided by $4 \cdot 15$ or 60. The standard deviation of the sampling distribution (usually called the **standard error of the mean**) is related to the standard deviation of the population by the following formula:

STANDARD ERROR OF THE MEAN

The set of all samples of size n drawn from a population of size N with mean μ and standard deviation σ has the mean $\mu_{\bar{x}} = \mu$ and standard deviation $\sigma_{\bar{x}}$ given by

$$\sigma_{\bar{x}} = \frac{\sigma}{\sqrt{n}} \sqrt{\frac{N - n}{N - 1}}$$

Note that if N is very large, or infinite, the term $\sqrt{(N - n)/(N - 1)}$ approaches one. As a rule of thumb, if the sample is drawn from an infinite population or constitutes less than 5% of the population, this **finite population correction factor** is not used and we have $\sigma_{\bar{x}} = \sigma/\sqrt{n}$.

Although this should be used only for theoretically infinite populations (such as sampling with replacement), in practice this is generally used, and use of the correction factor is fairly rare.

In the guinea pig litter example, if we ignore the fact that this is really a sample to begin with and used the formula $\sigma = \sqrt{E(x - \mu)^2}$, we have $\sigma \doteq 1.708$. Then $\sigma/\sqrt{4} \doteq 0.954$ while $(\sigma/\sqrt{4})(\sqrt{\frac{2}{5}}) \doteq 0.54$, and $\sigma_{\bar{x}}$ obtained from the sampling distribution is 0.54. Since the sample includes most of the population, the correction factor must be used. Populations this small are extremely rare.

A powerful tool for dealing with problems in inference is the **Central Limit Theorem**, which states that the theoretical sampling distribution of the means of all samples of size n is approximately a normal distribution with mean μ and standard deviation $\sigma_{\bar{x}}$. We can then use

$$z = \frac{\bar{x} - \mu}{\sigma_{\bar{x}}} = \frac{\bar{x} - \mu}{\sigma/\sqrt{n}}$$

as approximated by the *standard* normal distribution. The approximation becomes increasingly accurate as n increases in size.

EXAMPLE 1 A large population is normally distributed with a mean of 50 and a standard deviation of 12. A random sample of 36 is selected from the population. What is the probability that the mean of the sample is greater than 52?

SOLUTION Assuming that the population is larger than 720 (so that 36 is less than 5%), according to the Central Limit Theorem, the sample means will be normally distributed with mean 50 and standard error $12/\sqrt{36}$ or 2. The corresponding standard score is then given by

$$z = \frac{52 - 50}{2} = 1.00$$

so that the area under the normal curve to the right of $z = 1.00$ is equal (from Table 2) to 0.5000 − 0.3413 or 0.1587. Thus, the probability that the sample mean will be greater than 52 is 0.1587.

EXAMPLE 2 A sample of 100 pieces of copper tubing is examined for defects. If the process is in control, there will be a mean of 3.000 defects per tube, with a standard deviation of 0.400. If the sample contains a mean of 3.100 defects, or more, the entire shipment of 1,000 pieces will be refused on the assumption that the process is out of control. Assuming that the process is in control and the shipment is representative, what is the probability of refusing the shipment by error?

SOLUTION The shipment will be refused if the mean number of defects in the sample is greater than 3.100. If the mean of the shipment is actually 3.000 with a standard deviation of 0.400, the probability can be found with the aid of the standard score,

$$z = \frac{3.100 - 3.000}{\sigma_{\bar{x}}}$$

Since 100 is 10% of 1,000, it will be necessary to use the correction factor,

$$\sigma_{\bar{x}} = \frac{0.400}{\sqrt{100}} \sqrt{\frac{1{,}000 - 100}{1{,}000 - 1}} = \frac{0.400}{10} \sqrt{\frac{900}{999}}$$

$$\sigma_{\bar{x}} = \frac{(0.400)(30)}{(10)(31.61)} \doteq 0.038$$

Thus, $z = 0.100/0.038 \doteq 2.63$, and from Table 2 we see that the probability is $0.5000 - 0.4957$ or 0.0043. Thus, the sampling procedure seems to be a good one.

Problems

1. Suppose that a population is infinite, with standard deviation 100. Calculate the standard error of the mean for random samples of size $n =$

   (a) 100
   (b) 1,000
   (c) 10,000
   (d) 36
   (e) 144
   (f) 64
   (g) 128
   (h) 1024

2. A traffic engineer keeps track of the waiting time between automobiles passing an intersection for one minute. He obtains the following times in seconds: 3, 7, 5, 6, 7, 1, 1, 0, 2, 3, 4, 3, 5. Determine the mean, standard deviation, and standard error of the mean for these values.

3. A population of 10,000 has a mean of 187 with a standard deviation of 31. A sample of 100 is taken. What is the probability that the mean of the sample is greater than 190 if

   (a) no correction for finite population is made?
   (b) a correction for finite population is made?

4. Repeat problem 3 if the population is 2,000 and the sample 200.

5. A sample numbering 100 is taken from houseflies hovering around a garbage dump. Bacterial count on the feet of the flies is calculated and found to have mean of 10,000 per fly with the standard error of the mean for the sample 156. What was the standard deviation for the sample?

6. A researcher reports that he has a sample of persons for whom the mean number of jobs held in the last five years is 7.6. He reports the standard error of the mean for the sample is 0.925 and the standard deviation for the sample is 3.7. He neglects to report the sample size, however. From the information given, can you reconstruct his sample size? If so, how large was his sample?
7. An elevator bears a plate which states that the load limit is 2,640 pounds, and that no more than 16 persons may occupy the elevator at one time. Assume that the (large) population which rides the elevator is composed of men whose weights have a mean of 156 pounds with a standard deviation of 12 pounds. Assuming that the Central Limit Theorem applies (even though the sample has only $n = 16$), what is the probability of overloading the elevator with exactly 16 passengers?
8. A random sample of 100 is taken from an infinite population with mean 72.0 and standard deviation 8.0. What is the probability that the difference between the sample mean and the population mean is less than 0.50, that is, the probability that the value of $\bar{x}$ will be more than 71.5 but less than 72.5?

7.4 SUMMARY

When a sample is selected, care must be taken to use a method of selection which will produce a sample which represents the population. One method, which is widely used today, is to select the sample randomly from the entire population. A sample is considered random if each sample of that size had an equal chance to be included. Generalizations made from samples to a population must take into consideration the degree of certainty that a sample is truly random or otherwise representative of that population.

As we stated previously, characteristics describing a population are called **parameters**; characteristics describing a sample are called **statistics**. Two important sample statistics are the mean and standard deviation which are defined here again.

MEAN AND STANDARD DEVIATION OF A SAMPLE

The mean $\bar{x}$ and standard deviation s of a sample containing n pieces of data are given by

$$\bar{x} = \frac{\sum x}{n}$$

$$s = \sqrt{\frac{\sum (x - \bar{x})^2}{n - 1}}$$

Alternate formulas are also useful, and the most widely used formula for the standard deviation makes use of raw data only, and if the following:

RAW DATA FORMULA FOR SAMPLE STANDARD DEVIATION	The standard deviation of a sample of n measures is given by $$s = \sqrt{\frac{n \sum x^2 - (\sum x)^2}{n(n-1)}}$$

If all possible samples of a given size n are theoretically drawn from a population, the probability distribution for the means is approximately a normal distribution with mean μ and standard deviation $\sigma_{\bar{x}}$ (standard error of the mean). For very large populations, or infinite populations, with standard deviation σ, $\sigma_{\bar{x}} = \sigma/\sqrt{n}$, where n is the sample size. If the sample constitutes five per cent or more of the population, then

$$\sigma_{\bar{x}} = \frac{\sigma}{\sqrt{n}} \sqrt{\frac{N-n}{N-1}}$$

where N is the size of the population, and n is the size of the sample. In cases of doubt, it is preferable to use the correction factor than to omit it.

Problems

1. A student pays volunteers to participate in a study, then generalizes his results to the population as a whole. Comment on his procedure.
2. Ten years ago, students at Hay University had a mean score of 18.3 on a manual dexterity test, with a standard deviation of 2.4. Assume these statistics are representative of the entire population. A researcher wishes to discover whether there has been any change during the past ten years. He tests 64 students (of the university's 3,762) and obtains a mean score of 19.1 with a standard deviation of 2.2. What is the probability of obtaining a mean at least this high by chance, if the true situation remains as it was ten years ago?
3. Estimate the mean and standard deviation of each of the following samples:

   (a) 2.004; 2.012; 1.986; 2.006; 2.010; 1.992; 2.030; 1.986; 1.994; 1.998; 2.000
   (b) 56; 63; 74; 66; 61; 72; 68; 60
   c(c) 88; 96; 72; 63; 81; 74; 66; 88; 99; 73; 47; 61; 90; 76; 79; 83; 66; 75; 82; 71; 63; 84; 91; 76; 82; 69; 88; 73; 74; 82; 66; 91; 90; 86; 74; 70
4. A random sample of 40 aluminum stampings is checked each day to insure that the process is in control. If it is, a day's output of 6,000 stampings can be expected to have a mean index of 80.32 with a standard deviation of 3.16. If the process is

out of control, retooling must be accomplished. To this end, if the sample mean is greater than 81.62 or less than 79.02, retooling will begin. What is the probability of retooling by mistake?

5. Some students in a particular college favor co-ed dorms, while others do not. Suppose the dean of the college decided to conduct a survey to determine the proportion of students favoring co-ed dorms. The dean sends out two assistants to interview students at random. One assistant goes to the campus coffee shop and finds that 75 of 125 students favor co-ed dorms. The other assistant goes to the college gym and finds that 65 of 150 students favor co-ed dorms. If the results obtained by each interviewer are combined, would the results be a better representation of campus views, or should just one or the other of the samples be used? Do you feel that any of the results can be used, or should another poll be taken. If so, what procedure would you suggest?

6. The Ford Motor Company runs a quality check weekly on precision parts. The first week in November a sample of 25 was found to have a mean tolerance of 21 microns with a standard error of 0.38 microns. The second week in November a sample of 25 was again taken, but 4 were discarded. The remainder had a mean tolerance of 21.8 microns with a standard error of 0.25 microns. Determine the mean and standard error of the mean for the combined samples. This problem is fairly difficult and requires some recourse to the basic formulas. The time spent with it can be rewarding since research studies are often printed without raw data and it may be necessary to reconstruct at least some of the intermediate steps to acquire information about a set of data.

7. A sample of 60 delinquency letters is selected randomly from a day's output of 564. The average amount due on a delinquent account for the day is known to be \$80.54. Assume that the standard deviation for the amounts due is known to be \$13.66. What is the probability that the sample mean will fall between \$75 and \$85?

8. The sampling distribution for the standard deviation of a sample of size n from an infinite or very large population is approximately normal with mean σ and standard deviation $\sigma/\sqrt{2n}$. A sample of 50 company memorandums is examined and the number of words in each memorandum is recorded. Suppose that it is known that the company memorandums generally have a mean of 212.2 words with a standard deviation of 21.8.

   (a) What is the probability that this sample will have a mean of anywhere from 205 to 215 words?

   (b) What is the probability that the sample standard deviation will be between 17.5 and 25.5?

Chapter VIII

Estimation of Parameters

8.1 POINT ESTIMATION

In many cases it is impossible to calculate the mean or other population parameter directly and it must be estimated. One of the simplest ways to do this is to take a random sample, calculate the statistic for the sample which corresponds to the population parameter we are interested in, and then use the sample statistic as an estimate of the population parameter. This kind of estimate is called a **point estimate**.

For example, suppose that we wish to gauge the average take-home pay of New York secretaries. If we take a random sample of 100, we may find that this sample has a mean of \$90.40 and a standard deviation of \$13.60. In the absence of better information we may infer that these values approximate those of the corresponding population parameters. We also know, however, that we would be likely to obtain different values for both $\bar{x}$ and s if we were to take a different sample. The **Central Limit Theorem** tells us that the sample means of repeated samples drawn from the same population are approximately normally distributed with mean μ and standard deviation $\sigma_{\bar{x}}$ (the standard error of the mean) where μ is the population mean and $\sigma_{\bar{x}}$ depends upon the standard deviation of the population and the size of the sample. For a sample of one, the standard error is σ. This is sometimes symbolized by σ_x, the standard error of x.

In section 6.2 we investigated the properties of the standard normal distribution. Table 2 gives the area under this curve between the mean (0) and the given value of z. Using the Central Limit Theorem we can relate the distribution of the sample means to the standard normal distribution by using the formula:

$$z = \frac{\bar{x} - \mu}{\sigma_{\bar{x}}}$$

where $\bar{x}$ is the mean of a particular sample.

If we obtain a point estimate, $\bar{x}$, for μ the difference between $\bar{x}$ and μ is called the **error of estimation** incurred by using $\bar{x}$ in place of the unknown μ. The difference between $\bar{x}$ and μ is written $|\bar{x} - \mu|$ and it is the absolute value of $(\bar{x} - \mu)$, which is the positive difference between these two values. This is the value of $(\bar{x} - \mu)$ if $\bar{x}$ is greater than μ and the value of $(\mu - \bar{x})$ if $\bar{x}$ is less than μ.

Suppose we wish to know the probability that we will miss estimating correctly the population mean by, say, 3 or less, if we use the mean of a sample of 100 drawn from a population of 100,000 with a known standard deviation of 15. Here $\sigma_{\bar{x}} = 15/\sqrt{100} = 1.50$, so we have the following sampling curve.

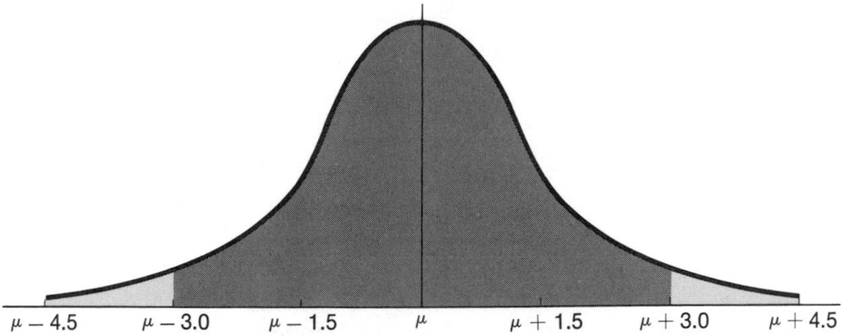

Since $3 = 2\sigma_{\bar{x}}$, it follows that we are looking for the probability that the sample mean will fall between $\mu - 2\sigma_{\bar{x}}$ and $\mu + 2\sigma_{\bar{x}}$, that is, between $z = -2$ and $z = +2$. From Table 2, we have 0.4772 of the area under the curve between 0 and 2.00, so the area under the curve between $z = -2.00$ and $z = 2.00$ is 0.9544. The probability that the error is 3 or less is 0.9544. This means that the probability is 0.9544 that a sample of 100 drawn from this population will have a mean which differs from the true population mean by no more than 3. Thus the maximum error with probability 0.9544 is 3. If the sample mean turned out to be 51, say, then we would know that probability is 0.9544 that the true mean will be no less than 48 and no more than 54. If we represent the maximum error by E, we then have $P(E = 3) = 0.9544$. We cannot assert anything definitely, but only with a given probability. This matter will be pursued a bit further in the next section. Now using the formula

$$z = \frac{\bar{x} - \mu}{\sigma_{\bar{x}}}$$

we have $|z| = E/\sigma_{\bar{x}}$ so that $E = |z|\sigma_{\bar{x}} = |z|\sigma/\sqrt{n}$ if n represents less than 5% of the population. Thus, we could also compute the maximum error from a given probability.

Since it is unlikely that we will know σ, the population variance, we use our best point estimate, s, the sample variance, obtaining

$$z = \frac{E}{s/\sqrt{n}} \quad \text{or} \quad E = z\left(\frac{s}{\sqrt{n}}\right)$$

The quantity $s/\sqrt{n}$ is usually denoted $s_{\bar{x}}$ and is an estimate for $\sigma_{\bar{x}}$ obtained from the sample variance. It is generally used as a substitute for $\sigma_{\bar{x}}$ if the samples are relatively large. In practice "relatively large" means different things to different people, but many statisticians accept the substitution for samples of size greater than 30. In reality substituting $s_{\bar{x}}$ for $\sigma_{\bar{x}}$ involves exchanging the standard normal distribution for a probability density called the t-distribution which will be discussed in section 8.3. For values of $n > 30$, however, the t-distribution is practically identical to the normal curve. In using $s_{\bar{x}}$ remember that the finite population correction factor still must be used for samples constituting 5% or more of the population if $n > 30$.

EXAMPLE 1 A sample of 100 New York secretaries has a mean take-home pay of \$90.40 with a standard deviation of \$13.60. What is the probability that the sample mean is an estimate of the population mean correct within \$2.00. That is, what is the probability that the true mean is between \$88.40 and \$92.40?

SOLUTION Here $E = 2$, $s = 13.6$, $n = 100$, so $s_{\bar{x}} = 13.6/\sqrt{100} = 1.36$. Then $|z| = 2/1.36 \doteq 1.47$. The associated area under the normal curve (from Table 2) is 0.4292. Thus, the probability that the sample mean is less than two units *above* the population mean is 0.4292; the probability that the sample mean is less than two units *below* the population mean is 0.4292; therefore, the probability that the sample mean is no more than two units *from* the population mean is 0.8584.

EXAMPLE 2 A researcher wishes to poll a sample in order to get public opinion on a certain issue. He will ask the people in his sample to rate their confidence in the administration on a scale from 0 to 100. Assuming an underlying continuous distribution, and assuming that measures of this type can be safely assumed to have standard deviations no greater than 20, how many people should he poll in order to estimate the population mean within 5 units, with a probability of 0.90?

SOLUTION Here $E = 5$, $\sigma = 20$ is a safe estimate, and $z \doteq 1.64$, so we have

$$5 = 1.64 \frac{20}{\sqrt{n}}$$

Solving for n we have

$$n = \left[\frac{(1.64)(20)}{5}\right]^2 = [(1.64)(4)]^2 = (6.56)^2 \doteq 43.03$$

Since we cannot poll a fractional number of persons, and 43 is too few, we must poll 44.

DISCUSSION Upon solving $E = |z|\sigma/\sqrt{n}$ for n, we have

$$n = \left[|z|\frac{\sigma}{E}\right]^2$$

Note that we cannot use this formula unless we know the standard deviation of the population. If, however, we can approximate it or estimate a *maximum* for it, the formula will depend only upon the maximum allowable error, E, and $|z|$, which can be computed from the degree of certainty (given by the probability) which is desired.

Problems

1. A sample of size n is drawn from a population of size N with a known standard deviation σ. Calculate the maximum error with a probability of 0.9500 if the mean of the sample is used to estimate the population mean, where

   (a) $N = 10{,}000$, $n = 100$, $\sigma = 13.4$.
   (b) $N = 500$, $n = 64$, $\sigma = 103.4$.
   (c) N is infinite, $n = 200$, $\sigma = 0.040$.
   (d) $N = 5{,}000$, $n = 400$, $\sigma = 8.62$.
   (e) $N = 1{,}000$, $n = 36$, $\sigma = 12.136$.

2. A sample of size 100 is drawn from a very large population with known standard deviation 11.60. If the mean of the sample is used to approximate the mean of the population, calculate the probability that the error of estimation is less than

   (a) 1.64
   (b) 0.80
   (c) 3.81
   (d) 0.78
   (e) 2.45

c3. A sample of 36 drawn from a population of about 1500 has values as follows: 8.6; 11.3; 9.4; 6.3; 8.0; 9.7; 10.8; 9.4; 12.2; 5.4; 7.6; 10.3; 9.4; 7.7; 9.6; 6.9; 9.3; 6.8; 9.1; 11.4; 8.8; 9.3; 12.4; 10.2; 10.5; 8.7; 9.4; 8.3; 9.1; 9.5; 10.4; 7.2; 13.1; 8.8; 6.7; 11.1. Calculate the maximum error if the mean of the sample is used to estimate the mean of the population. The desired probability level is 0.90. Use s as an estimate of σ.

4. Using the data of problem 3, determine the probability that the error will be less than 0.8.

5. To estimate the mean family income, a group of 100 incomes, chosen at random from the population to be sampled, was selected. This sample had a mean family income of \$11,443 with a standard deviation of \$3,762. What is the maximum error with a probability of 0.95 incurred by using these values? Suppose it is known that the standard deviation of the population is closer to \$3,000 than the value obtained from the sample. How does this change the maximum error calculated previously?

6. The **probable error of estimate** is the value for which 50% of the sample statistics will lie within that value of the population parameter. If a large population has a standard deviation of 10.84, what is the probable error of estimate of the mean of a sample of 64?

7. A biologist is searching for a method of controlling the common flour beetle. Using a spray recently developed, he discovers that the average number of eggs per laying which hatch from 50 beetles that have been sprayed is 743. What is the probable error of estimate if we use 743 as an estimate for the population mean and the known population standard deviation of 35?

8. The sampling distribution for the standard deviation of a sample of size n taken from a large population can be closely approximated by a normal curve with mean σ and standard deviation $\sigma/\sqrt{2n}$, where σ is the population standard deviation. Calculate the probability that the standard deviation of a sample of size n drawn from a population with standard deviation σ will be less than E if

   (a) $n = 32$, $\sigma = 10.4$, $E = 2.4$
   (b) $n = 128$, $\sigma = 1.08$, $E = 0.12$
   (c) $n = 100$, $\sigma = 134.2$, $E = 15.0$
   (d) $n = 50$, $\sigma = 1089$, $E = 200$

9. A sociologist wishes to sample a population to estimate the validity of a test given several years ago. The standard deviation of that test, 14.8, is used to estimate the current standard deviation and he wants to arrive at the current mean within 2.0 units. What sample size should he use to be sure he is correct with a probability of

   (a) 0.90?
   (b) 0.95?
   (c) 0.98?
   (d) 0.99?

10. A psychologist is dubious of the standardization technique used on a new test. He decides to make his own determination of the probable mean score on the test. He gives the test to 200 subjects and finds that the mean for the sample is 118.4. If he uses the published standard deviation of 22.6 for his computations, what is the maximum error in using 118.4 as the mean for the test? He is willing to take only a 1% chance of being wrong, so he wants the probability that the error is a maximum to be 0.99.

8.2 CONFIDENCE INTERVALS FOR THE POPULATION MEAN

A metallurgist wishes to determine the melting point of a new alloy. He takes 36 pieces of the alloy and records the melting point of each piece. The mean of the 36 numbers is found to be 2,356.0°C, and the standard deviation is 3.6°. What can he say about the melting point of the alloy? He may use 2,356° as a point estimate, but he is also aware that a different sample of pieces may have a mean which differs from the mean of this sample. He may also define a *range* of values within which the population mean is likely to fall, because he *can* assert that the melting point of the alloy is, in all likelihood, in the neighborhood of 2,356°. In fact, if 3.6° is σ, the standard deviation of the population (of the melting point of all pieces of the alloy), the Central Limit Theorem allows us to use the normal distribution to calculate the maximum error for a given probability that the population mean does fall within certain limits for a given value of z:

$$E = |z| \frac{3.6}{\sqrt{36}} = |z|(0.6)$$

If the standard deviation of the population is unknown, and it usually is, the normal distribution is replaced by the *t-distribution*, which is discussed in the next section. If the sample contains at least 30 measures, however, s is usually used to estimate σ, and

$$s_{\bar{x}} = \frac{s}{\sqrt{n}} \quad \text{or} \quad s_{\bar{x}} = \frac{s}{\sqrt{n}}\sqrt{\frac{N-n}{N-1}}$$

depending upon whether or not the sample of n measures is more than 5% of the population of N measures, a point discussed in section 7.3.

In this case, if we wanted to have the probability less than 0.05 that we are making an error in saying that the population mean falls within E of 2,356°, we choose $z = 1.96$, since $P(-1.96 < z < 1.96) = 0.95$. Then $E = 1.96(0.6) = 1.176 \doteq 1.2$, so we can assert that the probability that the sample mean is within 1.2° of the true mean is 0.95. Thus, in this case, we have a probability of 0.95 that the true melting point of the alloy is between 2,354.8° and 2,357.2°. The interval between these extremes is called a **confidence interval** for the mean, the end points are called the **confidence limits**, and the probability that the interval contains—or covers—the parameter is called the **degree of confidence**. In a way, the term *confidence* is a bit misleading. It is difficult to think about the concept of being 95% confident. Our degree of confidence is determined theoretically by the fact that if repeated samples of a given size are drawn from a population and their means determined, then 95% (or some other percentage) of all confidence intervals determined from these samples will cover the mean. The specific degree of confidence is chosen in advance and determines the specific value of z to be used. The probability that the true mean will lie outside a particular confidence interval is given the greek letter α (alpha), while the probability that it lies in the interval is $1 - \alpha$. This notation will be used extensively later on. In the case of the melting point of the alloy we have used a 95% or 0.95 degree of confidence, so the interval 2,354.8° to 2,357.2° is called a 95% or 0.95 **confidence interval for the population mean**. The most common values for the degree of confidence are 0.95, 0.98, and 0.99. In any case, the confidence interval is determined by its end points, $\bar{x} - E$ and $\bar{x} + E$, where $\bar{x}$ is the sample mean and E is the maximum error; that is, the true mean μ satisfies

$$\bar{x} - E < \mu < \bar{x} + E$$

where

$$E = 1.96\sigma_{\bar{x}} \quad \text{(if the degree of confidence is 0.95)}$$

$$E = 2.33\sigma_{\bar{x}} \quad \text{(if the degree of confidence is 0.98)}$$

$$E = 2.58\sigma_{\bar{x}} \quad \text{(if the degree of confidence is 0.99)}$$

in that proportion of the cases.

Thus, if we wish to be "more confident" and have a greater proportion of confidence intervals covering the mean, we must accept a greater maximum error. Conversely, if we wish a smaller interval, we must accept a lower confidence level and consequently a higher likelihood of the true mean lying outside the interval.

EXAMPLE 1 A sample of 40 loaves of bread is weighed and found to have a mean weight of 20.24 ounces with a standard deviation of 0.34 ounce. If the sample is a small portion of the daily output, determine 0.95 and 0.99 confidence intervals for the mean weight of the entire daily output.

SOLUTION Here $s_{\bar{x}} = 0.34/\sqrt{40} \doteq 0.054$, so the maximum error with a probability of $0.95 = 1.96(0.054) \doteq 0.106 \doteq 0.11$. The end points are $20.24 - 0.11 = 20.13$ and $20.24 + 0.11 = 20.35$ and the 0.95 confidence interval for μ is 20.13 to 20.35. The maximum error with a probability of 0.99 is $2.58(0.054) \doteq 0.139 \doteq 0.14$; the end points are $20.24 - 0.14 = 20.10$ and $20.24 + 0.14 = 20.38$. Thus, the 0.99 confidence interval for μ is 20.10 to 20.38.

DISCUSSION The foregoing shows that the more confident we wish to be, the wider the latitude that must be allowed. Thus, we are 95% sure that the true mean is between 20.13 and 20.35 and 99% sure that the true mean is between 20.10 and 20.38. In the first case, the probability that the mean is outside the confidence interval is 0.05, while in the second case it is 0.01.

EXAMPLE 2 A sample of 64 fuses is found to have a mean peak load of 30.840 amperes with a standard deviation of 0.420 amperes. Estimate 0.95 and 0.98 confidence intervals for the mean peak load for the entire shipment of 1,025 fuses.

SOLUTION Since 64 is greater than 5% of 1,025, the finite population correction factor must be used, so

$$s_{\bar{x}} = \frac{0.420}{\sqrt{64}}\sqrt{\frac{1{,}025 - 64}{1{,}025 - 1}} = \frac{0.420}{8}\sqrt{\frac{961}{1{,}024}}$$

$$= 0.00525\left(\frac{31}{32}\right) \doteq 0.0509$$

$$\doteq 0.051$$

Thus, the maximum error with a probability of 0.95 is 1.96(0.051) or 0.100; with a probability of 0.98 is 2.33(0.051) or 0.119. Thus, the 0.95 confidence interval is 30.740 to 30.940, and the 0.98 confidence interval is 30.721 to 30.959.

Problems

1. A random sample of 81 copper tubes is drawn from a shipment of 442 tubes, and the tubes are found to have a mean interior diameter of 2.080 cm with a standard deviation of 0.030 cm. Estimate 0.95 and 0.99 confidence intervals for the mean interior diameter of the shipment.
2. A sample of 100 "ten pound" sacks of sugar from a large shipment is found to have a mean weight of 9.83 pounds with a standard deviation of 0.70 pound. Calculate 0.90 and 0.98 confidence intervals for the mean weight of the shipment.
3. An elevator holds 10 men on a full load. If the weight of a sample of 36 of the 325 regular riders of the elevator shows a mean weight of 158.0 pounds with a standard deviation of 11.4 pounds, compute a 0.95 confidence interval for the weight of a full load on the elevator. (Hint: Find the 0.95 confidence interval for the mean weight of a sample of 10, then multiply the limits by 10.)
4. A load of 2,500 sacks of grain is inspected by weighing a random sample of 50 sacks and determining a 0.95 confidence interval for the mean weight of the entire load. If the mean of the sample is less than 150 pounds and the confidence interval does not contain 150 pounds, the shipment will be rejected. On a particular day, the mean is 149.0 pounds with a standard deviation of 5.2 pounds. Will the load be rejected?
5. A random sample of 50 grapefruit from a grove has weights as follows (in ounces): 10.7; 13.8; 9.7; 10.2; 9.6; 11.4; 11.9; 13.2; 10.8; 12.3; 10.3; 12.8; 9.1; 8.6; 13.1; 10.6; 11.4; 10.8; 9.1; 11.2; 7.6; 11.6; 13.2; 10.4; 9.9; 11.3; 10.6; 10.3; 10.9; 11.0; 9.3; 9.9; 13.0; 8.8; 10.6; 10.7; 12.4; 12.0; 13.1; 10.5; 9.8; 12.1; 11.8; 10.8; 11.2; 9.4; 8.9; 12.2; 10.7; 11.3. The mean is 10.92 and $s \doteq 1.38$. (Verify these values if a calculator is available.) Estimate the 0.95 and 0.99 confidence intervals for the mean weight of all grapefruit in the grove.
6. A water company wishes to discover the mean water consumption for the month of July in all homes in a certain area. If there are 618 homes in the area and a random sample of 30 homes shows a mean of 11,644 gallons, with a standard deviation of 1,206 gallons, estimate the 0.95 confidence interval for the population (a) without the correction factor and (b) with the correction factor. Compare the results.
7. A total of 400 castings is selected at random from a very large shipment and examined for flaws. The mean number of flaws is found to be 13.40, with a standard deviation of 3.60. What can be said about the possible size of the error,

with a probability of 0.98, if 13.40 is used to estimate the mean number of errors in the shipment?

8. Income data for a set of 100 incomes, chosen at random from a large population, showed a mean family income of \$11,443 with a standard deviation of \$3,762. Using s as an estimate for the population standard deviation construct a 0.95 confidence interval for the mean family income of the population. Suppose it is known that the standard deviation of the population is actually quite close to \$3,000. What would be the 0.95 confidence interval in that case? If you were to use one of these, which would be more nearly accurate?

9. Symptoms from a new influenza virus have a mean duration of 4.7 days with a standard deviation of 1.2 days. The data are gathered from all available evidence. A new vaccine has been given to 100 sufferers of the virus and the disease's mean duration with the vaccine for this sample was 3.2 days. Construct a 0.99 confidence interval for the mean duration of the symptoms of the disease if the new vaccine is used.

10. Confidence intervals for the standard deviation σ of a population as estimated from the standard deviation s of sample of size n are given by

$$\frac{s}{1 + \dfrac{z}{\sqrt{2n}}} < \sigma < \frac{s}{1 - \dfrac{z}{\sqrt{2n}}}$$

for an appropriate z determined by the degree of confidence, if $n > 30$ and does not constitute more than 5% of the population. Calculate 0.95 and 0.99 confidence intervals for σ if

(a) $s = 11.4$, $n = 32$.
(b) $s = 137.6$, $n = 50$.
(c) $s = 0.012$, $n = 120$.
(d) $s = 10.030$, $n = 68$.

8.3 CONFIDENCE INTERVALS OBTAINED FROM SMALL SAMPLES

The methods of the previous section can be used only if the population standard deviation is known or if the samples are of size $n > 30$. If $n \leq 30$, the sampling distribution of the mean is not accurately approximated by a normal distribution; the proportion of the area under the curve is greater in the tails than that of a normal curve, and the curve is consequently flatter. In fact, the smaller the sample, the flatter the curve. This sampling distribution is called the **t-distribution** (or *Student t-distribution,* since it was first investigated by W. S. Gosset under the pseudonym "Student"), and it may be used if the population from which the sample is drawn is approximately normal.

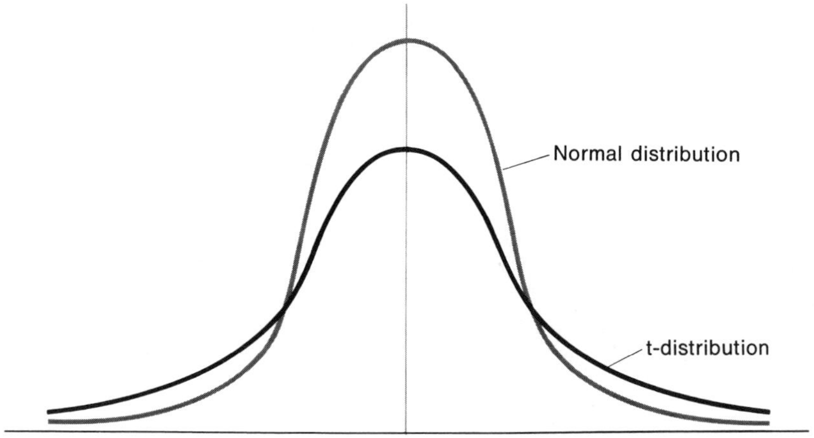

The preceding figure shows a t-distribution for a sample of size 6 compared to a normal distribution with the same mean and standard deviation. Each n, where n corresponds to the size of the sample, generates a different t-distribution; as n increases to infinity, the t-distribution approaches the normal distribution and for practical purposes can be considered as such for $n > 30$.

Thus, for each sample of size $n \leq 30$ drawn from a normally distributed population, we can use the statistic

$$t = \frac{\bar{x} - \mu}{s/\sqrt{n}}$$

in place of the z score as described in the previous section. Each sample of size n would generate a t-distribution which could have its areas tabulated as the normal curve does in Table 2. Such tables are available, but for our purposes we need only those values of t which correspond to 0.95, 0.98, and 0.99 confidence intervals. As with Table 2, only the positive points for t are given; the t-distribution is symmetric.

In cases in which the sample size is 30 or less, then, the maximum error E would be calculated by $E = |t| s/\sqrt{n}$ for an appropriate t depending upon the degree of confidence desired.

As we mentioned in section 8.2, whenever s is used in place of σ, the t values should be used, but t and z are practically identical for large values of n.

In Table 3, the numbers at the top give the area under the curve to the right of the given t, while the numbers at the side give the **degrees of freedom**,

$n-1$, for a sample of size n.* For instance if a sample contains 17 pieces of data, there are 16 degrees of freedom. A 0.95 confidence interval for μ corresponds to a probability that 0.025 of the sample means lie in *each* tail of the curve, thus the associated value of t, for 16 degrees of freedom, will be (from Table 3) 2.120, rather than 1.96 as it would be if the normal curve were used. Note that $t = z$ for infinite degrees of freedom.

EXAMPLE 1 A sample of 25 numbers is found to have a mean of 32.60 and a standard deviation of 1.30. Estimate 0.95 and 0.99 confidence intervals for the population mean.

SOLUTION As in section 8.2, the confidence interval is determined by the confidence limits; $\bar{x} - E$, and $\bar{x} + E$. The only difference is that t, not z, is used in finding E. Here

$$E = |t| \frac{1.3}{\sqrt{25}} = |t|(0.26)$$

for appropriate values of t. For a 0.95 confidence interval, 0.05 of the area under the curve is in the tails of the curve, hence 0.025 is in each tail. Specifically, 0.025 of the area under the curve is to the right (for 24 degrees of freedom) of t when $t = 2.064$. Thus for the 0.95 confidence interval for μ we have

$$E = 2.064(0.26) \doteq 0.54$$

so that the confidence limits are $32.60 - 0.54 = 32.06$ and $32.60 + 0.54 = 33.14$. The 0.95 confidence interval for the mean, then, is 32.06 to 33.14. Similarly for the 0.99 confidence interval 0.005 lies in *each* tail so that the corresponding t-value (for 24 degrees of freedom) is 2.797. Thus we have

$$E = 2.797(0.26) \doteq 0.73$$

so the confidence limits are $32.60 - 0.73 = 31.87$ and $32.60 + 0.73 = 33.33$. Thus, the 0.99 confidence interval for μ is 31.87–33.33.

* In some other uses of t- and other distributions, degrees of freedoms are defined in various ways. Generally speaking, the degrees of freedom are the number of values that can be assigned more or less arbitrarily within a framework. Thus, if there are 10 numbers which add up to 100, we could choose 9 of them to be any number we please, but the tenth must be that number necessary to make the sum 100. In this case, then, we would have 9 $(n-1)$ degrees of freedom.

EXAMPLE 2 A candy bar manufacturer has machines which supposedly make 2.30-ounce bars. It is known, however, that the weights vary slightly but are normally distributed with standard deviations varying slightly from machine to machine and often from day to day. To test quality, a random sample of 4 bars is drawn from the day's output, and the 4 bars are found to have weights of 2.24 ounces, 2.34 ounces, 2.28 ounces, and 2.30 ounces. Estimate a 0.98 confidence interval for the mean weight of the day's output.

SOLUTION For the 4 weights, $\bar{x} = 2.29$, while $s = \sqrt{0.0052/3} \doteq 0.04$. For the 0.98 confidence interval, 0.01 of the area under the curve lies in each tail, so with 3 degrees of freedom we have $t = 4.541$. Hence, the maximum error, E, with a probability of 0.98 is

$$E = 4.541\frac{0.04}{\sqrt{4}} \doteq 0.09$$

Thus, our confidence limits are $2.29 - 0.09 = 2.20$ and $2.29 + 0.09 = 2.38$, so then we are 0.98 certain (or confident) that the population mean lies between 2.20 and 2.38.

Problems

1. A new alloy is subjected to 9 determinations of hardness, with the result that the mean value on Moh's scale is 0.630 with a standard deviation of 0.081. Estimate 0.95 and 0.99 confidence intervals for the true hardness of the alloy.

2. A super-ball is dropped and the height of the bounce is measured. The proportion of the original height to which the ball returns is called the **coefficient of restitution** for the substance. Four determinations of the coefficient of restitution for this super-ball yield values of 0.84, 0.78, 0.86, and 0.81. Estimate the 0.95 and 0.99 confidence intervals for the coefficient of restitution of this ball.

3. On 15 tests, the reaction time of a volunteer to a given stimulus had the following values in seconds: 0.12, 0.14, 0.09, 0.13, 0.11, 0.12, 0.11, 0.11, 0.09, 0.13, 0.14, 0.10, 0.11, 0.10, 0.12. Assuming the tests were performed under similar conditions, and that the assumptions underlying the t-distribution apply, estimate a 0.99 confidence interval for the mean reaction time of this individual to the stimulus.

4. Sociologists classify kin groups by number of persons living in the same household. A person living together with spouse and 3 children is a kin group of size 5. A person living alone is a kin group of size 1. Suppose that 20 households are examined and the mean kin group size for the sample is found to be 3.14 with a standard deviation of 0.83. Construct a 0.90 confidence interval for the mean kin group size of the population from which the sample is taken.

5. A study of mating calls of moose was conducted and the mean duration of the calls for 12 different bull moose was recorded. The mean duration of these calls was 5.16 seconds with a standard deviation of 1.12 seconds. Determine a 0.95 confidence interval for the mean duration of the mating calls for the population of bull moose.

6. *If the population standard deviation is known* and the population is approximately normally distributed, the methods of section 8.2 apply; that is, we can use z rather than t. Suppose that a sample of 11 is drawn from a population with a known standard deviation of 11.62. If the sample mean is 38.44, estimate a 0.95 confidence interval for the population mean. If the sample standard deviation is 11.04, compute a 0.95 confidence interval using t and compare to the one obtained earlier. Thus, if σ is known, we can estimate more closely than if s must be used, even if s is less than σ.

8.4 CONFIDENCE INTERVALS FOR PROPORTIONS

Most of the variables we have been dealing with to this point have been measurable. A very important class of variables, called **attributes**, cannot be measured but can only be described. These variables are such things as sex or marital status of individuals, occupation, preferences, vacant or occupied status of an apartment, and so on. It is usual to describe attributes in terms of a *proportion* of a sample or a population.

In a manner similar to that described in section 8.2, population proportions can be estimated from sample proportions.

If P represents the probability that a member of a population possesses an attribute, P also represents the proportion of the population with that attribute. For instance, if the probability of drawing a red marble from an urn is 0.20, we can conclude that 0.20 of the marbles in the urn are red. Conversely, if we know that 3/5 of the apartments in a building are vacant we know that the probability of a randomly selected apartment being vacant is 3/5.

Because of this fact, proportions constitute a binomial population and, in most cases the binomial distribution can be used. Actually, it is more usual to use the normal approximation to the binomial (see section 6.4). Thus, if the actual population proportion is P, the number of "successes" calculated from the set of all samples of size n will be approximately normally distributed with mean nP and standard deviation $\sqrt{nP(1-P)}$, if both nP and $n(1-P)$ are at least 5. Then if we let x represent the number of those in a sample of size n with the attribute, the statistic

$$z = \frac{x - nP}{\sqrt{nP(1-P)}}$$

will relate our sampling distribution to the normal distribution. Actually, if p represents the sample proportion, then $x/n = p$, or $x = np$, so that by dividing numerator and denominator by n we have

$$z = \frac{np - nP}{\sqrt{nP(1-P)}} = \frac{p - P}{\sqrt{\dfrac{P(1-P)}{n}}} = \frac{p - P}{\sigma_p}$$

The quantity σ_p is called the standard error of proportion and is given by

STANDARD ERROR OF PROPORTION	$\sigma_p = \sqrt{\dfrac{P(1-P)}{n}}$

The standard error of a proportion is the standard deviation of the sampling distribution in question.

Again we note that the difference between the actual population proportion, P, and the sample proportion, p, is the error of estimation, E, and so we have

$$|z| = \frac{E}{\sigma_p}$$

or

$$E = |z|\sigma_p$$

Inasmuch as σ_p depends upon P, and P is often not known, it is usual in such cases to estimate σ_p by using

$$s_p = \sqrt{\frac{p(1-p)}{n}}$$

The values for z remain those discussed in section 8.2, so the confidence interval for a population proportion P as arrived at from the sample proportion p has endpoints $p - E$ and $p + E$ where E is the maximum error. That is, the true proportion satisfies

$$p - E < P < p + E$$

where

$$E = 1.96\sigma_p \text{ (if the degree of confidence is 0.95)}$$

$$E = 2.33\sigma_p \text{ (if the degree of confidence is 0.98)}$$

$$E = 2.58\sigma_p \text{ (if the degree of confidence is 0.99)}$$

in that proportion of the cases.

EXAMPLE 1 A sample of 100 fuses from a large shipment is found to have 10 which are defective. Construct 0.95 and 0.99 confidence intervals for the proportion of defective fuses in the shipment.

SOLUTION Here $n = 100$, $p = 0.1$ so we approximate σ_p by

$$s_p = \sqrt{\frac{(0.1)(0.9)}{100}} = \sqrt{\frac{0.09}{100}} = \frac{0.3}{10} = 0.03$$

The maximum error with a probability of 0.95 is given by $E = 1.96(0.03) \doteq 0.06$, so we have confidence limits of 0.04 and 0.16 and a 0.95 confidence interval of 0.04 to 0.16.

The maximum error with a probability of 0.99 is given by $E = 2.58(0.3) \doteq 0.08$, so the confidence limits are 0.02 and 0.18, and the 0.99 confidence interval is 0.02 to 0.18.

The problem of deciding upon the size of a sample, which is so important in experimental design, was dealt with briefly at the end of section 8.1. Where proportions are involved the problem is similar, but there is one difference.

If a maximum allowable error is specified for some probability level, we have

$$E = |z|\sigma_p = |z|\sqrt{\frac{P(1-P)}{n}}$$

Upon solving for n, we have

$$n = \frac{z^2(P)(1-P)}{E^2}$$

If we can estimate the population proportion, we can arrive at some estimate of the sample size needed. If not, note that if $P = 0.5$, $P(1 - P) = 0.25$. For any other value of P, $P(1 - P)$ is less than 0.25. Thus, if we know absolutely nothing about P, we can still obtain the maximum needed sample size by using $P = 0.5$. For any other value of P the needed sample size would be less. If $z^2/E^2 = 10{,}000$, for example, and if we could estimate $P = 0.3$, then $(0.3)(0.7) = 0.21$ and we would have $n = 2{,}100$. For $P = 0.4$, we would have $n = 2{,}400$; for $P = 0.5$, $n = 2{,}500$; for $P = 0.6$, $n = 2{,}400$, and so on. Thus $P = 0.5$ yields the largest needed sample size and we would know that if we took a sample of that size, no matter what the true value of P, our sample would be adequate.

EXAMPLE 2 A sample of a certain hybrid strain of corn will be examined to see what proportion exhibits a certain genetic characteristic. What sample size should be examined in order to justify generalizing the result to the entire population accurate to within 0.01 with a probability of 0.95? Assume (a) it is reasonable to use a figure of 0.20 as an estimate for P and (b) nothing is known about P.

SOLUTION If $E = 0.01$, $z = 1.96$ (for a 0.95 degree of confidence), we have:

$$n = \frac{(1.96)^2(P)(1-P)}{(0.01)^2}$$

(a) Assuming $P = 0.20$, we have

$$n = \frac{(1.96)^2(0.20)(0.80)}{(0.01)^2} \doteq 6{,}146.6$$

so we need a sample of 6,147.
(b) Knowing nothing about P, we let $P = 0.5$ since then n is maximum and we have

$$n = \frac{(1.96)^2(0.5)(0.5)}{(0.01)^2} = 9{,}604$$

so we need a sample of 9,604.

Problems

1. To obtain data for his thesis, a psychology major plans to interview people to determine if their reaction to a certain situation is positive or negative. She wants to estimate the true proportion to within 0.02 with a probability of 0.95. If a pilot study showed 60 of 100 people with positive reactions, how many should she interview?
2. After a certain amount of training, 48 of 58 rats tested managed to negotiate a maze successfully without error. Give a 0.95 confidence interval for the probability that any rat, given the same amount of training, would be able to traverse the maze successfully without error.
3. A sociologist interviews 120 of 650 families in an apartment complex, and found that 48 of them supported certain legislation pending. Give 0.95 and 0.99 confidence intervals for the proportion of families in the complex supporting the legislation.

4. Of 164 persons interviewed, 52 preferred black coffee, 78 preferred cream in their coffee, while the remainder did not drink coffee. Assuming the sample was random, calculate 0.95 and 0.99 confidence intervals for the proportion of coffee drinkers who like cream in their coffee.

8.5 SUMMARY

Estimating parameters from sample statistics is possible if certain assumptions are made and the necessary conditions are met. Estimation of population parameters generally involves an interval called a **confidence interval**. We say that there is a stated probability, usually 0.95 or 0.99, that the interval contains the parameter. The end points of the interval are called the **confidence limits** and the stated probability the **degree of confidence**.

If the standard deviation, σ, of the population is known, and the population is approximately normally distributed, the population mean can be estimated from a sample mean to lie within a confidence interval whose limits are $\bar{x} - E$ and $\bar{x} + E$ where $\bar{x}$ is the sample mean:

$$E = |z| \frac{\sigma}{\sqrt{n}} \quad \text{or} \quad E = |z| \frac{\sigma}{\sqrt{n}} \sqrt{\frac{N-n}{N-1}}$$

where N is the size of the population, n is the sample size and $|z|$ depends on the degree of confidence desired. The second formula is used if n is at least 5% of N; otherwise the first formula is used. The most commonly used degrees of confidence with associated values of $|z|$ are the following:

Degree of Confidence	*0.95*	*0.98*	*0.99*
$\lvert z \rvert$	1.96	2.33	2.58

If the standard deviation of the population is unknown and the sample is greater than 30, s can be used for σ in the above formulas.

If $n \leq 30$, the formula for E is modified so that

$$E = |t| \frac{s}{\sqrt{n}}$$

where $|t|$ is the positive value of the appropriate t obtained from Table 3 for $n - 1$ degrees of freedom.

For proportions, a confidence interval can be established whose limits are $p - E$ and $p + E$ where p is the sample proportion and, for large n,

$$E = |z| \sqrt{\frac{P(1-P)}{n}}$$

In this formula P is the population proportion, n is the sample size, and $|z|$ depends upon the degree of confidence desired. If P is unknown, p is commonly used in the formula as an estimate.

The following formulas are useful to determine the approximate sample size needed for estimation with a desired degree of accuracy.

For estimating the mean of a population within E units with a desired degree of confidence the sample should be at least n where

$$n = \left[|z| \frac{\sigma}{E} \right]^2$$

and where σ is the population standard deviation and $|z| = 1.96, 2.33$, or 2.58, if the desired degree of confidence is 0.95, 0.98, or 0.99, respectively. If σ is unknown, it should be estimated, or at least an upper bound computed.

For estimating a proportion of a population within E units with a desired degree of confidence, the sample should be at least n where

$$n = \frac{z^2(P)(1 - P)}{E^2}$$

and where P is a reasonable estimate of the population proportion and $z = 1.96, 2.33$, or 2.58 as previously. If no estimate for P is available, use $P = 0.5$.

For large samples ($n > 30$) a confidence interval for the population standard deviation, σ, can be constructed from the sample standard deviation, s, by using as confidence limits

$$\frac{s}{1 + \dfrac{|z|}{\sqrt{2n}}} \quad \text{and} \quad \frac{s}{1 - \dfrac{|z|}{\sqrt{2n}}}$$

where n is the sample size and $|z|$ depends on the degree of confidence desired, as in the estimate of μ.

Problems

1. A random sample of 120 Graduate Record Examination scores is taken from those of a university's graduating class of 4,182. If the mean of the sample was 1,082 with a standard deviation of 108, estimate 0.95 and 0.99 confidence intervals for (a) the population mean and (b) the population standard deviation.

2. A newspaper believes that 0.75 of retired couples prefer apartment living to house living. It wishes to sample retired couples and obtain a 0.99 confidence

interval correct to within 0.05. What sample size should be taken (a) if the newspaper estimate is used as a basis and (b) if no estimate is used?

3. A test run of 25 tires showed a mean mileage of 28,642 with a standard deviation of 1,246 miles. Determine 0.95 and 0.99 confidence intervals for the mean.

4. Under conditions of stress, the mean heart beat of 64 persons was found to be 134 with a standard deviation of 12. Determine 0.95 and 0.99 confidence intervals for the population's mean number of heart beats under the conditions of the experiment.

5. Four subjects were randomly selected from a population and asked to make subjective judgments on a situation involving interpersonal relationships. These judgments took the form of a rating sheet with the ratings weighted and an overall score obtained. These 4 subjects obtained scores of 173, 217, 143, and 166. Assuming the necessary conditions are met, estimate the 0.95 confidence interval for the population mean.

6. If a researcher wishes to obtain the mean income of a certain area correct to within \$100 with a probability of 0.95, what size random sample should he take if \$500 is a reasonable estimate for the standard deviation?

7. A random sample of 16 coffee cans is taken from a day's large output. The weights of the contents are known to be normally distributed with a standard deviation of 0.75 ounce and means which vary slightly from day to day. The sample is found to have a mean weight of 16.02 ounces with a standard deviation of 0.81 ounce. Calculate 0.95 and 0.98 confidence limits for the mean weight of the day's output.

8. Sixteen mentally handicapped patients were given a task requiring a certain degree of mental dexterity. The mean time required to do the task was 11 minutes 33 seconds, with a standard deviation of 3 minutes 12 seconds. Construct 0.95 and 0.99 confidence intervals for the mean time required to do this task by this type of patient.

9. A random sample of 200 families of 750 in a certain apartment complex is interviewed, and the mean number of children per family was found to be 3.2 with a standard deviation of 0.8. Calculate the 0.95 confidence interval for the mean number of children per family in the apartment complex.

10. A random sample of 80 students interviewed on campus showed 56 in opposition to a bill before the student senate. Construct 0.95 and 0.99 confidence intervals for the true proportion of the student body opposed to the bill.

11. Efficiency ratings of 12 identical machines chosen at random from among a company's output of thousands over the past five years were used to estimate the mean efficiency rating of all machines produced by the company over the past five years. The ratings of the twelve were 0.820, 0.913, 0.764, 0.881, 0.902, 0.893, 0.663, 0.862, 0.812, 0.778, 0.932, 0.824. If the mean of this sample is used to estimate the population mean, what can be said with a probability of 0.95 about the possible size of the error?

12. A new strain of bacteria is introduced into the water supply of a group of 84 white mice with the result that 36 of the mice are affected adversely with muscular spasms and general debilitation. Give 0.95 and 0.99 confidence intervals for the proportion of white mice which will be affected adversely under like conditions.
13. A sample of 200 light bulbs from a large number showed exactly 24 to be defective. Construct 0.95 and 0.99 confidence intervals for the proportion of defectives in the total number.

DECISIONS

Part V Photo: The Metropolitan Museum of Art, Gift of Thomas F. Ryan 1910

Chapter IX
Hypothesis Testing

9.1 EXPERIMENTAL ERROR

In addition to estimation, one of the primary purposes of statistical inference is to test theories or hypotheses. Many sets of data are obtained for this express purpose. For instance, a sociologist may wish to obtain data about characteristics of an ethnic group in a certain city and compare it with data on same ethnic group in its native land. He examines a random sample of both groups, and draws conclusions on the basis of this sample. A businessman may wish to discover if enough potential customers pass a possible site for a shop to warrant renting the shop. He spends several days observing customer traffic in order to obtain a representative sample. A physicist wishes to see if a radioactive isotope behaves as predicted. She observes a sample of the material to see how it behaves. A teacher wants to know which one of two presentations is more effective in promoting learning. He gives the presentations to two groups and examines the result.

In each case described above there was a **hypothesis**, or assumption; tests of the validity of this hypothesis were carried out. Since a sample does not constitute all cases, it is entirely possible to make an error in the conclusion. Either one can reject a true hypothesis, or else one can accept a false hypothesis. These two types of error are called type I and type II errors, respectively. These types are presented graphically as follows:

Hypothesis	*Accept*	*Reject*
true	no error	type I error
false	type II error	no error

EXAMPLE 1 A firm wishes to test the hypothesis that the volume of business in a certain territory warrants the opening of a branch office. To test this they will take a sample of one week's orders and examine it. Under what conditions will they make a type I error? Under what conditions will they make a type II error?

SOLUTION If the firm concludes, on the basis of the sample, that they should not open a branch office, yet in fact the hypothesis is true, they have made a type I error. On the other hand, if they decide to open a branch office as a result of the data gathered from the sample, and in fact are mistaken, they have made a type II error.

EXAMPLE 2 A psychologist wishes to test the hypothesis that rats can negotiate a maze in less than one minute. He takes a sample of rats and proceeds to test his hypothesis. If rats actually are able, on the average, to get through the maze in less than one minute, what was the psychologist's hypothesis if he made (a) a type I and (b) a type II error?

SOLUTION A type I error is made by rejecting a true hypothesis. Since the rats can negotiate the maze in less than one minute, this must have been his hypothesis, and he rejected it.

A type II error is made by accepting a false hypothesis. Since the rats can get through the maze in less than one minute, he must have hypothesized that it would have required at least one minute to get through the maze, then accepted it.

Problems

1. A farmer hypothesizes that fertilizer A will do a better job than fertilizer B in raising corn. To test this hypothesis, he tries each fertilizer on a plot of corn and compares the results. If he makes a type II error, what is his conclusion, and what

is the true state of things as regards the two fertilizers in their contribution toward raising corn?

2. A businessman claims that his business is going to be better this month than last, and examines a sample of orders to verify or disprove his contention. If he makes a type I error, what did he decide, and will his business actually be better this month?

3. A set of data is cited in a study concerning the antiquity of implements found in a grave site. As a result of these data, the researcher has decided that the implements are not less than 24,000 years old. Unfortunately, he is mistaken, as the implements are left over from a previous expedition, and were purchased in Pittsburgh, only 40 years ago. What hypothesis did he test if he made a type I error? A type II error?

4. An oil firm is testing a gasoline additive which may or may not increase gasoline mileage. In order to discover the additive's properties, they test gasoline with the additive in a large sample of automobiles against gasoline without the additive in the same sample. As a result of the sample they draw conclusions about the effectiveness of the additive as a mileage increaser. There are two possible hypotheses which can be tested; give each hypothesis and the conclusions, based on the sample, which would yield type I and type II errors, respectively.

9.2 FORMULATING HYPOTHESES TO CONTROL ERROR

As we pointed out in the preceding section, the possibility of error is implicit in any experiment. The experimenter wishes to control the error to the best of his ability. To this end, it is customary to state the hypothesis being tested in a way such that, if it is correct, the hypothesis can be used to calculate the probability of a type I error. A hypothesis which is of this kind states that a population mean is equal to a certain number ($\mu = \mu_0$). Hypotheses stating that two population means are equal ($\mu_1 = \mu_2$) or several population variances are equal ($\sigma_1^2 = \sigma_2^2 = \sigma_3^2$) are also of this type.

If these hypotheses are true, it can be calculated, on the basis of a normal distribution, t-distribution, or some other statistical distribution, what theoretical proportion of the sample statistics would actually fall so far from the true parameter. Since this type of hypothesis often is used to hypothesize that there is no difference between or among population parameters, it is called the **null hypothesis** and is symbolized by $\mathbf{H}_0$.

For instance, if we wished to test the hypothesis that the students at a university scored an average of more than 900 on the Graduate Record Examination, we could take a sample of students' scores, compare their mean to 900 on the basis of a z- or t-test, and decide whether or not to conclude that the average score was actually more than 900. In order to have some basis for comparison, it would be necessary to test the null hypothesis,

H_0: $\mu = 900$, against some suitable alternative. Obviously, we cannot use a hypothesis that the mean is not 900, since an extremely high *or* extremely low sample mean might cause us to reject H_0. Thus, selection of more than one **alternate hypothesis** is also possible.

Now in this case, as in many cases, we have three possible alternatives, which we designate as $\mathbf{H}_1$, H_1: $\mu > 900$; H_1: $\mu \neq 900$; H_1: $\mu < 900$.

In the first and third of these statements the alternate hypothesis is called **one-sided**; the second is **two-sided**. Which of these types of hypothesis is used generally depends upon the **research hypothesis**, which supplies the basis for obtaining the null hypothesis. In this case the research hypothesis was that the students scored an average of more than 900 on the GRE. Since the research hypothesis could be formulated H_R: $\mu > 900$, the alternative hypothesis is one-sided.

Now we come to selection of the alternate hypothesis which is to be used. If we use H_1: $\mu > 900$, we will not conclude that the mean actually is greater than 900 unless we can show that there is little chance of being mistaken. If we use H_1: $\mu < 900$, however, we will not conclude that the mean is less than 900 unless we can show little chance of error. The emphasis is a matter of placing the burden of the proof.

In sections 5.4 and 6.4 simple examples of hypothesis testing were presented. In each case the alternate hypothesis was one-sided. In each case rejection of the null hypothesis would be made only if the mean of our sample was found at one extreme of the distribution. In the case of a two-sided hypothesis, rejection would come if the sample was extreme at *either* end. If we want to show, for example, that income levels have increased in a certain area, we would use the null hypothesis that income levels have remained the same. If per capita income was $2,750 five years ago and we want to show that it has increased, we test the null hypothesis that $\mu = \$2,750$ where μ represents the per capita income. Our alternate hypothesis is that $\mu > \$2,750$ since that is what we want to show. As another example, we may want to try to determine which of two drugs for treatment gives better results. If we let μ_A and μ_B represent the mean survival or recovery rates for the two drugs, A and B, then our alternate hypothesis will be $\mu_A \neq \mu_B$ since we are willing to accept either as being better.

A word of caution about two-sided tests. Rejection of the alternate hypothesis and consequent acceptance of the alternate hypothesis does not, in itself, tell us in which direction our results point. Quite often the data themselves, or the test, will tell us, but, often this is not the case. Once we do get significant results, we must then examine these results to see if we can draw any additional conclusions or if we must perform another test.

Before an experiment is performed, a research hypothesis should be formulated. The research hypothesis sums up what the experimenter hopes to find out from the experiment and forms the basis for the null and alternate hypotheses which are used in statistical hypothesis testing.

If a research hypothesis presents an either-or situation, the selection of the null and alternate hypotheses is easy. An example of this type of hypothesis is one in which the experimenter wishes to know which one of two methods is the best. For example, suppose a researcher wishes to know whether fertilizer A or B will have the best effect on the yield of his corn crop. His null hypothesis would be that both fertilizers would have the same effect on crop yield, or that the yield with fertilizer A is the same as that with fertilizer B. The null hypothesis is symbolized by H_0: $Y_A = Y_B$. His alternate hypothesis would be that one fertilizer would have a better effect on crop yield, or that the yields would be different with the different fertilizers. The alternate hypothesis is symbolized by H_1: $Y_A \neq Y_B$. In this case the researcher has no preconception about which fertilizer will have the better effect on crop yield. If the results of the experiment indicate that the null hypothesis should be rejected at the selected level of significance, then the researcher knows that one fertilizer is better. At this point, however, the statistical tests performed thus far do not allow the researcher to say which one is better. In some experiments this is obvious from the data and the nature of the experiment. For others, however, further statistical tests must be used, and there are a variety of other statistical tests which can be used to indicate the direction of the difference if it is not obvious from the data itself. This is particularly true in multiple comparisons.

The use of the null hypothesis for testing has distinct advantages. If we can assume that the population parameters have specific values, then we can calculate the actual consequences of these values. By assuming that the population mean is actually 900, we can compare our hypothetical population to a normal distribution and see if our results are acceptable in the light of our hypothesis. We can determine, for example, the probability that a sample of size 100 would have a mean of as much as 946 by chance if the actual population mean is 900. If we feel that this chance is less than we think is reasonable, we can conclude that the mean of our population is greater than 900. We cannot, on the other hand, draw any kind of meaningful statistical conclusions about some vague statement such as "the mean is greater than 900." Thus we generally set up a null hypothesis which can be tested; that is, we perform the actual arithmetic as if the null hypothesis were true. If we doubt the reasonableness of this result, we then accept our alternate hypothesis which is, in most cases, what we want to prove. The word "prove" is used here with caution. Statistics cannot *prove* statements, but they can be used to demonstrate a high degree of likelihood.

Suppose that the research hypothesis includes some other, more complicated assumptions as well. For instance the researcher might want to show that fertilizer *A* is better than fertilizer *B*. In this case her alternate hypothesis would be H_1: $Y_A > Y_B$. In this case she would not conclude that fertilizer *A* was better unless the proof was fairly conclusive. Other, more subtle distinctions or desires may well shape the alternate hypothesis, however. Perhaps the company is bringing out a new fertilizer to compete with another company's product. It may be enough to show that the other company's product is not statistically superior. In the case where it is desired to show a direction there are two alternate hypotheses possible. In the case of the fertilizer yields, two alternatives are possible: H_1: $Y_A > Y_B$ or H_1: $Y_B > Y_A$. The alternate to be used depends on her purposes. She may wish to market fertilizer *B* only if it is better than fertilizer *A*, or she may wish to market fertilizer *B* unless it is inferior to fertilizer *A*. It may be obvious from the statement of the research hypothesis which alternative to use, but often it is quite confusing. In general, we may look at the result acceptance of H_0 will cause. H_1 is then assigned to the other result. In the case where fertilizer *B* will be marketed only if it is better than fertilizer *A*, acceptance of H_0: $Y_A = Y_B$ will result in *not* marketing the fertilizer. Then H_1 must result in marketing the fertilizer. Since H_1 must result in marketing the fertilizer, it must be H_1: $Y_A > Y_B$. In this case the burden of proof is on fertilizer *B* since H_0 will be rejected only if the probability of making an error is small.

If the manufacturer is anxious to get fertilizer *B* on the market and will market it unless it is inferior to fertilizer *A*, the failure to reject H_0: $Y_A = Y_B$ should result in marketing fertilizer *B*, and nonmarketing will occur only if the alternate hypothesis, H_1: $Y_A > Y_B$, is accepted. In this case the burden of proof is on fertilizer *A*.

It should be noted that a one-sided alternate hypothesis implies that the null hypothesis is somewhat more encompassing than a statement of equality. For instance our null hypothesis might be that the means of three populations are all equal. Since a number of directional or semi-directional alternate hypotheses are possible in this case, the only suitable alternate hypothesis is that they are not all equal.

In general, the alternate hypothesis coincides with the research hypothesis. The alternate hypothesis is that hypothesis which the experiment is set up to verify.

In any case, there is a risk of error. The probability of making a type I error is usually designated α, while the probability of making a type II error is designated β. Thus we have

$$\alpha = P\ (\text{reject } H_0 \mid H_0 \text{ is true})$$

$$\beta = P\ (\text{accept } H_0 \mid H_0 \text{ is false})$$

In general, statisticians use values of 0.05 or 0.01 for α. This value is often called the **level of significance** for the experiment and corresponds to the researcher's willingness to take a chance of being wrong. The more serious the consequences of making an error, the smaller the level of significance which is used. A result is usually called significant if the researcher can reject the null hypothesis in favor of the alternate hypothesis, at the desired value of α.

Controlling for a type II error is very difficult. β cannot be calculated exactly unless the actual population parameter is known. Quite often the possibility of a type II error is eliminated by refusing to accept a null hypothesis, even though it cannot be rejected. Such a procedure is called **reserving judgment**, and is useful when consequences of a type II error are great, or its likelihood is high.

Values of β can be calculated, of course, if we assume that the parameter has a particular value. By determining values of β for possible values of the parameter we can plot a distribution approximating all possible values of β called the **operating-characteristic curve** (or O–C curve). Examples of calculations for β are given in sections 5.4 and 6.5. As we stated there, the only way to decrease β is to increase the sample size or allow α to increase. One use of the O–C curve is that it allows us to weigh the probability of a type II error. In the case of large samples β is usually relatively small, so reserving judgement is done less than when the sample is small.

EXAMPLE 1 A farmer takes a sample of grapefruit and weighs them. His standard requires a mean weight of 12.5 ounces for his crop, or he will not market them. Give suitable null and alternate hypotheses for his experiment if (a) he has high standards and will not market his crop if it does not average more than 12.5 oz. (b) he is anxious to market the crop unless it is demonstrably substandard.

SOLUTION In either case we have $H_0\colon \mu = 12.5$, where μ represents the mean weight of the crop. In case (a) however, if H_0 is true he will not market the crop; he will market it only if $\mu > 12.5$, so we have for (a),

$$H_0\colon \mu = 12.5$$
$$H_1\colon \mu > 12.5$$

In case (b), if H_0 is true, the crop is not substandard so he will market it. He will not market it only if $\mu < 12.5$, so we have, for (b),

$$H_0\colon \mu = 12.5$$
$$H_1\colon \mu < 12.5$$

EXAMPLE 2 Two applicants for a job agree to a competition in which each will perform the operation required for the job a number of times in order to have a sample of each applicant's performance on the job. The applicant whose performance is significantly better will be given the job. Set up suitable null and alternate hypotheses for the experiment.

SOLUTION The key word here is *significantly*. Differences may be slight and may not meet desired level of significance until after a number of trials. Hence, we are searching for true difference in their long-term performance. If P_1 and P_2 represent the two individuals' *true* performance on the job, we have

$$H_0: P_1 = P_2$$

$$H_1: P_1 \neq P_2$$

The alternate hypothesis is two-sided, since we are only interested in there being some difference in the two performances. We do not address ourselves here to the problem of what to do if we cannot reject H_0: perhaps they will both be hired, or some other criteria will be devised.

Problems

1. A sociologist draws a sample of families from an apartment complex to examine whether the number of children in families in the complex differs significantly from the national average. If the national average of children is given by μ_0, set up suitable null and alternate hypotheses for the experiment.
2. A French teacher is testing a new method of teaching verbs. He knows that the class would be expected to learn an average of approximately 32 new verbs per week using the old method, but wishes to use the new method if at all possible. (a) Give suitable null and alternate hypotheses for the experiment. (b) The teacher next door is watching his experiment with interest, and will use the new method if it is definitely superior to the old. Give suitable null and alternate hypotheses for his version of the experiment.
3. A firm wishes to decide whether to open a branch office on the basis of volume and samples its daily orders to estimate the proportion of orders going into that area. The crucial amount is 5%. Formulate suitable null and alternate hypotheses for this experiment if (a) the firm is cautious and does not wish to open a new office unless it must and (b) the firm is experiencing a period of expansion and wishes to open a branch office on the slightest justification.

4. A psychologist tests two random samples of children with the same test after giving each group some instruction; the samples were given the instruction by two different methods. Let μ_1 and μ_2 represent the hypothetical population means of all possible samples instructed by methods 1 and 2 respectively. Set up hypotheses for the experiment if (a) the psychologist has no reason to suspect that there actually is a difference and (b) theoretical considerations indicate that method 2 is actually better, and she wishes to place the burden of proof on method 2.

9.3 TESTS OF SIGNIFICANCE

With the foregoing as background, let us recall the businessman's dilemma of sections 5.4 and 6.5. On his store shelves he has two products, identical except for brand name. Since his space is limited, he has decided to eliminate one brand from his inventory. Brand *A* and brand *B* have had approximately equal sales over the past several months, but he feels that recently his customers have shown a preference for brand *B*. On the other hand, brand *A* is advertised more extensively than brand *B*. He would like to stock the brand preferred by his customers, but if there is no real preference he wants to take advantage of the advertising offered by brand *A*. He therefore decides to perform an experiment by observing the buying habits of a sample of customers buying the product. He is thus testing a hypothesis.

What hypothesis is he testing? He feels that brand *B* is actually preferred by his customers; this is his **research hypothesis**, the hypothesis he wishes to prove. To control for type I error he sets up his **null hypothesis** that the customers prefer the two brands equally. Thus he bases his assumptions on the hypothesis that the proportion of customers preferring brand *A* is equal to the proportion of customers preferring brand *B*. This is symbolized $P_B = P_A$ (or some other appropriate designation). His **alternate hypothesis** corresponds to the research hypothesis and is determined by the research hypothesis. In this case it would be $P_B > P_A$. That is, if $P_B > P_A$ he will stock brand *B* only, otherwise he will stock brand *A* only. Note that the two hypotheses must result in different courses of action. Either $P_B = P_A$ or $P_B < P_A$ will result in stocking brand *A* only, so it is not necessary to consider them separately.

After deciding on the null and alternate hypotheses, he must set the **level of significance**, which is his willingness to accept a chance of making a type I error. This is α, and we generally set $\alpha = 0.05$ or $\alpha = 0.01$.

Before completing the experiment we must determine a **test statistic**. This is the basis for our decision. The test statistic is the variable which is used to accept or reject the null hypothesis. In the businessman's case, we let our statistic be the number of customers choosing brand *B*. It is more usual to relate the data obtained to one of the standard sampling distributions and calculate a value of z or some other test statistic that is compared to **acceptance** and **rejection regions**. The points which divide these regions are called the **critical values** and are set forth in the **criteria**. In the businessman's example in section 6.5, our criteria, based on a sample size of 163, were to reject the hypothesis if 90 or more customers bought brand B, and accept it otherwise. More typically we would relate our sample immediately to the normal distribution and reject the null hypothesis if, in this case, $z > 1.28$. Referring to section 6.5, you will see that we used this value of z to obtain the figure of 90 customers. In practice, we use the standard score formula to determine the value of z which our sample yields and then compare it to the critical value. In this case the critical value is 1.28. The general procedure is to set up the criteria independently of the precise size of the sample (although it is usually necessary to know whether the sample is large or small) and state them prior to actual performance of the experiment. Further discussion of acceptance and rejection regions is given in section 10.1.

After the preliminary work is completed, the experiment is performed and the results tabulated. These results are compared with the criteria and a **decision** is made about whether or not to reject the null hypothesis. Based upon this decision, a **conclusion** will be reached which agrees with the actual meaning of the hypotheses. This five-step format is known as the **hypothesis-testing procedure** and may be described as follows.

1. **Statement of hypotheses**
   H_0: Statement of the null hypothesis.
   H_1: Statement of the alternate hypothesis to be accepted if H_0 is rejected.
2. **Level of significance**
   α: The acceptable probability of making a type I error, that is, of rejecting H_0 erroneously. This is usually 0.05 or 0.01, by common usage. See the comment on page 223.
3. **Criteria**
   These criteria are the basis for later studies, and are determined entirely by three conditions: the appropriate sampling distribution, the nature of the alternate hypothesis (one- or two-sided), and α, the level of significance. They generally give a statement of a rejection region, such as the following: reject H_0, and accept H_1 if $z > 1.96$ or $z < -1.96$. This particular case, as will be seen later, is used when there is a normal distribution, a two-sided hypothesis, and a significance level of 0.05.
4. **Results**
   In performing the experiment, the researcher must actually apply the procedures indicated in the criteria and obtain a result, such as $z = 1.37$.

5. **Conclusions**

   Two types of conclusions will be reached, formal and informal. The formal conclusion will always be either to reject H_0 or to fail to reject H_0, and will be completely determined by comparing the results with the criteria.

   The second type of conclusion, the informal conclusion, states what the formal conclusion means, such as: the firm will open a branch office, will not open a branch office, or will reserve judgment, perhaps pending the results of further tests.

This five-step procedure is applicable to most statistical tests, and, if followed, it will provide a safeguard against improper uses or conclusions.

Selection of the criteria, as noted before, is dictated entirely by the selection of α, by the nature of H_1, and the sampling distribution used. In general, this distribution will, in turn, be indicated by the problem itself, but in most cases there are additional assumptions which should be met in order to use the test. These will be discussed as each test is presented and used.

Determination of sample size is also an important aspect of hypothesis testing. Although a representative sample is wanted, the larger the representative sample the better. If there is a strong possibility of a sample being limited in size, determination of a maximum allowable error, E, will dictate the *minimum* sample size. The methods given in example 2, page 184, and example 2, page 197, demonstrate some methods used for deciding the minimum sample size.

In the businessman's problem, suppose that 101 of 163 customers buy brand B. In the proper format, the presentation of the experiment would appear as follows.

1. H_0: $P_B = P_A$
   H_1: $P_B > P_A$
2. $\alpha = 0.05$
3. Criteria: reject H_0 if $x \geq 90$.
4. Results: $x = 101$.
5. Conclusions: reject H_0; the proportion of customers preferring brand B is greater than the proportion preferring brand A. Stock only brand B.

At this point you should examine the chart on the back end paper of this book. The purpose of this chart is to provide a handy reference to the statistical methods discussed in chapters 10 through 13, and to indicate which of the methods are to be used in each situation listed. To determine which test to use, begin at the top with number of samples, and work your way down to the appropriate rectangle in the classification system. When you have classified the situation, you will find the type of test, section reference, and appropriate hypotheses. The only technique which has not been placed on this chart, but is covered in the book, is the binomial test. This was covered in section 5.4 and is applicable to one sample, one variable, count data, small sample size.

The chart should be used in conjunction with *prior planning* for an experiment. The **design of an experiment** should precede the actual experimentation itself. Every statistician is familiar with cases in which an experimenter submits a mass of data and asks for an interpretation of the data only to be told that the data is useless because it does not test what the experimenter wanted to find out.

It is very important, then, to determine *in advance* what hypothesis is to be tested, and then to design an experiment using appropriate techniques which will make it possible to use statistical inference to answer the experimental questions. It is helpful when looking at an experiment to be able to evaluate the design and tell whether or not the results actually say what we are told they say. For example, were the samples chosen in such a way that we can reasonably assume that they were random? Are the samples independent, if appropriate? Were the proper tests used? These questions are very important and must be answered if we want to interpret the results correctly.

For further information the reader should consult any of the available books on design and analysis of experiments, such as B. J. Winer, *Statistical Principles in Experimental Design* (McGraw-Hill) or C. R. Hicks: *Fundamental Concepts in the Design of Experiments* (Holt, Rinehart, & Winston). Another text, which is specialized for use by biologists but which is still quite useful, is *Biometry*, by R. R. Sokal and F. J. Rohlf (W. H. Freeman).

9.4 SUMMARY

Hypothesis testing and decision making constitute most of the rest of the text. Two types of experimental error are encountered: type I error occurs when a true hypothesis is rejected; type II error occurs when a false hypothesis is accepted.

Since it is difficult to control for a type II error, hypotheses to be tested are formulated positively, in such a way that the probability of a type I error, α, can be controlled. The hypothesis to be tested is called the null hypothesis, H_0. The hypothesis which will be accepted if H_0 is false is called the alternate hypothesis, H_1. The acceptable probability of error, α, is called the level of significance.

In testing hypotheses and reaching decisions, five steps should be used.

1. **Statement of hypotheses:**
   H_0: Null hypothesis.
   H_1: Alternate hypothesis.
2. **Level of significance:**
   α = some number, usually 0.05 or 0.01.
3. **Criteria:**
   Determined by H_1, α, and the actual nature of the population and sample.
4. **Results:**
   Actual values obtained by applying the appropriate test.
5. **Conclusions:**
   Formal; reject H_0 or fail to reject H_0. Informal; the consequences of the formal conclusion.

Problems

1. A farmer wishes to test whether his pigs will gain more weight on corn or slop. He takes two randomly selected groups, *A* and *B*, and proceeds to perform the experiment. He will be satisfied if his significance level is 0.05. Give the statements of steps 1 and 2 of the experimental method if (a) he simply wishes to find out if one method is actually better than the other (b) all things being equal he would prefer to feed them slop since corn is expensive, and (c) he has a surplus of corn and would prefer to feed them corn. In each case give the informal conclusion as related to each possible formal conclusion.
2. A researcher tests samples of two models of sewing machines, *A* and *B*, and wants to draw conclusions as to which model actually does the better job. As a result of his sample, he finds that the sample of model *A* has a better job performance than the sample of model *B*. His hypotheses have been formulated in accordance with the guidelines set forth here. As a result of his tests, he decides that the probability of type II error precludes his acceptance of the results as conclusive. (a) What were his hypotheses? (b) Was the sample of model *A* significantly better than that of model *B*? (c) What were his conclusions?
3. The quality control procedure in a canning plant is to select a random sample of 40 cans and take the mean weight. If the procedure is in control the population mean weight will be 16.42 ounces. The level of significance used is 0.05, and the shipment will be passed unless the sample is significantly below standard. Monday's sample is weighed and its mean tested against the norm. It is found to be below 16.42, such that only 1 sample in 15 would fall so low. Will the shipment be passed? Will the shipment be passed if the mean of only 1 sample in 25 would fall so low? Suppose the sample has a mean greatly in excess of 16.42. In terms of the criteria stated, would the shipment be passed if only 1 sample in 25 would weigh so much?

4. A biologist wishes to decide whether a particular microorganism has a bad effect on a desirable strain of antibiotic. The antibiotic, under certain conditions, has a deterioration rate of 40%, which means that 40% of all ampoules in a sample will be ineffective after 21 days. She introduces the microorganism into 100 ampoules of the antibiotic and observes that 52 of them are ineffective after twenty-one days. Using the experimental procedure and a 0.05 level of significance, test the hypothesis that the microorganism is detrimental to the antibiotic. The research hypothesis is that $P > 0.4$, where P is the proportion of ampoules in the sample which are ineffective after 21 days.

Chapter X

Tests Concerning Means and Proportions

10.1 MEANS OF ONE LARGE SAMPLE

In many cases the purpose of obtaining a sample is to test a hypothesis, rather than to estimate a parameter. The two procedures are related, however, as we shall see. In this section we shall study a method of testing the hypothesis that the mean of a population is equal to some given value.

Now suppose, for some reason, we have a research hypothesis which leads us to believe that a population has a mean which is different from some value, μ_0. By the procedures outlined in Chapter 9 we have

$$H_0: \mu = \mu_0$$

$$H_1: \mu \neq \mu_0$$

where μ is the population mean and μ_0 is the value we are testing against.

If the mean actually is μ_0, a distribution of all means of samples of the same size (sampling distribution of the mean) will have mean μ_0 with standard error $\sigma_{\bar{x}}$. (Notice that the term *standard error* is routinely used to indicate the standard deviation of some sampling distribution.) The level of significance, α, is the chance we are willing to take of making a type I error. If μ_0 is the actual population mean, α is the risk we are willing to run that the sample mean is so far from μ_0 that we will reject H_0 erroneously.

The sampling distribution of the mean is approximately normally distributed with mean μ_0 and standard error $\sigma_{\bar{x}}$ as illustrated.

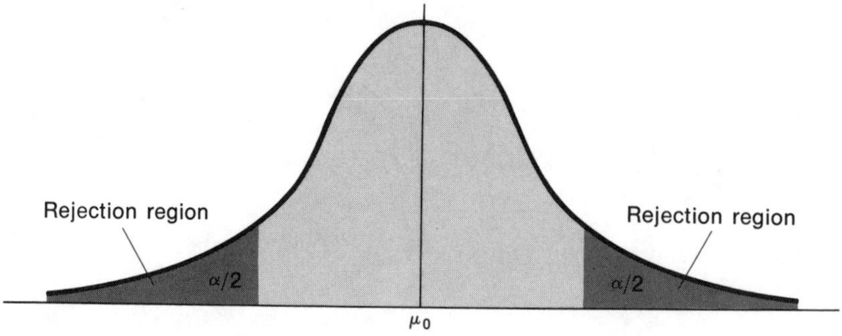

This curve can be related to the standard normal curve by the formula

$$z = \frac{\bar{x} - \mu_0}{\sigma_{\bar{x}}} \quad \text{or} \quad z = \frac{\bar{x} - \mu_0}{s_{\bar{x}}} \quad \text{if} \quad \sigma_{\bar{x}} \text{ is not known and } n > 30$$

where $\bar{x}$ is a particular sample mean. There is some value of z such that $\alpha/2$ of the area under the curve is to the right of z and $\alpha/2$ of the area under the curve is to the left of $-z$. These areas constitute the **rejection region**. If, for example, $\alpha = 0.05$, we know that 0.025 of the area under a standard normal curve is to the right of 1.96 and 0.025 of the area under the curve is to the left of -1.96. If the true population mean is actually μ_0, one of twenty (i.e., 0.05) of the sample means will fall in the tails *by chance alone*. Thus, if we reject the null hypothesis at a significance level of 0.05, we are still admitting that there is one chance in twenty that the null hypothesis is true and that it was pure bad luck that gave us a sample which led us to reject it. If we had specified a significance level of 0.01, we would be admitting only one chance in one hundred of making the same mistake. Of course, the **critical value** of z would be 2.58 rather than 1.96. That is, if the mean is actually μ_0, one sample mean in one hundred on the average will fall more than 2.58 standard errors either above or below μ_0 by chance alone. The critical value is dependent upon α, and upon the nature of the alternate hypothesis. If the alternate hypothesis is one-sided, it is (in this case) either $\mu > \mu_0$ or $\mu < \mu_0$. If we had

$$H_0\colon \mu = \mu_0$$

$$H_1\colon \mu > \mu_0$$

we would not reject H_0 if our sample mean was below μ_0, even far below, since we are only interested in the case in which μ is greater than the population mean. In this case we would reject H_0 only if our sample mean, $\bar{x}$, were so far above μ_0 that it occurred only α of the time, by chance.

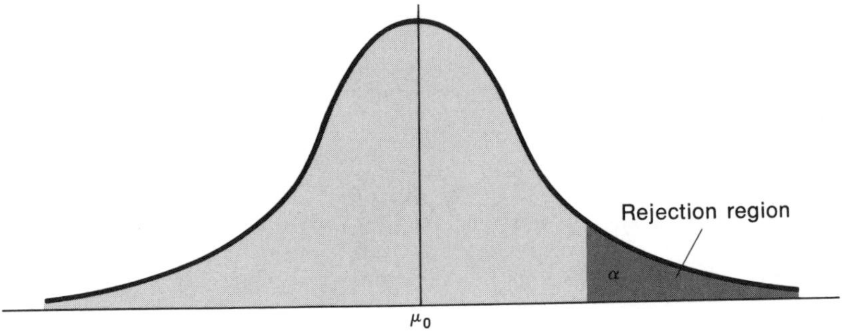

This type of test is called a **one-tailed** or directional test which may be contrasted to the **two-tailed** or nondirectional test described previously. In the case of $\alpha = 0.05$, we note that 0.05 of the cases would fall above z if $z = 1.64$ or below z if $z = -1.64$; this latter would be used in the cases where we had $H_1: \mu < \mu_0$. If $\alpha = 0.01$, the corresponding critical value for z would be 2.33 or -2.33.

The critical values will be incorporated into the criterion portion of the experimental format as appears later.

EXAMPLE 1 A firm wants to decide whether or not to purchase a new letter sorter for its office, since the new one is too expensive to buy if it will not be used. The company selling the sorter offers to let the firm try it for 6 weeks to make a decision. Since the firm works on Saturday, this means 36 working days. The office manager decides that she will order the machine if she can be 95% sure that it will sort more than 4,000 letters per day. Thus she performs an experiment, taking what she hopes to be a random sample of 36 days' performance of the machine and a significance level of 0.05. What will her decision be if this sample has a mean of 4,132 letters sorted per day with a standard deviation of 324?

SOLUTION Following the procedures outlined in Chapter 9, we have $H_0: \mu = 4{,}000$. Since the firm will not buy the sorter if this is so, but will if

$\mu > 4{,}000$, this becomes our alternate hypothesis. This makes the test one-tailed, and the critical value for z at 0.05 is 1.64; that is, 0.05 of the area under the standard normal curve is to the right of 1.64. Using the transformation formula, we determine the actual value of z in this case if $\mu = 4{,}000$, apply the criteria, and make the decision. The entire procedure is summarized here. Use of a continuity correction is unnecessary since the *means* of the set can be any rational value and so are not measured in integers only.

1. $H_0: \mu = 4{,}000$
   $H_1: \mu > 4{,}000$
2. $\alpha = 0.05$
3. Criteria: Reject H_0 if $z > 1.64$.
4. Results: Here $\bar{x} = 4{,}132$, $\mu_0 = 4{,}000$, and $s_{\bar{x}} = 324/\sqrt{36} = 54$

$$\text{hence } z = \frac{4{,}132 - 4{,}000}{54} = \frac{132}{54} = 2.44$$

5. Conclusions: Since $2.44 > 1.64$, we reject H_0 in favor of H_1. Thus the mean probably is actually greater than 4,000, and the firm will purchase the sorter.

DISCUSSION If the firm had decided to purchase the sorter only if it had been able to sort 4,000 letters or more per day, it would have been impossible to perform the test. $H_0: \mu = 4{,}000$ would have resulted in purchasing the sorter as would $H_1: \mu > 4{,}000$. Yet $H_1: \mu < 4{,}000$ would have caused the firm to purchase the sorter if the mean were not significantly below 4,000, which would have opposed the firm's desire to place the burden of proof on the machine. In such a case it is necessary to use judgment and note that an adjustment must be made in order to fulfill the burden of proof requirements. This would result in the problem being handled exactly as it was.

EXAMPLE 2 A sociologist examining a large apartment complex wishes to discover if the number of persons per family unit differs significantly from the national mean of 4.80. He interviews 100 of the 750 families in the complex, obtaining a mean of 4.70 with a standard deviation of 0.80. If he sets a significance level of 0.05, what conclusions does he draw?

SOLUTION His alternate hypothesis is two-sided, so his rejection region as stated in the criteria is two-tailed. His procedures are the following:

1. $H_0: \mu = 4.80$
   $H_1: \mu \neq 4.80$
2. $\alpha = 0.05$
3. Criteria: Reject H_0 if $z > 1.96$ or if $z < -1.96$
4. Results: Since 100 is more than 5% of 750, we have

$$s_{\bar{x}} = \frac{0.80}{\sqrt{100}}\sqrt{\frac{650}{749}} \doteq 0.073$$

   hence

$$z = \frac{4.70 - 4.80}{0.073} = \frac{-0.10}{0.073} \doteq -1.36$$

5. Conclusions: Since -1.36 is not less than -1.96, we cannot reject H_0. Thus the mean family unit of this complex does not differ significantly from the national average.

DISCUSSION The entire procedure is quite straightforward up to the informal conclusion. It is here that the possibility of a type II error must be considered. We know that H_0 cannot be rejected, but should it be accepted? We cannot know the risk of a type II error (i.e., the value of β) unless we know μ, the population mean. In this case it is the mean number in the 750 families. We can see from Table 2 in the Tables section, however, that only 0.0869 of all possible sample means would be as far as 1.36 standard errors below the mean by chance. If we must either accept or reject H_0, we have no choice. If, however, we have the option of reserving judgment, this is a way of avoiding a type II error. Another consideration involves the consequences of a type II error. If they are not particularly grave, we may accept H_0 with perhaps a passing comment, such as, "Although the risk of a type II error was fairly high, it was decided to use the national mean of 4.8, since the sample did not differ significantly (at the 0.05 level) from this value."

COMMENT It is usual to set significance levels based on all necessary considerations before actually performing the experiment. It can be tempting to change levels in order to achieve significance but this is poor practice since significance levels are based on other considerations. In addition, it is good practice in writing reports to give the actual significance level, such as 0.0869, one-tailed, in the previous example, so that the reader can be a judge. This is also something to watch for in reading a report. If you are reading a study and are concerned with the result, it is somewhat disconcerting to read "not significant at the 0.01 level" or "significant for $\alpha = 0.05$." It is possible that the first has a 0.03 level of significance and the second 0.04. The reader might be satisfied with $\alpha = 0.05$ in the first case but only with $\alpha = 0.01$ in the second. Thus it is courteous to the reader of a report for the author to be as precise as possible.

EXAMPLE 3 A psychologist hypothesizes that college students today sleep less during the night than college students of a generation ago. To test this hypothesis, he selects a random sample of 400 college students, asks them to record their nightly slumbers in the same way as a study of 25 years ago, and uses the mean for one quarter to test his hypothesis. The old study showed college students sleeping an average of 7.32 hours during the night. The 400 students used in the study had a mean of 7.21 hours per night with a standard deviation of 1.08 hours. What conclusions can you draw at a significance level of 0.01?

SOLUTION We have a one-sided hypothesis, and the procedures used are summarized as follows:

1. $H_0: \mu = 7.32$
   $H_1: \mu < 7.32$
2. $\alpha = 0.01$
3. Criteria: Reject H_0 if $z < -2.33$.
4. Results:

$$s_{\bar{x}} = \frac{1.08}{\sqrt{400}} = 0.054; \ \bar{x} = 7.21, \text{ so}$$

$$z = \frac{7.21 - 7.32}{0.054} = \frac{-0.11}{0.054} \doteq -2.04$$

5. Conclusions: Since -2.04 is not less than -2.33, we fail to reject H_0. However, the likelihood of a type II error is great, so we reserve judgment.

If the psychologist had chosen $\alpha = 0.05$, the result would have been significant since $-2.04 < -1.64$. Although he might not have been satisfied with the result he should include this fact in his report so that another person can draw his own conclusions.

In summary, this section has given criteria to use whenever a sample is used to test a hypothesis concerning the mean of the population from which the sample is drawn. These criteria are based on the sampling distribution for the mean as described in Chapter 7. The statistic relating the sample mean $\bar{x}$ to the hypothetical population mean μ_0 through the standard error of the mean $\sigma_{\bar{x}}$ is

$$z = \frac{\bar{x} - \mu_0}{\sigma_{\bar{x}}}$$

In each case the null hypothesis must be H_0: $\mu = \mu_0$, where μ is the actual population mean. The three possible alternate hypotheses and the associated criteria are given here.

Case I

1. H_0: $\mu = \mu_0$
   H_1: $\mu \neq \mu_0$

2. $\alpha = 0.05$	$\alpha = 0.01$
3. Criteria: Reject H_0 if $z > 1.96$ or if $z < -1.96$.	Reject H_0 if $z > 2.58$ or if $z < -2.58$.

Case II

1. H_0: $\mu = \mu_0$
   H_1: $\mu > \mu_0$

2. $\alpha = 0.05$	$\alpha = 0.01$
3. Criteria: Reject H_0 if $z > 1.64$.	Reject H_0 if $z > 2.33$.

Case III

1. H_0: $\mu = \mu_0$
   H_1: $\mu < \mu_0$

2. $\alpha = 0.05$	$\alpha = 0.01$
3. Criteria: Reject H_0 if $z < -1.64$.	Reject H_0 if $z < -2.33$.

In each case, the rejection region given by the criteria corresponds to the probability of α that a sample mean would fall into the rejection region by chance if H_0 is true.

Problems

1. A County Farm Agent wishes to determine the effect of a new fertilizer on the yield of corn per acre, as compared to that of the most popular brand. The popular brand is known to yield an average of 2.12 tons per acre. He persuades 64 farmers to each plant an acre of corn and treat it with the new fertilizer. These 64 acres had an average yield of 2.18 tons per acre with a standard deviation of 0.30 tons. Assuming the yield is attributed solely to the fertilizer and that the sample is random or representative, what can be said about the new fertilizer at a significance level of 0.05?

2. Specifications for a candy bar machine call for it to make bars weighing a mean of 2.30 ounces with a standard deviation of 0.12 ounces. A daily sample of 100 bars is weighed *in toto* to decide if the machine is in control. If the bars are significantly above or below the preset amount, the machine should be retooled. If a significance level of 0.05 is used, above and below what 100 bar weights should the machine be retooled?

3. A shipment of 10,000 copper rods must be accepted or rejected on the basis of a sample of 100. The shipment will be accepted unless there are significantly more than an average of 13.7 blemishes per bar. The sample of 100 showed an average of 13.9 blemishes with a standard deviation of 1.4. Using a significance level of 0.01, will the shipment be accepted or rejected? You cannot reserve judgment.

4. A new drug is hailed by its manufacturer as an excellent treatment for a disease. Using standard treatment, patients take an average of 9.2 days to recover. Using the new drug, 50 patients take an average of 9.3 days with a standard deviation of 1.1 days. The manufacturer claims that this shows that the old treatment is not significantly better than the new, so they will market it. Can their claim be substantiated at a 0.05 level of significance?

5. A disgruntled golfer claims that a new set of golf clubs has upset his game. He points to the fact that he used to average 93.4 strokes per round, but in the 36 rounds he has played since obtaining the new clubs, his average has jumped to 96.2 with a standard deviation of 8.1. Test his claim with $\alpha = 0.01$. Remember that low scores are desired in golf.

6. A group of 50 tenth-grade students is given a paragraph and asked to memorize as much as possible; the retention is measured on a scale from 1 to 99. The researcher is testing the hypotheses that the average tenth-grade student will score above 70 on the scale. This group scored a mean of 68.3 with a standard deviation of 11.62. Test the hypothesis at a significance level of 0.05.

7. A type of new tires is run through an endurance test to see whether or not they can average 20,000 miles under the worst conditions. A sample of 100 sets is tested and lasts an average of 20,226 miles, with a standard deviation of 863 miles. If the company does not wish to market the tires unless they have been shown to wear significantly more than 20,000 miles under these conditions, will they be marketed? A significance level of 0.01 is used.

10.2 MEANS OF ONE SMALL SAMPLE

The procedures of the preceding section are not conservative enough to be useful if the sample is small. When the sample is small H_0 may be rejected more often than it should be for $n \leq 30$. Incidentally, this number is relatively arbitrary and some statisticians may wish to use a larger number. As n increases, the approximation becomes better and better.

The method itself is valid, but if the sample is small, it is customary to base the criteria on the statistic

$$t = \frac{\bar{x} - \mu}{s/\sqrt{n}}$$

which was described in Chapter 8. This yields slightly higher critical values for the rejection region as you may see in Table 3. We use $n - 1$ degrees of freedom.

If the alternative hypothesis is two-sided, our rejection region is two-tailed; that is, *half* of the rejection region is located in each tail. For $\alpha = 0.05$, for example, having a two-sided alternative hypothesis means that we use the column headed 0.025. For ten degrees of freedom, for example, the critical value for $\alpha = 0.05$ and a one-sided alternative is 1.812 *or* -1.812, depending on the direction of the alternative; for a two-sided alternative, the critical values are 2.228 *and* -2.228.

It is easy to find the correct value to use since the last entry in each column is the critical value for z. If we are able to determine the appropriate value for z in the case of a large sample, the correct value for t will be found in the same column opposite the correct number of degrees of freedom.

The relationship between z and t can be clarified by noting that the normal distribution is the limiting distribution for Student's-t. That is, for larger degrees of freedom, the critical values for t get closer to those for z. Since printing of tables for large values of degrees of freedom is impractical, statisticians have agreed to use z in cases where there are more than 30 degrees of freedom. Ideally, the t-distribution should be used in all cases where the sample standard deviation is used, but in the past this has not been practical. Some medium and low-priced calculators now available can determine critical values of t, and other test statistics, for large numbers of degrees of freedom.

If we know the population standard deviation and also know that the population is normally distributed, it is still possible to use the methods of section 10.1, but this is not often the case.

Thus, if we have a sample of size n ($n \leq 30$) drawn from a relatively normally distributed population, and we are testing a hypothesis about the mean of the population, we have three possibilities.

Case I

1. $H_0: \mu = \mu_0$
   $H_1: \mu \neq \mu_0$
2. $\alpha = 0.05$ | $\alpha = 0.01$
3. Criteria:

$\alpha = 0.05$	$\alpha = 0.01$
Reject H_0 if $t > t'$ or if $t < -t'$ where t' is obtained from Table 3, column 2, opposite $n - 1$.	Reject H_0 if $t > t'$ or if $t < -t'$ where t' is obtained from Table 3, column 4, opposite $n - 1$.

Case II

1. $H_0: \mu = \mu_0$
   $H_1: \mu > \mu_0$
2. $\alpha = 0.05$ | $\alpha = 0.01$
3. Criteria:

$\alpha = 0.05$	$\alpha = 0.01$
Reject H_0 if $t > t'$ where t' is obtained from Table 3, column 1, opposite $n - 1$.	Reject H_0 if $t > t'$ where t' is obtained from Table 3, column 3, opposite $n - 1$.

Case III

1. $H_0: \mu = \mu_0$
   $H_1: \mu < \mu_0$
2. $\alpha = 0.05$ | $\alpha = 0.01$

Criteria:

$\alpha = 0.05$	$\alpha = 0.01$
Reject H_0 if $t < -t'$ where t' is obtained from Table 3, column 1, opposite $n - 1$.	Reject H_0 if $t < -t'$ where t' is obtained from Table 3, column 3, opposite $n - 1$.

EXAMPLE 1 A chemist is testing the hypothesis that the boiling point of certain substances is 846°C. He makes four determinations and obtains values of 844°, 847°, 845°, 844°. What conclusions can be drawn at a significance level of 0.05?

SOLUTION Assuming that such determinations are at least approximately normally distributed, the following procedure is used.

1. $H_0: \mu = 846$

   $H_1: \mu \neq 846$

2. $\alpha = 0.05$

3. Criteria: Reject H_0 if $t > 3.182$ or if $t < -3.182$. This is a two-tailed test with three degrees of freedom.

4. Results: Using the data, we have $\bar{x} = 845$, $s \doteq 1.4$, so

$$t = \frac{845 - 846}{1.4/\sqrt{4}} = \frac{-1}{0.7} \doteq -1.43$$

5. Conclusions: Since -1.43 is not greater than 3.182 nor less than -3.182, we fail to reject H_0. Thus he cannot say that the boiling point is not 846°. On the other hand, whether he is willing to say that it is 846° depends on many other things, such as other characteristics of the substance as compared to those of the substance he suspects it is, plus his willingness to make a type II error. This particular case does more to substantiate H_0 than to deny it.

EXAMPLE 2 A psychologist contends that young people of today are less irresponsible than their forebears. To prove this he shows evidence that nine young people selected at random and given an irresponsibility test showed a mean index of 0.63 as compared to 0.74 of the population of young people thirty years ago. If the data he gathered was found to have a standard deviation of 0.12, test his claim at a 0.05 significance level.

SOLUTION His research hypothesis is $\mu < 0.74$, where μ is the mean irresponsibility index of today's young people. Assuming normal distribution of the population, we apply the following method.

1. $H_0: \mu = 0.74$

   $H_1: \mu < 0.74$

2. $\alpha = 0.05$

3. Criteria: Reject H_0 if $t < -1.860$. The test is one-tailed with eight degrees of freedom.

4. Results:

$$t = \frac{0.63 - 0.74}{0.12/\sqrt{9}} = \frac{-0.11}{0.04} = -2.75$$

5. Conclusions: Reject H_0 since $-2.75 < -1.860$. We can conclude at a significance level of 0.05 that today's young people are less irresponsible than those of thirty years ago. This is assuming, of course, that the testing instrument was actually valid, and that one sample was truly random.

EXAMPLE 3 A drug manufacturer claims that their new drug causes faster red cell buildup in anemic persons than the drug currently used. A team of doctors tested the drug on 6 persons and compared the results with the current build-up factor of 8.3. The 6 persons had factors of 6.3, 7.8, 8.1, 8.3, 8.7, and 9.4. Test the manufacturer's claim at a significance level of 0.01.

SOLUTION Assuming the factors to be normally distributed in the population, the t-test is appropriate. We have

1. $H_0: \mu = 8.3$

   $H_1: \mu > 8.3$ (since this is the claim)

2. $\alpha = 0.01$

3. Criteria: Reject H_0 if $t > 3.365$.

4. Results: We have $\bar{x} = 8.1$, $s \doteq 1.04$, so

$$t = \frac{8.1 - 8.3}{1.04/\sqrt{6}} \doteq -0.47$$

5. Conclusions: Since $-0.47 < 3.365$, we fail to reject H_0. Since t is negative the claim is patently false and we conclude that the new drug is no better than the old one.

Problems

1. In nine tests under prescribed conditions, an automobile averaged 17 miles per gallon with a standard deviation of 1.7 miles. Test the hypothesis (at $\alpha = 0.05$) that the true mean is above 16 miles per gallon.

2. Sixteen tests were made of the effervescing time of a certain combination of ingredients. The purpose of the tests was to discover if this combination was the same as that of a rival product, which has a mean effervescing time of 16.2 seconds. The sample mean was 14.8 seconds with a standard deviation of 1.6 seconds. What was the conclusion at a 0.01 level of significance?

3. A cigarette manufacturer claims that one of their brands contains no more than 18 mg of "tar." Five cigarettes are smoked and found to contain, respectively, 21, 17, 19, 18, and 21 mg of "tar." Can the manufacturer's claim be rejected at a significance level of 0.05?

4. Coffee cans are filled to a "net weight" of 16 ounces, but there is considerable variability. In fact, a random sample of 8 cans of a particular brand showed net weights of 15.4, 16.1, 15.8, 16.1, 16.2, 15.5, 16.0, 15.1 ounces. Test (at $\alpha = 0.05$) the hypothesis that the net weight of this brand of coffee is actually 16 ounces.

5. A government testing agency routinely tests food products to see if they meet government requirements at 0.01 level of significance. In order to be labelled "ice cream," the product must maintain at least 10% butterfat content. What conclusions would they draw if a random sample of ten half-gallons of a certain brand had butterfat percents of 9.6, 10.1, 9.9, 9.7, 10.0, 9.9, 9.8, 10.1, 9.9, and 9.7?

6. Injections of a drug must be properly maintained. A certain firm manufactures ampoules labelled 400 units, and takes quality control samples using a 0.05 level of significance to reject lots which may not be maintaining the level of quality. Daily, a sample of 10 ampoules from each of 6 lots is analyzed for content, and each lot is accepted or rejected on the basis of the sample. Which samples will be accepted and which rejected if the following are the results for a particular day?

Lot A: $\bar{x} = 400.7$, $s = 3.4$
Lot B: $\bar{x} = 401.1$, $s = 1.8$
Lot C: $\bar{x} = 398.7$, $s = 3.2$
Lot D: $\bar{x} = 401.3$, $s = 0.6$
Lot E: $\bar{x} = 401.7$, $s = 2.4$
Lot F: $\bar{x} = 399.2$, $s = 0.8$

7. A biologist wishes to determine whether an insect population found in a particular location of a forest belongs to a particular species. The only morphological characteristic which appears different from that of known members of this species is wing length. The mean wing length of the species is 15.4 mm. The biologist measures wing length for 25 insects from the forest location and finds that the mean wing length is 17.4 mm with a standard deviation of 2.3 mm. Can he conclude, at the 0.05 significance level, that the insects are of a different species?

8. Suppose that a second set of 8 coffee cans showed the same mean and standard deviation as the set given in problem 4. Combine the two samples and test the same hypothesis. What conclusions do you reach this time? Note the effect of sample size on results and conclusions.

9. A traffic survey claims that the mean length of time required to go through a tunnel at rush hour is 3.5 minutes. A skeptical motorist decides to test this hypothesis. He times his trips through the tunnel for a total of 22 trips during rush hour. He finds that his mean time was 4.4 minutes with a standard deviation of 0.8 minutes. Can he conclude that the traffic survey claim was too low? Use the results of his trips and a 0.05 level of significance.

10. Administration of a sedative causes physiological changes in individuals which last for a period of time. A researcher wishes to test the accepted mean duration of effects of 1.3 hours by claiming that the duration is actually greater. She uses twelve subjects and finds the sample to have a mean duration of 1.8 hours with a standard deviation of 0.7 hours. Can she substantiate her claim at the 0.05 level of significance?

10.3 PROPORTIONS IN A SAMPLE

As we stated in section 8.4, attributes are measured by proportions. Hypotheses concerning the population proportion, P, can be tested on the basis of the sample proportion, p, and the statistic z, given by

$$z = \frac{p - P_0}{\sigma_p}$$

where σ_p is the standard error of proportion and P_0 is the assumed value of P. All the procedures outlined in the previous sections apply. It is not necessary to use s_p since σ_p is based on P_0, a hypothetical value which is always known.

EXAMPLE 1 A real estate promoter claims that at least 75% of the retired couples in an apartment village of over 2,200 couples prefer apartment living to single-unit living. Try to disprove this assertion at the 0.05 level of significance if a random sample of 100 couples interviewed showed that 63 of them did, in fact, prefer apartment living.

SOLUTION The experimental procedure outlined in Chapter 9 applies.

1. H_0: $P = 0.75$
   H_1: $P < 0.75$
2. $\alpha = 0.05$
3. Criteria: Reject H_0 if $z < -1.64$

4. Results:

$$\sigma_p = \sqrt{\frac{(0.75)(0.25)}{100}} \doteq 0.043, \text{ so}$$

$$z = \frac{0.63 - 0.75}{0.043} = \frac{-0.12}{0.043} \doteq -2.79$$

5. Conclusions: $-2.79 < -1.64$, so H_0 is rejected. The promoter's claim is exaggerated.

EXAMPLE 2 Of 400 samples of hybrid corn studied, 79 were found to have the recessive characteristic under study. By the Mendelian Law of Inheritance, 0.25 of the variety of corn the stock was taken from would exhibit the characteristic. A researcher hypothesized that a new hybrid is different than the parent stock variety. Test this hypothesis at a 0.01 level of significance.

SOLUTION If this were the old stock, the proportion would be 0.25, so any deviation would indicate that there was a difference. Thus we have

1. H_0: $P = 0.25$
   H_1: $P \neq 0.25$
2. $\alpha = 0.01$
3. Criteria: Reject H_0 if $z > 2.58$ or if $z < -2.58$.
4. Results:

$$\sigma_p = \sqrt{\frac{(0.25)(0.75)}{400}} \doteq 0.022, \text{ and } p = \frac{79}{400} \doteq 0.198, \text{ so}$$

$$z = \frac{0.198 - 0.250}{0.022} = \frac{-0.095}{0.022} \doteq -2.36$$

5. Conclusions: Since -2.36 is not less than -2.58 nor greater than 2.58, we fail to reject H_0. However, the probability of obtaining a z-score so far from zero, purely by chance, is less than 0.02. Thus, it would be best to reserve judgment and probably a larger sample is in order.

Problems

1. The businessman's problem, discussed at length in sections 5.4, 6.5, and 9.3, was given in terms of number of customers buying brand B. Restate the problem in the terms of this section (in proportions) if 101 of 163 customers bought brand B.

2. A television manufacturer claims that at least 80% of its color picture tubes last at least two years. A random sample of 200 tubes sold over two years ago showed that 144 of them lasted at least two years. Test the manufacturer's claim at the 0.01 significance level.
3. A supply house has ordered a large shipment of dingbats and is willing to accept no more than 3% defectives. If a sample of 200 contains 8 defectives, should the shipment be rejected at 0.05 level of significance?
4. The Canter Poll claims that 0.60 of the electorate supports a certain proposition to be placed on the ballot. The Stringer Poll decided to test this contention at the 0.05 level. What are their conclusions if a random sample of 1,563 people revealed 884 who supported the proposition?
5. To test a hypothesis that, on the average, more than 4 out of 10 persons buying soft drinks still prefer bottles, a soft drink firm uses one day's sales of drinks to test the hypothesis at $\alpha = 0.01$. Stretching the assumptions of a representative sample quite a bit, what conclusions can they draw if the day's sales showed 687 bottle sales and 876 can sales?
6. In repeated tests with a pair of dice, seven was rolled 138 times in 540 tries. Does this support or deny (at the 0.05 significance level) a contention that the dice are loaded?
7. In a study relating student seating (proximity to the teacher) to grades, three classes with a total of 97 students were randomly rearranged with respect to seating arrangement. If a student moved closer to the teacher and grades increased, or the student moved further from the teacher and grades decreased, this was considered a success. A random relationship between proximity to the teacher and grades would lead to a probability of success of 0.5. Sixty-four students were considered to be successes. Can we conclude that the hypothesis $P > 0.5$ is substantiated at the 0.01 level of significance? This would show that proximity to the teacher is a factor in success in school as indicated by grades.
8. Use the data of problem 5, section 7.4, to test the hypothesis that at least one-third of the students on campus favor co-ed dorms. Combine the two samples and use the 0.01 level of significance.

10.4 DIFFERENCES BETWEEN MEANS

One important application of statistical inference is answering questions posed by research hypotheses which ask which of two methods yields the best result or whether two samples belong to the same or different populations.

The simplest test of this kind of hypothesis is called a **correlated test**. In this test, two related data sets are compared with each other. Two paired samples may receive different treatments and the results of the treatments may be compared, or before and after measurements may be taken from the same sample after the application of a single treatment. For example, suppose that ten people took part in a diet experiment. Their weights were measured at the beginning and the end of the experiment. The results are presented in the following Table.

Subject	*Starting Weight*	*Ending Weight*	*Net Difference*
A	183	177	−6
B	144	145	+1
C	151	145	−6
D	163	162	−1
E	155	151	−4
F	159	163	+4
G	178	173	−5
H	184	185	+1
I	142	139	−3
J	137	138	+1

EXAMPLE 1 Using the data given previously, discover whether or not the diet is effective. Use a 0.05 level of significance.

SOLUTION Letting $\bar{x}_d$ represent the individual difference scores, μ_d the (hypothetical) mean weight loss for the population, SE*, the appropriate standard error, we have the following procedure.

1. $H_0: \mu_d = 0$

   $H_1: \mu_d < 0$

2. $\alpha = 0.05$

3. Criteria: With a sample of 10, the t-test would be appropriate with 9 degrees of freedom; thus, we reject H_0 if $t < -1.833$.

4. Results: We have $\bar{x}_d = -1.8$, $s \doteq 3.49$, and $SE = 3.49/\sqrt{10} \doteq 1.10$. Then $t = -1.8/1.10 \doteq -1.64$.

5. Conclusions: We do not reject H_0; we reserve judgment in this case. Either get a larger sample, or try another week of the diet.

If more than thirty pieces of data are involved, the appropriate z-score rather than t-score would be used.

---

* This standard error could be symbolized as $s_{\bar{x}_d}$, using our previous notational methods, but this is quite clumsy, so it will be written as SE.

A somewhat more complicated case occurs when two *independent* samples are involved. Suppose that we take two samples, obtain the same statistic, such as the sample mean, and wish to decide if there is a statistically significant difference between the two means. We may measure the weight of menhaden taken from Escambia Bay and the weight of the same variety of fish taken from Pensacola Bay, and wish to see if there is any significant difference between the weights of these two samples which may be attributable to some cause we are interested in. If μ_1 represents the mean weight of all fish from Escambia Bay and μ_2 represents the mean weight of all fish from Pensacola Bay, we are testing the null hypothesis $\mu_1 = \mu_2$ against some alternative such as $\mu_1 \neq \mu_2$. Another way of phrasing it is to say that we are testing a hypothesis that the samples are from the same population. Now the differences between means of *all possible* pairs of samples of size n_1 and n_2, respectively, from the same population are normally distributed with $u = 0$ and standard deviation σ_d **(standard error of the difference)** where

STANDARD ERROR OF THE DIFFERENCE

$$\sigma_d = \sigma \sqrt{\frac{1}{n_1} + \frac{1}{n_2}}$$

and σ is the known population standard deviation. If we know σ, we can base our decision on the statistic

$$z = \frac{\bar{x}_1 - \bar{x}_2}{\sigma_d}$$

Unfortunately, as usual, it is unlikely that the population standard deviation will be known. In such cases it is customary to estimate σ_d by means of the **pooled variance** of the samples which is given by

$$s_d^2 = \frac{s_1^2}{n_1} + \frac{s_2^2}{n_2}$$

Then, we use $s_d = \sqrt{(s_1^2/n_1) + (s_2^2/n_2)}$ if both n_1 and n_2 are larger than 30. If not, we must use the *t*-distribution and we have

$$t = \frac{\bar{x}_1 - \bar{x}_2}{\sqrt{\dfrac{(n_1 - 1)s_1^2 + (n_2 - 1)s_2^2}{n_1 + n_2 - 2} \cdot \left(\dfrac{1}{n_1} + \dfrac{1}{n_2}\right)}}$$

The expression under the square root sign is the pooled variance for small samples. The number of degrees of freedom is given by $n_1 + n_2 - 2$.

Note that if n_1 and n_2 are both equal to n, the pooled variance in any case becomes $(s_1^2 + s_2^2)/n$. Note also that if we are calculating the pooled variance from raw scores we have

$$s_1^2 = \frac{\sum (x_1 - \bar{x}_1)^2}{(n_1 - 1)} \quad \text{and} \quad s_2^2 = \frac{\sum (x_2 - \bar{x}_2)^2}{(n_2 - 1)}$$

so that, for small samples

$$s_d^2 = \frac{\sum (x_1 - \bar{x}_1)^2 + \sum (x_2 - \bar{x}_2)^2}{n_1 + n_2 - 2} \cdot \left(\frac{1}{n_1} + \frac{1}{n_2}\right)$$

or

$$s_d^2 = \frac{\sum x_1^2 - \frac{(\sum x_1)^2}{n_1} + \sum x_2^2 - \frac{(\sum x_2)^2}{n_2}}{n_1 + n_2 - 2} \cdot \left(\frac{1}{n_1} + \frac{1}{n_2}\right)$$

EXAMPLE 2 Two samples of menhaden are drawn, one from Escambia Bay, one from Pensacola Bay. The 37 fish from Escambia Bay had a mean weight of 11.6 ounces with a standard deviation of 1.30 ounces. The 52 fish from Pensacola Bay had a mean weight of 12.1 ounces with a standard deviation of 2.1 ounces. Test the hypothesis that there is no weight difference between Escambia Bay and Pensacola Bay menhaden. Use the 0.05 level of significance.

SOLUTION The standard procedure is as follows:

1. $H_0: \mu_1 = \mu_2$
   $H_1: \mu_1 \neq \mu_2$
2. $\alpha = 0.05$
3. Criteria: Reject H_0 if $z > 1.96$ or if $z < -1.96$.
4. Results: The given data are summarized here.

$$n_1 = 37 \qquad n_2 = 52$$
$$\bar{x}_1 = 11.6 \qquad \bar{x}_2 = 12.1$$
$$s_1 = 1.3 \qquad s_2 = 2.1$$

Then

$$s_d = \sqrt{\frac{(1.3)^2}{37} + \frac{(2.1)^2}{52}} = \sqrt{0.1305} \doteq 0.36$$

and

$$z = \frac{11.6 - 12.1}{0.36} = \frac{-0.5}{0.36} = -1.39$$

5. Conclusions: Since -1.39 is not greater than 1.96 nor less than -1.96, we fail to reject H_0. In this case, z is relatively large and we can afford to take another sample.

EXAMPLE 3 A teacher tries two methods of teaching verbs to his French class. He has divided the class randomly into two samples of 14 students each. With method 1, the pupils learn an average of 43.8 verbs each, with a standard deviation 4.6. The other sample, using method 2, learns an average of 38.6 verbs each, with standard deviation 5.2. Theoretical considerations suggest that method 1 is superior to method 2. Estimate, at $\alpha = 0.01$, if method 1 is actually superior to method 2.

SOLUTION With the obvious correspondence we have

1. $H_0\colon \mu_1 = \mu_2$
   $H_1\colon \mu_1 > \mu_2$
2. $\alpha = 0.01$
3. Criteria: Reject H_0 if $t > 2.479$. NOTE: We are hypothesizing $\mu_1 - \mu_2 > 0$ with 26 degrees of freedom.
4. Results:

$$n_1 = 14,\ \bar{x}_1 = 43.8,\ s_1 = 4.6$$
$$n_2 = 14,\ \bar{x}_2 = 38.6,\ s_2 = 5.2$$

Our pooled variance can be found by

$$s_{\bar{d}}^2 = \frac{(14 - 1)(4.6)^2 + (14 - 1)(5.2)^2}{14 + 14 - 2} \cdot \left(\frac{1}{14} + \frac{1}{14}\right)$$

but is more easily found by

$$s_{\bar{d}}^2 = \frac{(4.6)^2 + (5.2)^2}{14} \doteq 3.44 \quad \text{since } n_1 = n_2 = 14$$

Then

$$t = \frac{43.8 - 38.6}{\sqrt{3.44}} \doteq \frac{5.2}{1.85} \doteq 2.81$$

5. Conclusions: Since $2.81 > 2.479$, we reject H_0 and conclude that method 1 is better than method 2 at 0.01 level of significance. That is, we are sure there is no more than one chance in a hundred that we are mistaken.

EXAMPLE 4 A psychologist gives a test to four blue-collar workers and five white-collar workers to see if there is any significant difference (at $\alpha = 0.05$) in their scores as groups. The blue-collar workers made scores of 23, 18, 22, 21, and the white-collar workers made scores of 17, 22, 19, 18, 20. What conclusions can be drawn from this study?

SOLUTION

1. H_0: $\mu_1 = \mu_2$
   H_1: $\mu_1 \neq \mu_2$
2. $\alpha = 0.05$
3. Criteria: Reject H_0 if $t > 2.365$ or if $t < -2.365$
4. Results: We must compute $\bar{x}$ in each sample. We can use a raw score formula for the pooled variance. Now $\bar{x}_1 = 21.0$, $\bar{x}_2 = 19.2$, and we have

	x_1	$x_1 - \bar{x}_1$	$(x_1 - \bar{x}_1)^2$	x_2	$x_2 - \bar{x}_2$	$(x_2 - \bar{x}_2)^2$
	23	2	4	17	−2.2	4.84
	18	−3	9	22	2.8	7.84
	22	1	1	19	−0.2	0.04
	21	0	0	18	−1.2	1.44
				20	0.8	0.64
Total	84		14.00	96		14.80

so

$$s_d^2 = \frac{14 + 14.80}{4 + 5 - 2} \cdot \left(\frac{1}{4} + \frac{1}{5}\right) = \frac{28.80}{7} \cdot \frac{9}{20} \doteq 1.8514.$$

Then $\bar{x}_1 = 21.00$, $\bar{x}_2 = 19.20$, and

$$t = \frac{21.00 - 19.20}{\sqrt{1.8514}} \doteq \frac{1.80}{1.36} \doteq 1.32$$

5. Conclusions: Since 1.32 is neither greater than 2.365 nor less than -2.365, we fail to reject H_0. This value of t is fairly small, so it is probably safe to accept H_0 and conclude that there is no difference between the two groups on this test.

NOTE: Table 3 only gives values of t for up to 29 degrees of freedom, but tables of t exist for values above this number of degrees of freedom.

Problems

1. The department of Institutional Research at The University of West Florida wishes to determine how seniors there do on the Graduate Record Examination compared to seniors at Florida State University. A random sample of 50 UWF seniors was found to have a mean of 1,183 on the GRE with a standard deviation of 137. The FSU sample of 100 had a mean of 1,168 with a standard deviation of 146. Test (at $\alpha = 0.05$) the hypothesis that there is no difference in the population.

2. A group of 10 subjects was given tests before and after being subjected to a hypothetical learning situation. If the table to follow shows the results of the experiment, test the hypothesis that learning actually took place at a significance level of 0.05. Assume that an increase in score indicates that learning actually took place.

Subject	*Before*	*After*
1	27	29
2	21	32
3	34	29
4	24	27
5	30	31
6	27	26
7	33	35
8	31	30
9	22	29
10	27	28

3. To determine which of two brands of golf balls is better for his long game, a professional golfer drives two samples of 50 balls from the tee, using the same club for each shot, alternating brands at random. For brand 1 the mean yardage is 231 yards with a standard deviation of 24 yards. For brand 2 the mean yardage is 226

yards with a standard deviation of 20 yards. Can he conclude that there is a difference in the balls? Use $\alpha = 0.01$.

4. An amateur chemist discovered some residue left in a test tube and hypothesizes that it is the same as the residue left in a different tube last week. He subjects it to four determinations of a melting point, arriving at 1,543°C, 1,540°C, 1,542°C and 1,543°C. The residue last week was tested three times for its melting point, getting 1,545°C, 1,544°C, and 1,544°C. Do these results tend to substantiate or deny the hypothesis that the residues are the same substance? Use a 0.05 level of significance.

5. To test the effect of two different fertilizers on corn yield, 6 one-acre plots of corn are fertilized with fertilizer 1 and 6 one-acre plots of corn are fertilized with fertilizer 2. The plots fertilized with fertilizer 1 showed a mean yield of 1.32 tons per plot with a standard deviation of 0.22 tons; those fertilized with fertilizer 2 showed a mean yield of 1.41 tons per plot with a standard deviation of 0.31 tons. Do these results support the manufacturer's claim that fertilizer 2 is superior to fertilizer 1? Use a 0.01 level of significance.

6. A hospital is considering two suppliers for ampoules of a wonder drug. Both suppliers deliver ampoules which are supposed to contain 500,000 units of the drug. An analysis was made of a sample of 6 ampoules from each company. Brand A had a mean of 510,000 units with a standard deviation of 20,000, while brand B had a mean of 490,000 units with a standard deviation of 15,000. Can you conclude there is probably no difference between the brands? Use a 0.05 level of significance.

7. The correlated test presented at the beginning of this section can also be used for two different samples if the scores or observations are *paired* using some criterion; that is, you could reasonably expect the score in each pair to be the same. Two varieties of corn are planted in paired plots to see if there is a difference in the yield. Calculate whether or not there is a significant difference at a 0.05 level of significance. Yields are given in tons per plot. Paired plots are of the same size, though different pairs are not necessarily of a similar size.

Pair	*Variety A*	*Variety B*
1	123	108
2	91	76
3	102	103
4	64	60
5	144	135

8. The methods of this section may also be used when it is hypothesized that differences between two samples are greater than a fixed amount. A survey was made which tested the hypothesis that college graduates read at least two books per year *more* than non-college graduates. As a result the null hypothesis was H_0: $\mu_1 = \mu_2 + 2$, against a suitable alternate hypothesis, where μ_1 is the number of books

per year read by college graduates and μ_2 is the corresponding value for non-college graduates. What were their conclusions at a 0.05 level of significance if their results are given as follows?

College graduates: $n_1 = 66$, $\bar{x}_1 = 4.3$, $s_1 = 4.6$

Non-college grads: $n_2 = 78$, $\bar{x}_2 = 1.9$, $s_2 = 3.1$

10.5 DIFFERENCES BETWEEN PROPORTIONS

In many cases it is necessary to decide whether the difference in the proportions of two independent* samples is significant. If, for instance, we find that 75 of 100 retired couples in Miami Beach are happy while only 62 of 100 retired couples in Chicago are happy, we may test the hypothesis that there is a difference in proportion between these two samples, in the same manner as in section 10.5. In this case, however, the **standard error of the difference for proportions** is estimated by

STANDARD ERROR OF THE DIFFERENCE FOR PROPORTIONS

$$s_{dp} = \sqrt{\frac{p_1(1-p_1)}{n_1} + \frac{p_2(1-p_2)}{n_2}}$$

where p_1 and n_1 are the sample proportion and size for one sample, while p_2 and n_2 are the sample proportion and size for the second sample. We can then base the test on the statistic

$$z = \frac{p_1 - p_2}{s_{dp}}$$

if $n_1 p_1$, $n_1(1 - p_1)$, $n_2 p_2$, and $n_2(1 - p_2)$ are at least 5.

EXAMPLE 1 On a certain campus a poll shows that 45 of 60 underclassmen are in favor of a proposed judicial reform while 48 of 80 upperclassmen are in favor of it. Do these results support the contention (at $\alpha = 0.05$) that a greater percentage of underclassmen than upperclassmen support the reform?

* Independence is important since, for instance, husbands and wives would probably have related opinions.

SOLUTION Since our research hypothesis here is that the proportion for underclassmen (P_1) is greater than that for upperclassmen (P_2), our test is one-tailed. We have

1. $H_0: P_1 = P_2$
   $H_1: P_1 > P_2$

2. $\alpha = 0.05$

3. Criteria: Reject H_0 if $z > 1.64$.

4. Results: $p_1 = 0.75$, $p_2 = 0.60$, so we have

$$\begin{aligned} s_{dp} &= \sqrt{\frac{(0.75)(0.25)}{60} + \frac{(0.60)(0.40)}{80}} \\ &= \sqrt{\frac{0.1875}{60} + \frac{0.2400}{80}} \\ &= \sqrt{0.003125 + 0.003} \\ &= \sqrt{0.006125} \\ &\doteq 0.078 \end{aligned}$$

   Then

$$z = \frac{0.75 - 0.60}{0.078} = \frac{0.150}{0.078} \doteq 1.92.$$

5. Conclusions: Reject H_0; underclassmen do support the reform more than upperclassmen.

DISCUSSION The apparently great difference (0.15) between the two sample proportions is barely significant. In fact, it would not be significant at $\alpha = 0.01$. In proportions it is important to obtain a fairly large sample size to be sure of getting good results.

NOTE: If the samples are fairly close in size, it is not unusual to estimate s_{dp} by using a pooled sample proportion, p, obtained simply by combining the two samples together. Then we have

$$s_{dp} = \sqrt{p(1 - p)\left(\frac{1}{n_1} + \frac{1}{n_2}\right)}$$

In this example $p = 45 + 48/60 + 80 = 93/140 \doteq 0.66$, so the estimate would be

$$s_{dp} = \sqrt{(0.66)(0.34)\left(\frac{1}{60} + \frac{1}{80}\right)}$$
$$= \sqrt{0.006545}$$
$$\doteq 0.081$$

Then we have $z = 0.150/0.081 \doteq 1.85$. The use of this estimate tends to be a bit conservative. As a general rule, particularly with modern calculators available, it is probably best to just use the basic formula for s_{dp} since the resulting estimate is a slightly better approximation to σ_{dp}, the true standard error of the difference for proportions.

Problems

1. Two companies submit bids and a sample of ball bearings to gain a contract. Firm *A* has 82 acceptable ball bearings out of 90 while firm *B* has 96 acceptable out of 110. Firm *B*'s price is slightly lower than firm *A*'s, so that unless it can be shown that firm *A* has a higher percentage of acceptable bearings (with $\alpha = 0.01$), firm *B* will get the contract. Who should get the contract?
2. Two moving companies are applying for the job of moving a major concern across country. The office manager views completion of the move within the promised time as the major criteria. She feels that she can flip a coin and decide which to use on the basis of evidence that company *A* has completed 344 out of 388 moves on time during the last year while company *B* has completed 217 of 232 on time. Test her hypothesis at the 0.05 level of significance.
3. A psychologist tests men and women for mental dexterity, grading each subject pass or fail on a task requiring such activity. If 83 of 124 men and 72 of 103 women passed the test, can she conclude that men and women differ significantly ($\alpha = 0.01$) on mental dexterity as measured by this test?
4. A class is randomly divided into two sections of 18 each, and one section is given instruction in assembling a puzzle. Then each subject is tested to see if he can assemble the puzzle within one minute. The results in the test group (given instructions) showed that 11 of the students could do the task while the results in the control group (given no instructions) showed that only 9 of the students could do the task. Do these results bear out at the 0.05 significance level the contention that instruction in assembling a puzzle aids in actually assembling the puzzle?
5. Use the data of problem 5, section 7.4, to test the hypothesis that the samples represent different populations; that is, that the differences between the proportions are real. Use the 0.01 level of significance. How does this affect a decision to pool the data?

6. A biologist wishes to determine whether two insect populations found in different locations in a forest are actually two different species. The prime morphological characteristic which differentiates the two populations appears to be the incidence of white eyes. The biologist observes that 32 of 100 from one population and 23 of 100 from the other population have white eyes. Can he conclude that the two populations are different species? Use the 0.05 level of significance.
7. A recent campus poll showed 40% of students living on campus opposed to pets on campus. Feeling that there are perhaps differences depending upon where the students live, the campus newspaper polls students living in dormitories and students living in fraternity and sorority houses separately. The newspaper claims on the basis of the poll that more students living in fraternity and sorority houses are opposed to pets on campus than those living in dormitories. If the poll showed that 21 of 39 fraternity/sorority house residents were opposed to pets on campus, but 37 of 94 dorm residents, test the newspaper's claim at the 0.05 level of significance.

10.6 SUMMARY

This chapter dealt with various hypotheses about sample means and proportions. The techniques used were outlined in Chapter 9. The criteria are based upon specific hypotheses and sampling distributions depending upon the parameter in question. These criteria are summarized here. If assumptions of a normal population are not met, alternative procedures may be used. Some of these are given in Appendix B.

Tests concerning means

1. Mean of one sample

$$H_0: \mu = \mu_0$$

$$H_1: \mu > \mu_0\,, \mu \neq \mu_0\,, \mu < \mu_0$$

Sampling statistic—criteria based upon the value of

$$z = \frac{\bar{x} - \mu_0}{\sigma_{\bar{x}}} \quad \text{(for large samples)}$$

or

$$t = \frac{\bar{x} - \mu_0}{s/\sqrt{n}} \quad \text{(for small samples with } n - 1 \text{ degrees of freedom)}$$

where

$$\sigma_{\bar{x}} = \frac{\sigma}{\sqrt{n}} \quad \text{or} \quad \sigma_{\bar{x}} = \frac{\sigma}{\sqrt{n}}\sqrt{\frac{N-n}{N-1}}$$

depending upon whether or not the sample size, n, is at least 5% of the population size, N. If σ is not known, s is used.

2. Two sample means

Case I Two means from one sample (before and after) or paired observations from two samples. If x_d represents the difference score for each individual, we have

$$H_0: \mu_d = 0 \text{ (or } \mu_1 = \mu_2 \text{, for paired observations)}$$

$$H_1: \mu_d > 0,\ \mu_d \neq 0, \text{ or } \mu_d < 0$$

Sampling statistic—criteria based upon the value of

$$z = \frac{\bar{x}_d}{SE} \quad \text{(for large samples)}$$

and

$$t = \frac{\bar{x}_d}{SE} \quad \text{(for small samples, with } n - 1 \text{ degrees of freedom)}$$

where SE is calculated as $s_{\bar{x}}$.

Case II Means from two different samples.

$$H_0: \mu_1 = \mu_2$$

$$H_1: \mu_1 > \mu_2,\ \mu_1 \neq \mu_2, \text{ or } \mu_1 < \mu_2$$

Sampling statistic—criteria based upon the value of

$$z = \frac{\bar{x}_1 - \bar{x}_2}{\sigma_d} \quad \text{(for large samples)}$$

or

$$t = \frac{\bar{x}_1 - \bar{x}_2}{\sqrt{\frac{(n_1 - 1)s_1^2 + (n_2 - 1)s_2^2}{n_1 + n_2 - 2}\left(\frac{1}{n_1} + \frac{1}{n_2}\right)}}$$

(for small samples with $n_1 + n_2 - 2$ degrees of freedom)

Since it is generally unknown, σ_d is usually approximated by s_d where $s_d = \sqrt{(s_1^2/n_1) + (s_2^2/n_2)}$. If $n_1 = n_2 = n$, then $\sqrt{(s_1^2 + s_2^2)/n}$ can be used in the denominator for both z and t.

Tests concerning proportions

1. Proportions of one sample

$$H_0\colon P = P_0$$

$$H_1\colon P > P_0,\ P \neq P_0,\ \text{or } P < P_0$$

Sampling statistic—criteria based upon the value of

$$z = \frac{p - P_0}{\sigma_p}$$

where p is the sample proportion and $\sigma_p = \sqrt{P_0(1 - P_0)/n}$.
NOTE: This is valid only if both nP_0 and $n(1 - P_0)$ are at least 5.

2. Proportions in two samples

$$H_0\colon P_1 = P_2$$

$$H_1\colon P_1 > P_2,\ P_1 \neq P_2,\ \text{or } P_1 < P_2$$

Sampling statistic—criteria based upon the value of

$$z = \frac{p_1 - p_2}{\sqrt{\dfrac{p_1(1 - p_1)}{n_1} + \dfrac{p_2(1 - p_2)}{n_2}}}$$

and should be used only if $n_1 p_1$, $n_1(1 - p_1)$, $n_2 p_2$, $n_2(1 - p_2)$ are all at least 5.

Relation between confidence interval and rejection region

1. Two-tailed tests
Using the appropriate z or t value for the significance level α, the rejection region can be found. But if the probability is α that a sample falls in the rejection region by chance, the probability is $1 - \alpha$ that it does not. Therefore, the complement of the rejection region is the $1 - \alpha$ confidence interval for the parameter. Hence the rule

> A null hypothesis concerning means or proportions will be rejected in favor of a two-sided alternate hypothesis at a significance level α if and only if the hypothetical parameter (μ_0, P_0) is *not* included in the $1 - \alpha$ confidence interval obtained from the sample.

2. One-tailed tests
   In this case we have α in one tail so that the appropriate confidence interval is $1 - 2\alpha$. Thus the rule

> A null hypothesis concerning means or proportions will be rejected in favor of a one-sided alternate hypothesis if and only if the hypothetical parameter is *not* included in the $1 - 2\alpha$ confidence interval obtained from the sample, and the sample does not contradict the alternate hypothesis.

The purpose of the latter clause is to rule out one tail.

Problems

1. A firm wishes to examine whether or not it should expand its offerings and is considering increasing the output of type 7 widgets if necessary. It decides to increase the output if the daily demand volume exceeds 27,500, on the average. A sample of 40 days during the past 6 months is examined and it is found that the mean daily demand is 27,654 with a standard deviation of 384. If the level of significance chosen is 0.05, does this sample tend to indicate that the mean daily demand volume exceeds 27,500?
2. In a study of 100 patients admitted to a hospital suffering from appendicitis, the mean stay was found to be 9.33 days with a standard deviation of 2.83 days. Does this support the contention of the hospital administrator that this is significantly less than the national average of 10.17 days? Use the 0.05 level of significance.
3. In a high school, students are interviewed regarding the desirability of having a prom this year. Of 317 girls interviewed, 208 wanted a prom; 107 of 203 boys wanted a prom. Does this sample show (at $\alpha = 0.01$) that there is a difference in the proportion of boys wanting a prom as opposed to the proportion of girls?
4. A traffic engineer tries two different settings for traffic signals in an attempt to find the most efficient setting, using number of traffic delays per week as his criterion. Using setting *A*, he obtained a mean of 26.1 traffic delays per week, for 11 weeks, with a standard deviation of 5.6 delays. Using setting *B*, he obtained a mean of 22.4 traffic delays per week for 13 weeks, with a standard deviation of 6.3 delays. What can he conclude (at $\alpha = 0.05$) about the two settings?
5. Under the usual treatment, a certain disease produces undesirable side effects in 25% of the victims. A new drug has been developed which is claimed to be effective in reducing the side effects with this disease. A sample of 225 patients with the disease is treated with the new drug. Test the hypothesis that the drug is effective in reducing the proportion of patients developing side effects at the 0.05 level of significance if 43 patients develop the side effects.

6. Testing for tensile strength of a new alloy, a metallurgist obtains readings, on separate tests, of 117.6, 122.4, 119.8, 118.8, 121.6, and 123.4. The theoretical tensile strength is 120.0. Do these tests tend to confirm or deny the theory? Use $\alpha = 0.05$.

7. Forty percent of the geldings racing at a certain track ran in the money, while only thirty-two percent of the fillies did so. If the sample is based on the performance of 105 geldings and 75 fillies, test the jockey's hypothesis (at the 0.05 significance level) that geldings run better than fillies.

8. A hospital administrator introduces a new accounting system designed to reduce the average amount of unpaid bills. Under the old system, the average unpaid bill was \$217.00 with a standard deviation of \$34.00. Under the new system, the first 50 unpaid bills showed an average of \$203.00 with a standard deviation of \$47.00. Is this good enough evidence to assert that the new system has reduced the amount of the average unpaid bill? Use the 0.01 level of significance.

9. Two companies submit samples of light bulbs for bids. On the average, each of company *A*'s sample of 300 bulbs has a mean life of 1,065 hours, with a standard deviation of 133 hours. Company *B*'s sample of 300 exhibits a mean life of 1,047 hours with a standard deviation of 56 hours. If mean bulb life is the prime criterion for awarding the contract, does the sample yield sufficient information to decide, using $\alpha = 0.05$, which company should get the contract? If so, which one?

10. Use the data of problem 9, but the criterion is that the company which will produce the greatest proportion of bulbs lasting at least 1,000 hours will get the contract.

11. A medical researcher studies the effect of two drugs on guinea pigs infected with pneumococcus bacillus. One group of 44 animals has a mortality of 17 while the second group of 52 animals has a mortality of 24. Can he conclude, at the 0.01 level of significance, that there is a difference in the effectiveness of the two drugs?

12. A random sample of a day's output of a chemical revealed that the ester level for 20 vials was 11.6% with a standard deviation of 0.8%. Specifications require a minimum of 12% but allow the output to pass if a sample is not significantly lower with $\alpha = 0.05$. Will the output be passed?

13. An antipyretic is being tested as a replacement for aspirin. A total of nine experimental animals are artificially given high temperatures and the drug administered. Given the before and after temperatures, test whether or not the drug can be considered effective at the 0.05 level of significance.

Before	*107.2*	*111.4*	*109.3*	*106.5*	*113.7*	*108.4*	*107.7*	*111.9*	*109.3*
After	106.1	111.7	105.4	107.2	109.8	108.8	106.9	109.6	110.5

14. To test the efficiency of a procedure designed to increase the daily output of a machine, five machines are tested with and without the procedure, with the following results:

Machine	*With*	*Without*
A	15.6	14.4
B	18.2	16.7
C	14.3	13.1
D	14.9	14.4
E	16.7	14.7

Test the hypothesis that the procedure is effective. Use $\alpha = 0.05$.

15. Repeat problem 14 assuming that the data were obtained from two independent random samples.

16. In a study of attitudes of military personnel toward women the Attitude Toward Women Scale (ATWS) was used to assess general attitude. The questionnaire was administered to 172 Naval personnel, of which 48 were Marines and 124 were in the Naval Air branch. The Marines had a mean score of 86.4, with a standard deviation 18.5, and the Naval Airmen had a mean of 99.7, with a standard deviation 18.32. Use the 0.01 level of significance to test the hypothesis that attitudes toward women differ between these two branches of military service.

17. A study of a group of twelve juvenile delinquent boys and twelve nondelinquent boys from similar backgrounds to those of the delinquents was conducted to assess differences in self-concept. The following results were obtained on the Osgood Semantic Differential test. The nondelinquent boys had a mean of 96.4 with a standard deviation of 8.1; the delinquent boys had a mean of 98.5 with standard deviation 10.8. Are the obtained differences significant at the 0.05 level of significance?

18. A study of personality differences among LSD users and nonusers was conducted using 40 users and 40 nonusers as subjects. The Minnesota Multiphasic Personality Inventory was administered. Some of the results are given here:

	Nonusers		*Users*	
Scale	*Mean*	*SD*	*Mean*	*SD*
Depression	55.94	13.60	57.10	11.83
Hysteria	58.05	7.92	62.08	8.31
Paranoia	55.85	10.12	59.35	9.69
Schizophrenia	55.05	12.15	61.25	12.82

Using the 0.05 level of significance, in which cases would the hypothesis that users score higher than nonusers be substantiated?

Chapter XI

The Chi-Square Distribution

11.1 PROPORTIONS

The methods of Chapter 10 apply when the null hypothesis is a statement that a population parameter is equal to a particular value or that two population parameters are equal. Quite often more than two samples are involved in an experiment so that these methods do not apply. In cases where the hypotheses apply to attributes, a sampling distribution called Chi-square is useful. This distribution is useful whenever we compare observed data with expected outcomes. For example hypotheses that proportions are equal in several populations from which samples are obtained can be handled in this way.

For instance, suppose that we have taken a poll on a campus to determine if left-wing speakers should be allowed on campus and we have obtained the results presented in the following table.

	Freshmen	*Sophomores*	*Juniors*	*Seniors*	*Total*
Should be allowed	99	98	98	97	392
Should not be allowed	121	112	92	83	408
Totals	220	210	190	180	800

We would like to decide whether the proportion which approves of the proposition is the same for each class. To do this we must decide if the observed differences in proportions are due to chance or to the samples having been drawn from populations with different proportions of opinions. To test this idea, we hypothesize that all the samples were drawn from the same population and that any differences in proportions observed are due to chance. If this is true, then we should be able to predict the proportion of students in each class who hold each opinion by knowing the proportion of the total number of students (of all classes) who hold that opinion.

In this case, 392 of 800 students, or 0.49, feel that left-wing speakers should be allowed on campus. If this proportion is an estimate of the proportion of students in each class who hold this opinion, then we can obtain an *expected* value for the number of students in each class who should hold this opinion. In the freshman class, for example, 0.49 of 220 is 107.8. This number is the expected number of freshman who would hold this opinion if freshman hold this opinion in the same proportion as the general school population.

We now use a statistic called **Chi-Square** (χ^2) to evaluate the differences between the expected and the observed values for each class. This statistic is calculated by taking the difference between the observed value and the expected value, squaring it, and dividing it by the expected value. This is done for each class, or cell, and the sum of the calculations is called Chi-Square, or the sample value of Chi-Square. Thus we have

CHI-SQUARE $$\chi^2 = \sum \frac{(\text{observed} - \text{expected})^2}{\text{expected}}$$

If the proportions in each class were equal to the proportions predicted and the samples reflected this, then the observed and expected values would be equal and χ^2 would be equal to zero. However, even if the proportions in each class were equal to the predicted proportions, we would expect to find some variation in the proportions just by chance. The Chi-Square statistic allows us to decide whether the variation is due to chance or to true differences in proportions.

The Chi-Square distribution, like the *t*-distribution, is determined by the number of degrees of freedom. Each number of degrees of freedom gives rise to a different Chi-Square curve. In the example we have been discussing, there are four classes and three degrees of freedom. There are three degrees of freedom for the following reason. If we know the total number of students in the "should be allowed" category, we can put in any numbers we want in this category under freshmen, sophomores, and juniors (as long as the total does not exceed the total number in this category). But the number of seniors in this category is then fixed. For a Chi-Square test involving k samples with two proportions per sample, there are $k - 1$ degrees of freedom. Problem 3 describes how to determine the number of degrees of freedom where there are several proportions per sample.

Table 4 gives critical values of Chi-Square for $\alpha = 0.05$ and $\alpha = 0.01$ to thirty degrees of freedom. For six degrees of freedom, for instance, these values are, respectively, 12.592 and 16.812.

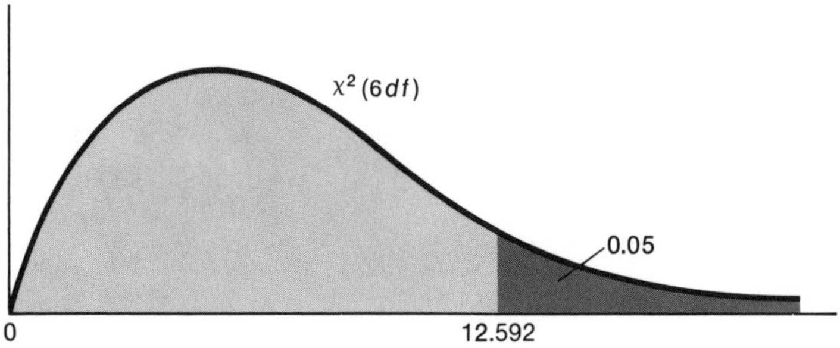

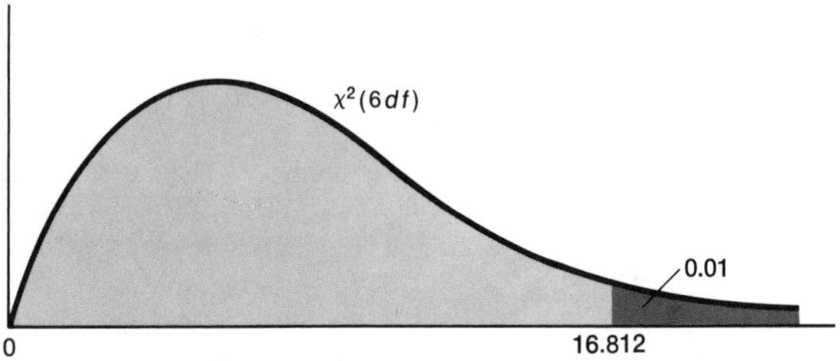

Thus if a hypothesis of equal proportions is true about a sample with six degrees of freedom, we would expect a χ^2 greater than 12.592 no more than one time in twenty by chance, and greater than 16.812 no more than one time in one hundred.

A Chi-Square test, by its nature, is one-tailed, since $\chi^2 = 0$ indicates that all observed outcomes are exactly as expected and we would be able to accept the null hypothesis in this case.

A few cautions are necessary in the use of the Chi-Square procedure. One necessary assumption is that the samples are independently obtained, so that bias is avoided. Another is that each observation falls into *exactly one* cell. A third is that the samples are large enough so that the expected number in each cell is at least five. If this is not so, categories or samples should be combined or eliminated in order to obtain this number. For instance, if we had not had a sufficient sample size of seniors to meet this assumption, juniors and seniors could have been combined into a category of "upperclassmen" or even "juniors and seniors."

EXAMPLE 1 Using the data given, test the hypothesis that responses do not differ according to class; that is, that the proportion in each category will be the same, regardless of class, in the total population of students. Use a 0.05 level of significance.

SOLUTION As before, we found that 0.49 of the total sample belonged to the "should" category. We can determine the expected number in each cell and calculate χ^2, comparing it with the critical value. The appropriate procedure is outlined as follows:

1. H_0: $P_1 = P_2 = P_3 = P_4$ (where these represent the true proportions)
   H_1: Proportions are not all equal
2. $\alpha = 0.05$
3. Criteria: There are 3 degrees of freedom. From Table 4, the critical value of χ^2 for 3 degrees of freedom and $\alpha = 0.05$ is 7.815. Thus we will reject H_0 if $\chi^2 > 7.815$.
4. Results: The table is summarized here with the expected number in each cell in parentheses below the observed number. The expected numbers are found in each case by taking the pooled proportion as an estimate of each sample proportion. Expected numbers are calculated to one decimal.

	Freshmen	*Sophomores*	*Juniors*	*Seniors*	*Total*
Should be allowed	99	98	98	97	392
	(107.8)	(102.9)	(93.1)	(88.2)	
Should not be allowed	121	112	92	83	408
	(112.2)	(107.1)	(96.9)	(91.8)	
Totals	220	210	190	180	800

Then

$$\chi^2 = \frac{(99 - 107.8)^2}{107.8} + \frac{(121 - 112.2)^2}{112.2} + \frac{(98 - 102.9)^2}{102.9} + \frac{(112 - 107.1)^2}{107.1}$$
$$+ \frac{(98 - 93.1)^2}{93.1} + \frac{(92 - 96.9)^2}{96.9} + \frac{(97 - 88.2)^2}{88.2} + \frac{(83 - 91.8)^2}{91.8}$$
$$\doteq 4.093$$

5. Conclusions: Since 4.093 is not greater than 7.815, we cannot reject H_0 and we conclude that our sample does not show a significant difference among the proportions of the samples.

Problems

1. A survey of business failures during the past year showed that 44 of 110 small businesses failing had been in business for less than a year. The same survey showed that 24 of 75 medium size businesses and 14 of 50 large businesses failing had been in business for less than a year. Estimate (at $\alpha = 0.05$) if the difference in observed proportions is attributed to chance.

2. A poll was conducted to discover whether citizens were more concerned with the problem of inflation or corruption in government. The questions asked were: which of the two issues they considered more pressing and their party affiliation or preference. Using the results given here, test the hypothesis that party preference does not affect the proportion of people concerned with pollution (or war) as the primary issue. Use $\alpha = 0.01$.

	Republican	*Democrat*	*Independent*
Inflation	187	316	191
Corruption	154	187	97

3. Chi-Square analysis can also be extended to cases where several proportions are involved. The procedure is the same as outlined except that the number of degrees of freedom is calculated as $(r - 1)(c - 1)$ where r is the number of proportions (rows) and c is the number of samples (columns). Using Chi-Square analysis, estimate whether the proportion of people preferring each type of television program differs significantly at the 0.01 level among the various cities. Each of the samples is a random sample of viewers in that city.

	PREFERENCE BY CITY			
TYPE OF PROGRAM	*New York*	*St. Louis*	*Atlanta*	*San Francisco*
Variety	80	60	75	25
Drama	40	50	80	50
Comedy	80	80	50	60
Other	50	30	25	65

4. Long-term data on frequency of catching fish was accumulated regarding catch distribution by quarters on days which were clear and on days which were rainy. Test the hypothesis that distribution of catch does not differ according to weather conditions. Use $\alpha = 0.05$.

	NUMBER OF CATCHES			
Weather Condition	*Midnight–6 AM*	*6 AM–Noon*	*Noon–6 PM*	*6 PM–Midnight*
Rainy	1,013	3,143	2,965	987
Clear	5,550	16,567	13,787	8,341

11.2 CONTINGENCY TABLES

The Chi-Square distribution is also used to work with a hypothesis that count data are in two independent classifications. For instance, in a year's statistics classes at a university, grades were distributed as follows by category of prior course work.

	GRADE					
PRIOR COURSE WORK IN	*A*	*B*	*C*	*D*	*F*	*Total*
1. College algebra, but not logic	15	20	40	5	0	80
2. Logic, but not college algebra	10	15	70	20	5	120
3. Both college algebra and logic	10	20	25	5	0	60
4. Neither college algebra nor logic	15	15	75	30	55	190
Totals	50	70	210	60	60	450

We want to analyze the data to discover whether there is a relation between prior course work and grade in the course, or whether these factors are independent.

In category 1, 80/450 of the students had prior work in college algebra, but not logic and 50/450 of the students received a grade of "A" in the course. If a student is chosen at random from among the 450 in the sample, these numbers also give the probability that he will belong to one of these categories, respectively. If the classifications are *independent* (see section 3.4), the probability that a student chosen at random will belong to *both* categories is their product. Then the *expected number* of students who will belong to both categories is that probability times the number of students in the sample or 450. The expected number of students in that cell is $(80/450)\cdot(50/450)\cdot(450)$, or simply $(80)\cdot(50)/(450)$. Each of the other expected values for the cells can be found in the same manner, χ^2 calculated, and the result compared with the critical value. The number of degrees of freedom is found to be equal to $(r - 1)(c - 1)$ where r is the number of rows and c is the number of columns.

In this case there are $(4 - 1)(5 - 1)$ or 12 degrees of freedom.

A table which sets forth count data in this fashion, classifying observations from a sample into two categories, is called a **contingency table**. The use of Chi-Square in this case has assumptions similar to those with proportions.

EXAMPLE 1 Test the hypothesis that the grade achieved in the course is independent of prior course work in college algebra or logic in the table given. Use $\alpha = 0.01$.

SOLUTION Considering the data as an independently obtained random sample of all students ever having had the course and using the experimental procedure, we have

1. H_0: Grade achievement in statistics and prior course work in college algebra and logic are independent.

   H_1: They are not independent.

2. $\alpha = 0.01$

3. Criteria: Since $df = (4 - 1)(5 - 1) = 12$, reject H_0 if $\chi^2 > 26.217$.

4. Results: If it were necessary, some columns could be combined, but it would be more difficult to combine rows. Each expected frequency is calculated (e.g., the first cell is found to be $(80)(50)/(450) \doteq 8.9$), and the data follows with the expected frequency in parentheses.

	GRADE				
CATEGORY	*A*	*B*	*C*	*D*	*F*
1.	15	20	40	5	0
	(8.9)	(12.4)	(37.3)	(10.7)	(10.7)
2.	10	15	70	20	5
	(13.3)	(18.7)	(56.0)	(16.0)	(16.0)
3.	10	20	25	5	0
	(6.7)	(9.3)	(28.0)	(8.0)	(8.0)
4.	15	15	75	30	55
	(21.1)	(29.6)	(88.7)	(25.3)	(25.3)

Then

$$\chi^2 = \frac{(15-8.9)^2}{8.9} + \frac{(20-12.4)^2}{12.4} + \frac{(40-37.3)^2}{37.3} + \frac{(5-10.7)^2}{10.7}$$
$$+ \frac{(0-10.7)^2}{10.7} + \frac{(10-13.3)^2}{13.3} + \frac{(15-18.7)^2}{18.7} + \frac{(70-56)^2}{56}$$
$$+ \frac{(20-16)^2}{16} + \frac{(5-16)^2}{16} + \frac{(10-6.7)^2}{6.7} + \frac{(20-9.3)^2}{9.3} + \frac{(25-28)^2}{28}$$
$$+ \frac{(5-8)^2}{8} + \frac{(0-8)^2}{8} + \frac{(15-21.1)^2}{21.1} + \frac{(15-29.6)^2}{29.6}$$
$$+ \frac{(75-88.7)^2}{88.7} + \frac{(30-25.3)^2}{25.3} + \frac{(55-25.3)^2}{25.3}$$
$$\doteq 106.59$$

5. Conclusions: Since $106.59 > 26.217$, we reject H_0 and conclude that there is some ıelationship between taking one or another of the two courses, or both, and subsequent success in statistics.

EXAMPLE 2 A college follows up its graduating class and collects income data for students five years after graduation. Excluding students still in school, the data on income level as compared with grade-point average (GPA) follows. Test a hypothesis that there is no relationship between income level five years after graduation and undergraduate GPA, that is, that the two variables are independent. Use $\alpha = 0.05$.

	INCOME LEVEL (DOLLARS PER YEAR)				
GPA	*1* *Under 5,000*	*2* *5,000–9,999*	*3* *10,000–14,999*	*4* *15,000–19,999*	*5* *20,000 and up*
3.50–4.00	1	18	22	9	2
3.00–3.49	5	63	39	16	11
2.50–2.99	10	152	77	38	14
2.00–2.49	11	84	46	34	2

SOLUTION Since we are able to calculate expected frequencies in each cell and can meet the other assumptions, the Chi-Square test is appropriate. Using the experimental procedure, we have

1. H_0: Undergraduate grade achievement and income level five years after graduation are independent.

   H_1: They are not independent.

2. $\alpha = 0.05$

3. Criteria: Since $df = (4 - 1)(5 - 1) = 12$, reject H_0 if $\chi^2 > 21.026$.

4. Results: Calculating the expected frequencies (in parentheses) we obtain the following table.

	INCOME LEVEL CATEGORY					
GPA	*1*	*2*	*3*	*4*	*5*	*Total*
3.50–4.00	1 (2.1)	18 (25.2)	22 (14.6)	9 (7.7)	2 (2.3)	52
3.00–3.49	5 (5.5)	63 (65.0)	39 (37.7)	16 (19.9)	11 (5.9)	134
2.50–2.99	10 (12.0)	152 (141.1)	77 (81.9)	38 (43.2)	14 (12.9)	291
2.00–2.49	11 (7.3)	84 (85.8)	46 (49.8)	34 (26.3)	2 (7.8)	177
Totals	27	317	184	97	29	654

Since two of the expected frequencies are below 5, Chi-Square may not be used with this arrangement of data. However we could combine the columns on each end, or combine the top two rows. The latter course involves the least loss of data sensitivity, so we can combine "3.00–3.49" and "3.50–4.00" to obtain the following table.

GPA	*1*	*2*	*3*	*4*	*5*
3.00–4.00	6 (7.7)	81 (90.2)	61 (52.3)	25 (27.6)	13 (8.2)
2.50–2.99	10 (12.0)	152 (141.1)	77 (81.9)	38 (43.2)	14 (12.9)
2.00–2.49	11 (7.3)	84 (85.8)	46 (49.8)	34 (26.3)	2 (7.8)

Expected frequencies should be recalculated because of the possibility of a rounding error.

Note that this changes the number of degrees of freedom since the number of rows is changed from four to three. Degrees of freedom are then changed from 12 to 8 and the critical value of χ^2 is reduced from 21.026 to 15.507. Finally, we have

$$\chi^2 = \frac{(6-7.7)^2}{7.7} + \frac{(81-90.2)^2}{90.2} + \frac{(61-52.3)^2}{52.3} + \frac{(25-27.6)^2}{27.6}$$
$$+ \frac{(13-8.2)^2}{8.2} + \frac{(10-12.0)^2}{12.0} + \frac{(152-141.1)^2}{141.1} + \frac{(77-81.9)^2}{81.9}$$
$$+ \frac{(38-43.2)^2}{43.2} + \frac{(14-12.9)^2}{12.9} + \frac{(11-7.3)^2}{7.3} + \frac{(84-85.8)^2}{85.8}$$
$$+ \frac{(46-49.8)^2}{49.8} + \frac{(34-26.3)^2}{26.3} + \frac{(2-7.8)^2}{7.8}$$
$$\doteq 16.774$$

5. Conclusions: Since $16.774 > 15.507$, we reject H_0 and conclude that the variables are not independent. Thus there is some relationship between undergraduate grade-point average and income level five years later. Some further analysis is probably warranted using the methods of Chapter 13.

Problems

1. Using the data of example 2, test the hypothesis by combining columns rather than rows. Note that loss of degrees of freedom (sensitivity of data) caused a marked difference.
2. A survey of 450 randomly selected persons was made and two pieces of information—education level and income level—were collected. The results are as follows:

	INCOME		
EDUCATION	*High*	*Medium*	*Low*
Less than high school	30	50	70
High school graduate	80	100	120

Are income level and education level independent?

3. A sociologist contends that requirements for mortgage approval by lending institutions tend to keep lower income people from buying houses even if the houses

are actually in their price range. To back up this contention he cites the following data:

	INCOME LEVEL (DOLLARS PER YEAR)			
	Less than 5,000	*5,000–9,999*	*10,000–14,999*	*15,000–over*
Home owners	38	64	31	12
Renters	55	58	15	8

Assuming his sample was random, does this actually support his contention at a 0.05 level of significance?

4. A marriage counsellor kept records of the reasons given by husband and wife for marriage difficulty. NOTE: The stated reasons may not necessarily be the true reasons. He classified each couple into two categories—by husband's reason and by wife's reason—with the results which follow. Test the hypothesis at a 0.05 level of significance that husbands' and wives' reasons are independent.

WIFE'S REASON	HUSBAND'S REASON FOR MARITAL DISCORD			
	Money	*Children*	*Consideration**	*All Other*
Money	86	31	132	19
Children	17	64	43	13
Consideration*	54	39	132	33
All other	30	17	37	54

* Including "doesn't love me any more."

5. In a study of technological complexity of societies and the relative punitiveness of legal sanctions in twenty West and North African societies, the following results were obtained.

TECHNOLOGICAL COMPLEXITY	PUNITIVENESS	
	Low	*High*
Low	8	2
High	2	8

Use Chi-Square analysis to determine whether punitiveness of legal sanctions and technological complexity in these societies are independent. Use the 0.05 level of significance.

6. A study relating anxiety to success in Naval Aviation training was conducted using 258 Aviation Officer Candidates who were randomly selected from beginning aviation training classes. Each was given a test of anxiety and classified as very high, high, medium, or low. At the conclusion of the program, 189 students had passed while 69 had dropped out. The number in each anxiety level is given below:

Anxiety Level	*Pass*	*Drop*
Very high	8	11
high	54	25
medium	87	29
low	40	4

Test whether the tendency to drop out was independent of anxiety level. Use the 0.05 level of significance.

7. A study comparing dream recall vs. need for achievement as measured on a standard test gave the following results.

	DREAM RECALL CATEGORIES		
NEED TO ACHIEVE	*Non-Recallers*	*Moderate Recallers*	*Frequent Recallers*
Low	4	9	3
Medium	19	17	22
High	7	4	5

Use Chi-Square analysis to determine if any association exists between frequency of dream recall and need for achievement as measured on this test. Use the 0.05 level of significance.

11.3 CURVE FITTING

The Chi-Square distribution is also used to test hypotheses that the outcomes of an experiment can be predicted by use of some formula or curve. In such a case, the expected frequencies can be computed for each possible outcome and a value of χ^2 calculated. If this value is less than a predetermined critical value, it can be concluded (unless the probability of type II error appears to be too great) that the outcomes can be so predicted. Degrees of freedom are given by $n - 1$, where n is the number of outcomes.

EXAMPLE 1 A sociologist has obtained a good deal of data about a community by taking an extensive random sample which included over 6,000 respondents. She found, however, that one question was inadvertently left off the questionnaire: a question pertaining to income level. In order to avoid having to re-do the study, she examines the material closely and compares it with other communities. She decides that a population such as that from which the sample was drawn should have the family incomes distributed as follows:

Category	*Income*	*Proportion*
I	over \$15,000	0.40
II	\$12,000–\$14,999	0.20
III	\$9,000–\$11,999	0.20
IV	\$6,000–\$8,999	0.10
V	under \$6,000	0.10

She decides to test the validity of this distribution by taking a smaller sample of 500 and asking respondents into which category their family incomes fall. Her sample had the following results.

Category	*I*	*II*	*III*	*IV*	*V*
Number	166	97	134	61	42

Do these results substantiate her claim at the 0.05 level of significance?

SOLUTION The experimental procedure is as follows:

1. H_0: The outcomes conform satisfactorily to the predictions.
   H_1: The outcomes deviate from the predictions.
2. $\alpha = 0.05$
3. Criteria: Reject H_0 if $\chi^2 > 9.488$
4. Results: Since there are 500 responses, we can calculate the expected values in each category, and determine χ^2 as follows:

Category	*I*	*II*	*III*	*IV*	*V*	*Total*
Number	166	97	134	61	42	500
Expected	200	100	100	50	50	500

Then

$$\chi^2 = \frac{(166-200)^2}{200} + \frac{(97-100)^2}{100} + \frac{(134-100)^2}{100} + \frac{(61-50)^2}{50} + \frac{(42-50)^2}{50}$$

$$= \frac{1{,}156}{200} + \frac{9}{100} + \frac{1{,}156}{100} + \frac{121}{50} + \frac{64}{50}$$

$$= 21.13.$$

5. Conclusions: Since $21.13 > 9.488$, we reject H_0 and decide that this hypothetical distribution is not a good predictor of the larger sample income levels.

EXAMPLE 2 A poll on a campus asks students to name their favorite one of four soft drinks. The results are as follows:

Brand	*A*	*B*	*C*	*D*
No. preferring	190	198	187	225

(a) Test the hypothesis that differences in preference are due to chance.
(b) Test the hypothesis that brand D is preferred by more students than any other particular brand. Use $\alpha = 0.05$.

SOLUTION In (a) we would expect 200 people to choose each brand if there were only chance differences (since $P = 1/4$ in each case), so we would have

$$\chi^2 = \frac{(190-200)^2}{200} + \frac{(198-200)^2}{200} + \frac{(187-200)^2}{200} + \frac{(225-200)^2}{200}$$

$$= \frac{100}{200} + \frac{4}{200} + \frac{169}{200} + \frac{625}{200}$$

$$= 4.49$$

1. H_0: The differences observed are due to chance.
   H_1: There are real differences in the preferences.
2. $\alpha = 0.05$
3. Criteria: Reject H_0 if $\chi^2 > 7.815$.

4. Results: $\chi^2 = 4.49$

5. Conclusions: 4.49 is not greater than 7.815, so we cannot reject H_0 and may conclude that the differences are due to chance.

In (b) we have only two categories, "D" and "other." If D is not preferred, we may assume that $P = 1/4$ for D and would expect 200 for "D" and 600 for "other." Then

$$\chi^2 = \frac{(575 - 600)^2}{600} + \frac{(225 - 200)^2}{200} \doteq 4.168$$

Thus:

1. H_0: D is not preferred over the other brands.
   H_1: D is the preferred brand.

2. $\alpha = 0.05$.

3. Criteria: Reject H_0 if $\chi^2 > 3.841$.

4. Results: $\chi^2 = 4.168$.

5. Conclusions: $4.168 > 3.841$, so we reject H_0 and conclude that brand D is indeed preferred.

COMMENT Note the importance of taking the research hypothesis into consideration prior to analyzing the data. Pitting brand D against the others yielded significant results whereas examining all the data did not.

The student may wish to use the methods of section 10.3 and test the hypothesis that $P = 0.25$ (for D) against the alternate hypothesis that $P \neq 0.25$. The results will be the same since $z^2 = \chi^2$ for one degree of freedom. One of the hazards of using χ^2 in such cases, however, is that you cannot test the hypothesis that $P = 0.25$ against $P > 0.25$. If brand D had had 175 takers against the other brands' 625, the value of χ^2 would have been the same. Only the actual results would have indicated the direction of the preference.

Problems

1. In 600 throws of a pair of dice "7" was obtained 148 times and "11" 37 times. Would you conclude (at $\alpha = 0.01$) that the dice were loaded?

2. Four balls are drawn from an urn, with replacement, and the number of white balls in each set of four recorded. The experiment is repeated 100 times with the following results:

Number of White Balls	*Number of Occurrences*
0	11
1	16
2	38
3	33
4	2

Test the hypothesis (at $\alpha = 0.01$) that exactly 40% of the balls in the urn are white.

3. A student poll obtained data on the relative popularity of 3 statistics instructors. Of 150 students who had a preference, 62 preferred instructor *A*, 47 instructor *B*, and 41 preferred instructor *C*. Accepting this as a random sample of statistics students, test the hypothesis that the 3 instructors are equally popular. Use $\alpha = 0.05$.

4. Using the data of example 1, section 11.2, test the hypothesis that prior course work in college algebra and logic have equal effects on success in statistics as measured by the proportion of students with each background that each grade category contains.

c5. Chi-square tests for curve fitting are particularly useful in genetics to test whether a set of offspring has bred true-to-type or is mutant. Suppose, for example, that we examine the result of several litters of guinea pigs with respect to color of coat and length of hair. We have the following results.

	Expected	*Observed*
long hair/brown-white	53.1	61
long hair/black-white	26.5	33
long hair/brown-black	17.7	23
long hair/white	8.8	12
short hair/brown-white	79.6	72
short hair/black-white	53.1	49
short hair/brown-black	26.5	21
short hair/white	17.7	12
Total		283

Expected frequencies were calculated using genetic probabilities and a ratio of 12 : 6 : 4 : 2 : 18 : 12 : 6 : 4 was hypothetically determined for this situation. Use the 0.05 level of significance to test the hypothesis that observed frequencies conform satisfactorily to the prediction. Rejection of that hypothesis would lead us to a speculation that we have a mutation.

6. Prior to an election a public opinion poll claimed that 38% of the voters were for candidate *A*, 33% for candidate *B*, 12% for candidate *C*, and the remainder undecided. Candidate *A* received 5,132 votes, candidate *B* 4,376 votes, and candidate *C* 2,034 votes. Assuming that the undecided voters voted in the same proportion as the rest of the voters, would we say ($\alpha = 0.05$) that the poll reasonably predicted the outcome of the election?

11.4 SUMMARY

The **Chi-Square (χ^2) distribution** is often useful when comparing expected outcomes to obtained outcomes. This distribution is particularly useful in assessing if two classifications of one sample are independent, if proportions among several samples are equal, and if results can be predicted by use of a certain formula or curve.

In each case the statistic is obtained by use of the formula

$$\chi^2 = \sum \frac{(\text{expected} - \text{observed})^2}{\text{expected}}$$

The obtained value of χ^2 is then compared to a critical value arrived at by appropriate choice of α and calculation of degrees of freedom $[(r - 1) \cdot (c - 1)]$ for tables with r rows and c columns, and $n - 1$ for predictions with n outcomes.

If χ^2 is greater than the critical value we reject the null hypothesis of independence, or of equality of proportion, or of fit to a given curve by use of the selected criteria.

The Chi-Square statistic does not assume that the data are normally distributed. For this reason it is called a **nonparametric statistic.** See Appendix B for other such statistics.

Problems

1. In an apartment complex, a sociologist asked a sample of workers if they were satisfied with their work. He then classified the work into three categories. His results were as follows.

	White Collar	*Blue Collar*	*Menial*
Satisfied	81	124	44
Unsatisfied	49	76	56

Test the hypothesis, at $\alpha = 0.05$, that job satisfaction is not related to type of job or that the proportion of satisfied workers is the same in each category.

2. In a large firm, sales people were classified as aggressive, nonaggressive, and shy. Their sales for the month were classified as high, average, or low. Given the data which follows, test the hypothesis that relative aggressiveness and sales are independent. Use $\alpha = 0.01$.

	SALES		
	High	*Average*	*Low*
Aggressive	64	28	38
Nonaggressive	45	22	29
Shy	28	29	27

3. A coin is tossed 3 times and the number of heads recorded. The experiment is repeated 80 times, with the results given here.

Number of Heads	*Frequency of Occurrence*
0	7
1	24
2	35
3	14

Test the hypothesis that the coin is biased. Use $\alpha = 0.05$. If it is biased, in what direction is the bias?

4. A medical researcher is testing the hypothesis that there is a relationship between blood type and susceptibility to a certain liver condition. Now about 40% of the population has blood type O, 30% has type A, 20% has type B, and 10% has type AB. He obtains data on all patients admitted to the hospitals in his area during the past ten years who have had this condition and finds that there were 1,043 who had blood type O, 801 with blood type A, 861 with blood type B, and 745 with blood type AB. Can he safely conclude, at a significance level of 0.01, that there is a relationship?

5. A biologist exposes some culture plates to each of five different, but related, strains of bacillus. She wishes to test the hypothesis that strain makes no difference in the number of plates which will attain propagation level in 36 hours. Given the data below, decide whether or not the hypothesis is substantiated at the 0.05 level of significance.

	STRAIN				
	A	*B*	*C*	*D*	*E*
Number reaching propagation level	32	23	28	36	31
Number not reaching propagation level	68	77	72	64	69

6. To test whether susceptibility to disease A and prior contracting of disease B are related, a sample of 1,000 persons is randomly selected during an epidemic of disease A. The severity of each case of disease A in the sample and the prior severity of any case of disease B are recorded. The results are given in the table below. Test, at the 0.01 level of significance, to see if there is any relationship between disease A and prior experience with disease B.

	DISEASE A		
DISEASE B	*Did not contract*	*Light case*	*Severe case*
Did not contract	302	267	31
Light case	86	156	8
Severe case	12	77	61

Chapter XII

Analysis of Variance

12.1 BASIC ASSUMPTIONS FOR ANALYSIS OF VARIANCE

In Chapter 11, we discussed a method useful in testing the hypothesis that differences among several sample proportions are due to chance. It is also useful to have a method to test the hypothesis that differences among several sample *means* are due to chance or, alternatively, that the samples are drawn from the same population.

The procedure used for this is called **analysis of variance**. It can be used only if the samples are independent, the population (or populations) from which they are drawn are very nearly normally distributed, and the variances of the populations are approximately equal as measured by requiring close equality of the sample variances. The approximation required is rather loose. Experience indicates that analysis of variance is appropriate unless the normality assumption is violated badly by the samples, or unless sample variances differ greatly—for instance if one is ten times more than the other. In the former case, a distribution-free test known as the Kruskal-Wallis Analysis of Variance may be used. This test is discussed in Appendix B. Wide variations in cell variance are harder to handle, but techniques such as the arcsine transformation can be used to solve the problem. This technique will not be discussed in this text.

If these assumptions are met to the required degree, we can proceed to test the null hypothesis that all samples are drawn from populations with equal means, or from the same population. Thus we will have

$$H_0: \mu_1 = \mu_2 = \cdots = \mu_k$$

$$H_1: \text{The means are not all equal}$$

for k different samples.

If H_0 is true, and the samples are drawn from the same or equivalent populations, the population variance will be the same, and no matter how it is calculated from the samples, the estimates for the population variance should be approximately equal. Analysis of variance is a technique in which two estimates of the population variance are obtained from the samples and these estimates compared. If they do not differ too greatly, it is assumed that the difference is due to chance. If they differ so greatly that the probability is less than α (usually 0.05 or 0.01) that it is due to chance, H_0 can be rejected at the α level of significance.

12.2 ESTIMATING POPULATION VARIANCE

Suppose we have several samples, each containing n pieces of data. Each sample variance, $s_1^2, s_2^2, s_3^2, \ldots, s_k^2$ for k samples is an estimate of the population variance if all samples are drawn from the same population (or equivalent populations).

Since each is an estimate of the population variance, an even better calculation can be made by taking the mean of all these variances. This estimate is called the **mean square within samples**.

MEAN SQUARE WITHIN SAMPLES

$$MS_{\text{WITHIN}} = \frac{s_1^2 + s_2^2 + \cdots + s_k^2}{k}$$

if the k samples are equal in size.

If we know the actual population variance, σ^2, we can obtain the standard error of the mean by the formula $\sigma_{\bar{x}} = \sigma/\sqrt{n}$. If we have k samples of size n, with means $\bar{x}_1, \bar{x}_2, \ldots, \bar{x}_k$, we can calculate the variance of these numbers. Since this variance is an estimate of $\sigma_{\bar{x}}^2$, we can estimate σ^2 by noting that if $\sigma_{\bar{X}} = \sigma/\sqrt{n}$, $\sigma_{\bar{x}}^2 = \sigma^2/n$ and $\sigma^2 = n\sigma_{\bar{x}}^2$. This estimate for σ^2 is called the **mean square between samples.**

MEAN SQUARE BETWEEN SAMPLES

$$MS_{\text{BETWEEN}} = \frac{n}{k-1}[(\bar{x}_1 - \bar{x})^2 + (\bar{x}_2 - \bar{x})^2 + \cdots + (\bar{x}_k - \bar{x})^2]$$

where each of k samples is of size n with means $\bar{x}_1$, $\bar{x}_2, \ldots, \bar{x}_k$ and x is the overall mean.

If the samples truly represent the same populations, one estimate is probably as good as the other. If they are from different populations, however, the mean square between the samples will be appreciably larger as we shall show. Naturally, some differences will be expected among samples, even if they are from the same population. These differences generate a sampling distribution called the **F-distribution**, based on the statistic

F-RATIO

$$F = \frac{MS_{\text{BETWEEN}}}{MS_{\text{WITHIN}}}$$

If both estimates for σ^2 are reasonably close, F will be close to 1.00. If MS_{BETWEEN} is appreciably larger, of course, F will be larger. Each set of samples generates an F-distribution based on the null hypothesis of equality of means and the underlying hypothesis that the samples were drawn from the same population or equivalent populations. The distribution depends upon two parameters, the degrees of freedom between samples, $k - 1$, and the sum of the degrees of freedom of each of the samples. This latter number is equal to $N - k$ where N is the total number of cases in all the samples. NOTE: If there are k samples, each with $(n - 1)$ *df*, then $k(n - 1) = kn - k = N - k$. Just as with Chi-Square, some variation is expected by chance. With each F-distribution there is a critical value for $\alpha = 0.05$ and $\alpha = 0.01$ such that only that proportion would be greater than the critical value by chance. Tables 5 and 6 give the 0.05 and 0.01 critical values of F. Degrees of freedom between samples is given at the top of the table; degrees of freedom within samples is given at the side.

EXAMPLE 1 Suppose three dice are rolled four times and the results are recorded as four different samples:

I:	2	4	3
II:	3	6	3
III:	4	6	5
IV:	4	5	3

To calculate both MS_{BETWEEN} and MS_{WITHIN} we note $\sum x = 48$, so $\bar{x} = 4$. The sample means are $\bar{x}_1 = 3$, $\bar{x}_2 = 4$, $\bar{x}_3 = 5$, $\bar{x}_4 = 4$, so

$$MS_{\text{BETWEEN}} = \frac{3}{4-1}[(3-4)^2 + (4-4)^2 + (5-4)^2 + (4-4)^2]$$

$$= \frac{3}{3}[1 + 0 + 1 + 0]$$

$$= 2$$

Also,

$$s_1^2 = \frac{(2-3)^2 + (4-3)^2 + (3-3)^2}{3-1} = 1$$

$$s_2^2 = \frac{(3-4)^2 + (6-4)^2 + (3-4)^2}{3-1} = 3$$

$$s_3^2 = \frac{(4-5)^2 + (6-5)^2 + (5-5)^2}{3-1} = 1$$

$$s_4^2 = \frac{(4-4)^2 + (5-4)^2 + (3-4)^2}{3-1} = 1$$

so

$$MS_{\text{WITHIN}} = \frac{1 + 3 + 1 + 1}{4} = 1.5$$

Now suppose we add 2 units to each number in sample II, 4 units to each number in sample III, and 10 units to each number in sample IV. Now $\sum x = 96$ and $\bar{x} = 8$; $\bar{x}_1 = 3$, $\bar{x}_2 = 6$, $\bar{x}_3 = 9$, $\bar{x}_4 = 14$, so

$$MS_{\text{BETWEEN}} = \frac{3}{4-1}[(3-8)^2 + (6-8)^2 + (9-8)^2 + (14-8)^2]$$

$$= \frac{3}{3}[25 + 4 + 1 + 36]$$

$$= 66$$

From problem 3, section 4.3, recall that adding a constant to each term of a population (or sample) leaves the variance unchanged, so MS_{WITHIN} is still equal to 1.5. Since we constructed this second set of samples to be definitely *not* from the same population we note that $F = 66/1.5 = 44$. From Table 6, the 0.01 critical value for 3 and 8 degrees of freedom is 7.59. This means that only 0.01 of all such samples will have an F-ratio greater than 7.59 by chance if there really is no difference between the samples. In this case we could easily conclude that the differences were not due to chance.

EXAMPLE 2 An English teacher hypothesizes that simplification of paragraphs promotes understanding of content. He divides his class randomly into three groups of ten each. He gives a standardized test of paragraph understanding to the first group. The second and third groups are given the same test with the paragraphs modified in two different ways, each of which is designed to promote understanding. Each group was given the same questions and scored in the same way. The results follow. Test the hypothesis that there are real (nonchance) differences due to the modifications. Use the 0.05 level of significance.

I:	38,	54,	39,	52,	63,	54,	47,	52,	46,	25
II:	58,	44,	63,	94,	72,	42,	89,	68,	53,	47
III:	76,	51,	83,	84,	51,	67,	40,	89,	76,	53

SOLUTION The usual experimental procedure applies. We have

1. $H_0: \mu_1 = \mu_2 = \mu_3$

   $H_1: \mu_1, \mu_2, \mu_3$ are not all equal

2. $\alpha = 0.05$

3. Criteria: Taking it for granted that the assumptions for analysis of variance are met (and theory suggests that considerable variation is permitted), we can calculate F. Since $K = 3$ and $N = 30$, we have 2 and 27 degrees of freedom, so we will reject H_0 if $F > 3.35$.

4. Results: Arranging the data in tabular form we have

x_1	$x_1 - \bar{x}_1$	$(x_1 - \bar{x}_1)^2$	x_2	$x_2 - \bar{x}_2$	$(x_2 - \bar{x}_2)^2$	x_3	$x_3 - \bar{x}_3$	$(x_3 - \bar{x}_3)^2$
38	−9	81	58	−5	25	76	9	81
54	7	49	44	−19	361	51	−16	256
39	−8	64	63	0	0	83	16	256
52	5	25	94	31	961	84	17	289
63	16	256	72	9	81	51	−16	256
54	7	49	42	−21	441	67	0	0
47	0	0	80	26	676	40	−27	729
52	5	25	68	5	25	89	22	484
46	−1	1	53	−10	100	76	9	81
25	−22	484	47	−16	256	53	−14	196
470		1,034	630		2,926	670		2,628

$$\bar{x}_1 = 47 \qquad \bar{x}_2 = 63 \qquad \bar{x}_3 = 67$$

$$s_1^2 = \frac{1{,}034}{9} \doteq 114.89 \qquad s_2^2 = \frac{2{,}926}{9} \doteq 325.11 \qquad s_3^2 = \frac{2{,}628}{9} = 292$$

Then

$$MS_{\text{WITHIN}} = \frac{114.89 + 325.11 + 292}{3}$$

$$= \frac{732}{3} = 244$$

Since

$$\bar{x} = \frac{470 + 630 + 670}{30} = \frac{1{,}770}{30} = 59.0$$

$$MS_{\text{BETWEEN}} = \frac{10}{3-1}[(47 - 59)^2 + (63 - 59)^2 + (67 - 59)^2]$$

$$= \frac{10}{2}[(-12)^2 + (4)^2 + (8)^2]$$

$$= 5(144 + 16 + 64)$$

$$= 5(224)$$

$$= 1{,}120$$

Then

$$F = \frac{1{,}120}{244} = 4.59.$$

5. Conclusions: Since $4.59 > 3.35$, reject H_0. It appears that simplification of paragraphs does promote understanding. There is no evidence, however, that one method of simplification is clearly better than the other.

COMMENT A significant value of F must be interpreted with caution. It is usual to follow a significant F-test with a t-test between pairs of means. There are some hazards which should be noted, however, and the student is referred to articles by Ryan* and Tukey.†

There are many ways to make further tests following an F-test. These tests are usually called *post hoc* or *a posteriori* tests, from the Latin words meaning "to follow." Practically every statistician has his favorite test, so only one will be given here, as an example. This test, which was developed by Scheffé, has the advantages of simplicity and wide applicability, but it tends to be more conservative than some of the others.

For pairwise comparisons with equal sample sizes, a critical value for differences between means is obtained from the formula

$$CV_d = \sqrt{(k-1)(F^*)(MS_{\text{WITHIN}})(2/n)}$$

where k is the number of groups, F^* is the critical value used in the analysis of variance, MS_{WITHIN} is the value determined from the data, and n is the number in each sample. For unequal sample sizes, the n is replaced by $\tilde{n}$ where $\tilde{n} = k/\Sigma(1/n_i)$ and the n_i are the numbers in the k samples. If the difference between sample means exceeds the critical value, we conclude that there is a significant difference between the two samples at the given level of significance.

EXAMPLE 3 Use Scheffé's method to perform *post hoc* tests on the data of Example 2.

SOLUTION The means are 47, 63, and 67, $k = 3$, $F^* = 3.35$, $MS_{\text{WITHIN}} = 244$, $n = 10$, and the significance level is $\alpha = 0.05$. Thus

$$CV_d = \sqrt{(2)(3.35)(244)(2/10)} \doteq 18.08.$$

* Ryan, T. A., "Significance Tests for Multiple Comparison of Proportions, Variances, and other Statistics," *Psychological Bulletin* **57** (1960), 318–328.

† Tukey, J. W., "Comparing Individual Means in the Analysis of Variance," *Biometrics* **5** (1949), 99–114.

Since the means of groups I and III differ by 20, which is greater than 18.08, we conclude that the difference between these means is significant at the 0.05 level.

Problems

1. Students are assigned randomly to 4 different groups and 4 different methods are used to teach the same material to the groups, one method per group. Use analysis of variance to unearth any significant difference between the groups given means and standard deviations of each of the samples. Each group contained 21 students. Use the 0.05 level of significance. Note the effect of different groupings and of the size of the standard deviations.

   (a) $\bar{x}_1 = 78.4$, $\bar{x}_2 = 88.4$, $\bar{x}_3 = 71.6$, $\bar{x}_4 = 70.4$
   $s_1 = 14.7$, $s_2 = 14.4$, $s_3 = 21.6$, $s_4 = 9.8$

   (b) $\bar{x}_1 = 78.4$, $\bar{x}_2 = 88.4$, $\bar{x}_3 = 71.6$, $\bar{x}_4 = 70.4$
   $s_1 = 19.7$, $s_2 = 18.4$, $s_3 = 24.6$, $s_4 = 29.8$

   (c) $\bar{x}_1 = 78.4$, $\bar{x}_2 = 88.4$, $\bar{x}_3 = 81.6$, $\bar{x}_4 = 80.4$
   $s_1 = 14.7$, $s_2 = 14.4$, $s_3 = 21.6$, $s_4 = 9.8$

2. Anthropologists uncovered three burial mounds some distance apart. They found 4 adult skulls in each mound. The skulls measured as follows in each mound (in cm):

   I: 48.4, 46.2, 47.1, 46.3

   II: 52.8, 49.6, 50.5, 48.7

   III: 44.2, 44.7, 46.1, 45.4

   Try to estimate at the 0.05 level of significance if the skulls are representative of the same population.

3. Sociologists studying 6 different cultures did some research into the "waiting time" of an interview—that is, the time spent in amenities before getting down to business. Eleven interviews were timed in each culture. Waiting times are given in minutes. Use analysis of variance to decide whether there are differences among the cultures in regard to this aspect. Use the 0.01 level of significance.

Culture	*Waiting Times*
A	11, 7, 8, 4, 6, 9, 8, 5, 7, 9, 10
B	6, 4, 8, 2, 7, 6, 3, 5, 7, 6, 3
C	5, 9, 8, 7, 6, 10, 9, 11, 9, 8, 7
D	6, 6, 5, 4, 5, 7, 6, 3, 5, 4, 8
E	13, 14, 16, 12, 17, 11, 14, 16, 13, 11, 12
F	7, 6, 5, 9, 11, 3, 8, 14, 3, 7, 5

12.3 OTHER METHODS FOR ANALYSIS OF VARIANCE

The methods described in the previous section can be used only if all samples have an equal number of data points. These methods do not utilize, except indirectly, the raw scores. The methods developed in this section will be useful in these other cases.

Suppose that we wish to obtain the MS_{WITHIN} from samples of unequal size. In this case, we use a pooled variance similar to that used with the t-score between independent samples, namely

$$MS_{\text{WITHIN}} = \frac{(n_1 - 1)s_1^2 + (n_2 - 1)s_2^2 + \cdots + (n_k - 1)s_k^2}{n_1 - 1 + n_2 - 1 + n_3 - 1 + \cdots + n_k - 1}$$

$$= \frac{(n_1 - 1)s_1^2 + (n_2 - 1)s_2^2 + \cdots + (n_k - 1)s_k^2}{N - k}$$

where $n_1, n_2, \ldots, n_k$ are the sizes of the k samples and N is the total number of measures.

To use the raw data we note that

$$s_i^2 = \frac{\sum x_i^2 - \dfrac{(\sum x_i)^2}{n_i}}{n_i - 1}$$

or

$$(n_i - 1)s_i^2 = \sum x_i^2 - \frac{(\sum x_i)^2}{n_i}$$

for each sample, so

$$MS_{\text{WITHIN}} = \frac{\sum x_1^2 - \dfrac{(\sum x_1)^2}{n_1} + \sum x_2^2 - \dfrac{(\sum x_2)^2}{n_2} + \cdots + \sum x_k^2 - \dfrac{(\sum x_k)^2}{n_k}}{N - k}$$

The numerator of this last fraction is called the **sum of squares within samples**. Collecting terms and noting that $\sum x_1^2 + \sum x_2^2 + \cdots + \sum x_k^2 = \sum x^2$, we have

SUM OF SQUARES WITHIN SAMPLES

$$SS_{\text{WITHIN}} = (\sum x^2) - \left[\frac{(\sum x_1)^2}{n_1} + \frac{(\sum x_2)^2}{n_2} + \cdots + \frac{(\sum x_k)^2}{n_k}\right]$$

Similarly, we can determine a **sum of squares between samples**:

$$MS_{\text{BETWEEN}} = \frac{SS_{\text{BETWEEN}}}{k - 1}.$$

SUM OF SQUARES BETWEEN SAMPLES

$$SS_{\text{BETWEEN}} = \left[\frac{(\sum x_1)^2}{n_1} + \frac{(\sum x_2)^2}{n_2} + \cdots + \frac{(\sum x_k)^2}{n_k}\right] - \frac{(\sum x)^2}{N}$$

The sum of squares in each case is simply the estimate for the variance (MS) multiplied by the degrees of freedom. The overall or **total sum of squares** is $(N - 1)s^2$ or

TOTAL SUM OF SQUARES

$$SS_{\text{TOTAL}} = \sum x^2 - \frac{(\sum x)^2}{N}$$

This is also $SS_{\text{WITHIN}} + SS_{\text{BETWEEN}}$. A good check for errors is to determine all three independently and then make sure that

$$SS_{\text{TOTAL}} = SS_{\text{WITHIN}} + SS_{\text{BETWEEN}}$$

EXAMPLE 1 A psychologist tests how long it takes each of four groups of white mice to learn a maze. Three groups have been taught to traverse a maze. Group I had learned the maze six months ago, group II three months ago, and group III one month ago. Group IV had not been taught the maze at all and is designated the control group. The number of trials required for each mouse to learn the maze is given here. Test the hypothesis that there is no difference in mean learning time required among the groups at a 0.05 significance level.

I: 13, 11, 6, 18, 7, 15, 14

II: 9, 13, 15, 11, 10, 6, 7, 9

III: 6, 9, 13, 10, 9, 7

IV: 16, 10, 14, 17, 9, 11, 14

SOLUTION

1. $H_0: \mu_1 = \mu_2 = \mu_3 = \mu_4$
   $H_1: \mu_1, \mu_2, \mu_3, \mu_4$ are not all equal
2. $\alpha = 0.05$
3. Criteria: Reject H_0 if $F > 3.01$ (for 3 and 24 degrees of freedom).
4. Results: For convenience, the calculations are arranged in tabular form:

I		*II*		*III*		*IV*	
x	x^2	x	x^2	x	x^2	x	x^2
13	169	9	81	6	36	16	256
11	121	13	169	9	81	10	100
6	36	15	225	13	169	14	196
18	324	11	121	10	100	17	289
7	49	10	100	9	81	9	81
15	225	6	36	7	49	11	121
14	196	7	49			14	196
		9	81				

$\sum x_1 = 84$ $\quad \sum x_2 = 80$ $\quad \sum x_3 = 54$ $\quad \sum x_4 = 91$

$\sum x_1^2 = 1{,}120$ $\quad \sum x_2^2 = 862$ $\quad \sum x_3^2 = 516$ $\quad \sum x_4^2 = 1{,}239$

$n_1 = 7$ $\quad n_2 = 8$ $\quad n_3 = 6$ $\quad n_4 = 7$

$\bar{x}_1 = 12$ $\quad \bar{x}_2 = 10$ $\quad \bar{x}_3 = 9$ $\quad \bar{x}_4 = 13$

Then

$$\sum x = 84 + 80 + 54 + 91 = 309$$
$$\sum x^2 = 1{,}120 + 862 + 516 + 1{,}239 = 3{,}737$$
$$N = 28$$
$$k = 4$$

Thus, we have

$$SS_{\text{BETWEEN}} = \frac{(84)^2}{7} + \frac{(80)^2}{8} + \frac{(54)^2}{6} + \frac{(91)^2}{7} - \frac{(309)^2}{28}$$
$$= 3{,}477 - 3{,}410.04$$
$$= 66.96$$
$$SS_{\text{WITHIN}} = 3{,}737 - 3{,}477$$
$$= 260$$

$$SS_{\text{TOTAL}} = 3{,}737 - 3{,}410.04$$
$$= 326.96$$

$$MS_{\text{BETWEEN}} = \frac{66.96}{4-1} = \frac{66.96}{3} = 22.32$$

$$MS_{\text{WITHIN}} = \frac{260}{28-4} = \frac{260}{24} \doteq 10.83$$

$$F = \frac{22.32}{10.83} \doteq 2.06$$

Generally, these are summarized in a table such as the one that follows.

Source of Variation	*df*	*SS*	*MS*	*F*
Between Samples	3	66.96	22.32	2.06
Within Samples	24	260.00	10.83	
Total	27	326.96		

Critical value of F for $\alpha = 0.05$ is $F = 3.01$.

5. Conclusions: We fail to reject H_0. The differences are not significant.

EXAMPLE 2 Repeat example 2 of section 12.2 using the formulas of this section.

SOLUTION Using the tabular methods of the preceding example we have

I		*II*		*III*	
x	x^2	x	x^2	x	x^2
38	1,444	58	3,364	76	5,776
54	2,916	44	1,936	51	2,601
39	1,521	63	3,969	83	6,889
52	2,704	94	8,836	84	7,056
63	3,969	72	5,184	51	2,601
54	2,916	42	1,764	67	4,489
47	2,209	89	7,921	40	1,600
52	2,704	68	4,624	89	7,921
46	2,116	53	2,809	76	5,776
25	625	47	2,209	53	2,809

$$\sum x_1 = 470 \qquad \sum x_2 = 630 \qquad \sum x_3 = 670$$
$$\sum x_1^2 = 23{,}124 \qquad \sum x_2^2 = 42{,}616 \qquad \sum x_3^2 = 47{,}518$$
$$n_1 = 10 \qquad n_2 = 10 \qquad n_3 = 10$$
$$\bar{x}_1 = 47 \qquad \bar{x}_2 = 63 \qquad \bar{x}_3 = 67$$

Then

$$\sum x = 470 + 630 + 670 = 1{,}770$$
$$\sum x^2 = 23{,}124 + 42{,}616 + 47{,}518 = 113{,}258$$
$$N = 30, k = 3$$

So we have

$$SS_{\text{BETWEEN}} = \frac{(470)^2}{10} + \frac{(630)^2}{10} + \frac{(670)^2}{10} - \frac{(1{,}770)^2}{30}$$
$$= 106{,}670 - 104{,}430$$
$$= 2{,}240$$
$$SS_{\text{WITHIN}} = 113{,}258 - 106{,}670$$
$$= 6{,}588$$
$$SS_{\text{TOTAL}} = 113{,}258 - 104{,}430$$
$$= 8{,}828$$
$$MS_{\text{BETWEEN}} = \frac{2{,}240}{3 - 1} = 1{,}120$$
$$MS_{\text{WITHIN}} = \frac{6{,}588}{30 - 3} = 244$$
$$F = \frac{1{,}120}{244} = 4.59$$

In summary:

Source of Variation	*df*	*SS*	*MS*	*F*
Between Samples	2	2,240	1,120	4.59*
Within Samples	27	6,588	244	
Total	29	8,828		

* Significant at the 0.05 level.

Critical value of F for $\alpha = 0.05$ is $F = 3.35$.

Problems

1. Students at 5 schools are being studied to discover if there are differences in an "ambition level" index between students attending these schools. Students selected for the study at random from each school were given an instrument evaluating this index during their freshman and senior years, and the increase or decrease was calculated. Equal numbers were selected at the start, but attrition reduced the numbers to the data given here.

School	*Difference Scores*
A	+11, +32, +18, +16, +14, −3, +24, +13, +6
B	−4, +24, +16, −2, +13, +17, +22, +24, +16, +9, +12
C	+13, +31, +34, −7, +18, +14, +34, +28, +19
D	+16, +11, +2, +4, +13, +4, +9, +11, +7, +10
E	+8, −11, +4, +7, +12, −1, −3, +24, +17, +28, +9, +14, +22

Use $\alpha = 0.05$ to discover if the difference among means of the groups is significant.

2. A survey of businesses which have failed during the past year revealed that the mean length of time in business for 42 large businesses was 6.2 years with a standard deviation of 2.2 years; for 34 medium size businesses it was 3.9 years with a standard deviation of 1.6 years; and 47 small businesses had a mean life of 3.1 years with a standard deviation of 1.8 years. Test, at a 0.05 significance level, to see if a difference in duration prior to failure may be attributed to difference in size. (Hint: Refer to the formulas on pages 278–279 to determine the sums of squares).

3. Students were taught a lesson using one of 3 methods—programmed instruction, a tape recorder and visual aids, and television. A learning test was used to determine the amount retained. The scores are given here. Test the hypothesis of no difference between groups at a 0.05 level of significance.

Group	*Scores*
Programmed	10, 13, 3, 38, 11, 23, 36, 3, 61, 21, 5
Tape	8, 36, 61, 23, 36, 48, 51, 36, 48, 36
Television	36, 48, 23, 48, 61, 61, 23, 36, 61

12.4 SUMMARY

To test for differences between means of several groups, a method called analysis of variance is employed. Two estimates of population variance are obtained and their ratio is compared to the appropriate F distribution. If F

is greater than the critical value, we reject the null hypothesis of equality of means.

The procedure restated here is recommended for k samples containing $n_1, n_2, n_3, \ldots, n_k$ pieces of data where the total number of pieces of data is N. These procedures are given for raw data; if samples are equal in size, some special formulas given in section 12.2 may be used.

SUM OF SQUARES BETWEEN SAMPLES

$$SS_{\text{BETWEEN}} = \left[\frac{(\sum x_1)^2}{n_1} + \frac{(\sum x_2)^2}{n_2} + \cdots + \frac{(\sum x_k)^2}{n_k}\right] - \frac{(\sum x)^2}{N}$$

SUM OF SQUARES WITHIN SAMPLES

$$SS_{\text{WITHIN}} = \sum x^2 - \left[\frac{(\sum x_1)^2}{n_1} + \frac{(\sum x_2)^2}{n_2} + \cdots + \frac{(\sum x_k)^2}{n_k}\right]$$

MEAN SQUARE BETWEEN SAMPLES

$$MS_{\text{BETWEEN}} = \frac{SS_{\text{BETWEEN}}}{k - 1}$$

MEAN SQUARE WITHIN SAMPLES

$$MS_{\text{WITHIN}} = \frac{SS_{\text{WITHIN}}}{N - k}$$

Then

$$F = \frac{MS_{\text{BETWEEN}}}{MS_{\text{WITHIN}}}$$

is compared with the critical value from Table 5 or Table 6 with $k - 1$ and $N - k$ degrees of freedom.

Problems

1. Five companies submit samples of paint to a company which is considering the purchase of a large quantity. Six samples of each paint are tested for drying time. Drying times are given as follows. Determine whether the differences are significant at a 0.05 level.

Company	*Drying Times (Minutes)*
I	34, 36, 29, 38, 35, 32
II	30, 34, 30, 32, 31, 28
III	27, 32, 31, 30, 34, 30
IV	28, 35, 29, 29, 37, 33
V	34, 31, 36, 38, 40, 37

2. Stenographers trained in three different systems of stenography are given tests to measure their maximum dictation rate in words per minute. Assuming the stenographers are sampled at random, test the hypothesis that the systems are equally effective at a 0.05 level of significance.

System	*Maximum Dictation Rates*
A	147, 188, 162, 144, 157, 179, 165, 180
B	153, 161, 157, 155, 163, 160, 154
C	173, 152, 194, 186, 166, 194, 178, 192, 186

3. Four samples of thirty soldiers taken from different companies are given marksmanship tests. The mean number of points per ten shots per man are given as follows. Do the data substantiate, at a 0.01 level, the contention that the companies are equal in marksmanship?

Company	*Results (mean scored)*	*Standard Deviation*
A	$\bar{x}_A = 57.8$	$s_A = 11.3$
B	$\bar{x}_B = 62.3$	$s_B = 17.4$
C	$\bar{x}_C = 58.4$	$s_C = 13.5$
D	$\bar{x}_D = 48.9$	$s_D = 22.6$

4. A medical researcher is testing the effect of a drug in inhibiting the release of adrenalin. She wishes to use the 0.05 level of significance. She selects five samples each composed of ten male individuals and injects the members of the samples with 0 cc, 10 cc, 20 cc, 30 cc, and 40 cc, respectively. She then subjects each man to a stress situation and measures the stress level at which adrenalin is released by

each individual. Given the data below, test the hypothesis that there are non-chance variations among the groups which may be caused by the effect of the drug.

cc of drug injected	*Stress Level at Which Adrenalin Is Released*
0	14, 21, 16, 18, 23, 15, 19, 22, 26, 17
10	13, 17, 19, 18, 21, 16, 25, 18, 22, 23
20	16, 22, 18, 23, 22, 25, 19, 17, 21, 25
30	21, 19, 24, 22, 28, 23, 22, 28, 24, 20
40	26, 22, 23, 25, 27, 29, 22, 24, 27, 26

Chapter XIII

Correlation and Regression

13.1 THE COEFFICIENT OF CORRELATION

One of the most widely used—and misused—statistics is the **coefficient of correlation**, r. Roughly speaking, r measures the degree of association between two related sets of data. In most cases, this means that the sets of data result from classifying one set of subjects into two distinct variable classifications. Athletes may be classified according to height and weight. More loosely, we may have gross national product and national alcohol consumption for the same years. The essential factor in using the methods presented in this chapter is that the two sets of data are paired in some way, each member of one set matched with exactly one member of the other set.

A great deal of care must be exercised in the use of the correlation coefficient. It does *not* necessarily imply any true *causal relationship* between the two variables; it may simply be a reflection of one or more variables affecting both of the variables.

The correlation coefficient takes on values from $+1$ to -1. If two sets of data have $r = +1$, they are said to be perfectly correlated positively; if $r = -1$, they are perfectly correlated negatively; and if $r = 0$, they are uncorrelated; r takes on values from $+1$ to -1.

To illustrate the use of r, suppose we have the following sets of data (weights and systolic blood pressures) on a group of ten individuals.

Subject	*Weight*	*Systolic Blood Pressure*
1	188	140
2	231	160
3	176	130
4	194	130
5	244	180
6	207	160
7	198	140
8	217	150
9	181	140
10	194	150

If we plot these on a graph, we notice a general upward trend.

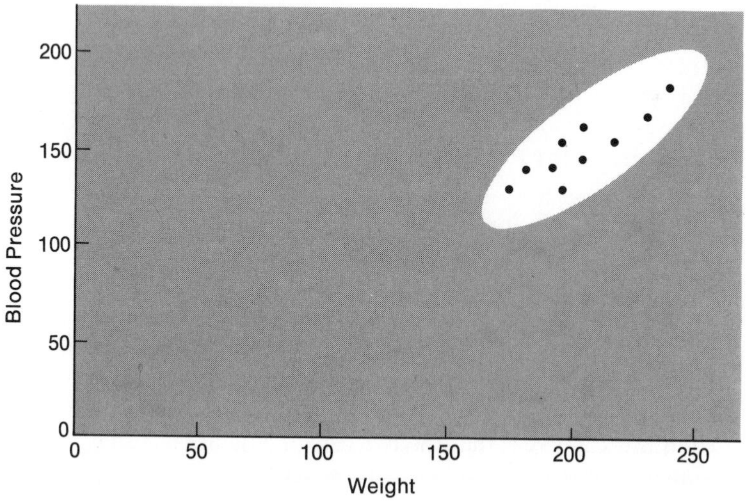

Concentrating on the circled area of the graph, and labelling each point, the upward trend becomes even more apparent and it is evident that there is a tendency for higher weights and higher blood pressures to go together. This does not indicate any *causal* relationship between them; it only indicates an association.

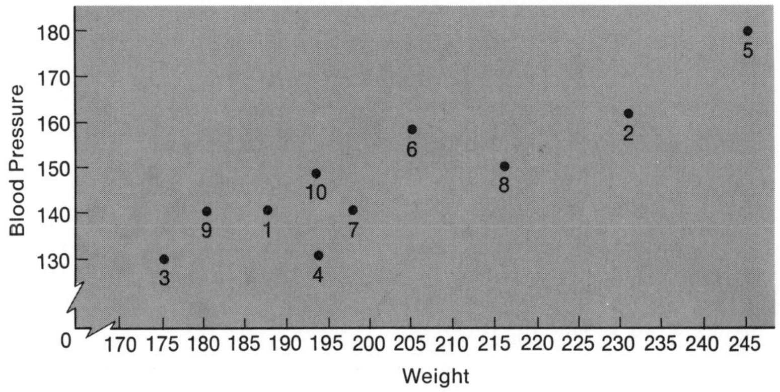

The **product-moment coefficient of correlation**, a statistic developed by Pearson, can be calculated in any of several forms. The most common form is the formula

$$r = \frac{n \sum xy - \sum x \sum y}{\sqrt{[n \sum x^2 - (\sum x)^2][n \sum y^2 - (\sum y)^2]}}$$

Although this formula looks formidable, it is easy to apply, particularly with desk calculators. It can be made even simpler if we note that the value of r is not affected by changes in the values of the variables. We can add and subtract any number we like, as long as every value of that variable gets the same treatment. We can even multiply or divide each value of a variable by the same number.

This will be illustrated by calculating, in two ways, the coefficient of correlation for the data already given. All we need to do is to obtain the five quantities needed for the formula and substitute them into the formula. If we let x represent the weight and y the blood pressure, then we have

Subject	x	y	x^2	y^2	xy
1	188	140	35,344	19,600	26,320
2	231	160	53,361	25,600	36,960
3	176	130	30,976	16,900	22,880
4	194	130	37,636	16,900	25,220
5	244	180	59,536	32,400	43,920
6	207	160	42,849	25,600	33,120
7	198	140	39,204	19,600	27,720
8	217	150	47,089	22,500	32,550
9	181	140	32,761	19,600	25,340
10	194	150	37,636	22,500	29,100
$\sum =$	2,030	1,480	416,392	221,200	303,130

Then

$$r = \frac{(10)(303{,}130) - (2{,}030)(1{,}480)}{\sqrt{[(10)(416{,}392) - (2{,}030)^2][(10)(221{,}200) - (1{,}480)^2]}}$$

$$= \frac{26{,}900}{\sqrt{929{,}232{,}000}}$$

$$\doteq 0.88$$

Using a calculator, the above computation is straightforward and not particularly difficult. If a calculator is not available, however, these computations may be very difficult. One method of simplifying the calculations is based on the observation that changing the values of a variable does not change the value of r. If you subtract 200 from each value of x, for example, this will make the numbers smaller and leave r unaffected. This technique, which is called *coding*, is similar to the procedure of finding the mean and standard deviation by using an assumed mean, except that no correction factor is necessary. If we divide each value of y by 10 and subtract 13 from the quotient, the value of r will remain unchanged. With these changes, the problem we solved before can be done again.

Subject	x	y	x^2	y^2	xy
1	−12	1	144	1	−12
2	31	3	961	9	93
3	−24	0	576	0	0
4	−6	0	36	0	0
5	44	5	1,936	25	220
6	7	3	49	9	21
7	−2	1	4	1	−2
8	17	2	289	4	34
9	−19	1	361	1	−19
10	−6	2	36	4	−12
$\sum =$	30	18	4,392	54	323

Then

$$r = \frac{10(323) - (30)(18)}{\sqrt{[10(4{,}392) - (30)^2][10(54) - (18)^2]}}$$

$$= \frac{2{,}690}{\sqrt{9{,}292{,}320}}$$

$$\doteq 0.88$$

The value of r in each case is the same. It would appear that the two variables are highly correlated. The significance of the value of r will be discussed in the next section.

One can often visualize the degree of correlation with a **scattergram** which is constructed by plotting the two variables on one pair of axes. A few hypothetical scattergrams are given below.

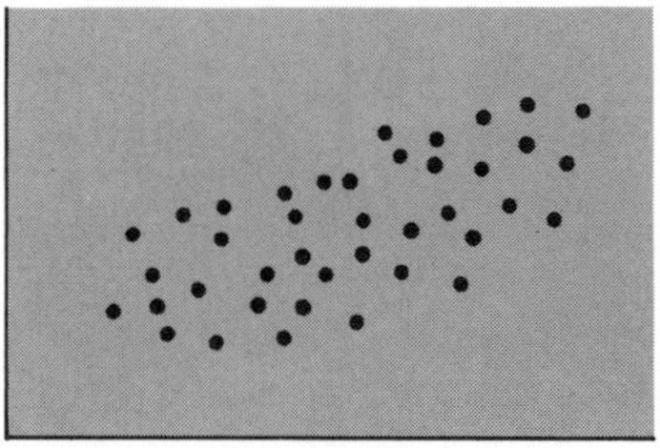

r = 0.82

(A)

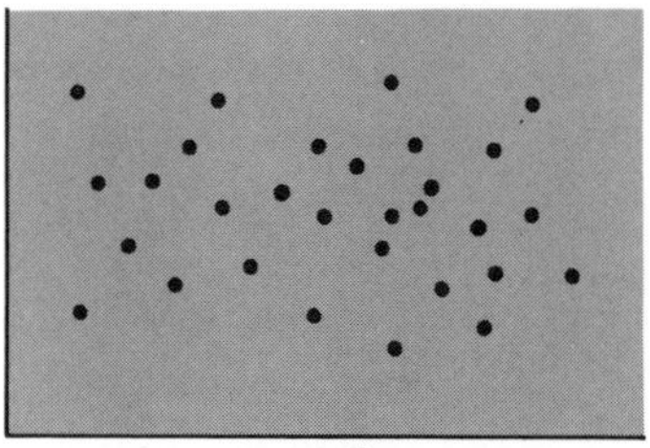

r = 0

(B)

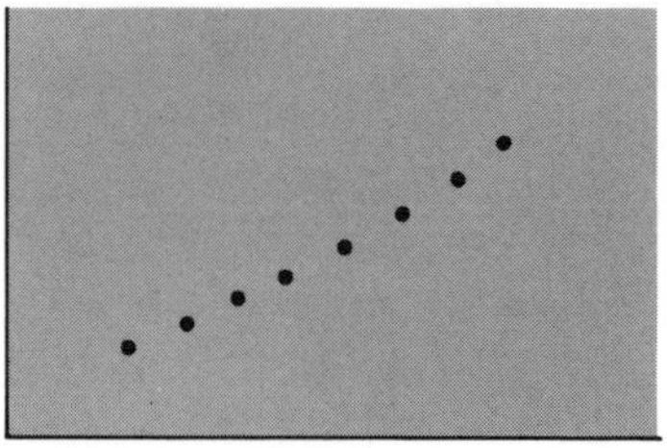

r = 1

(C)

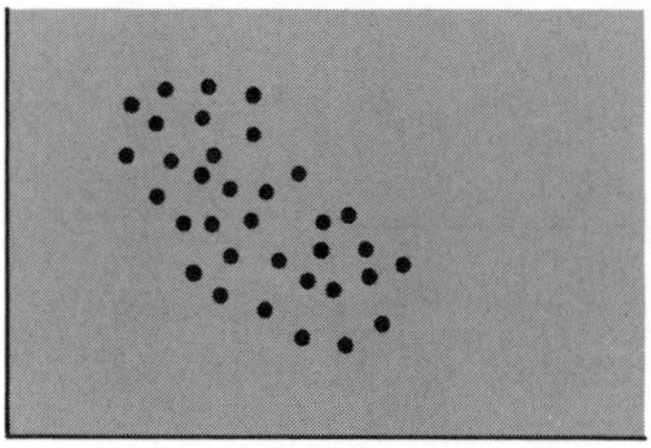

r = − 0.76

(D)

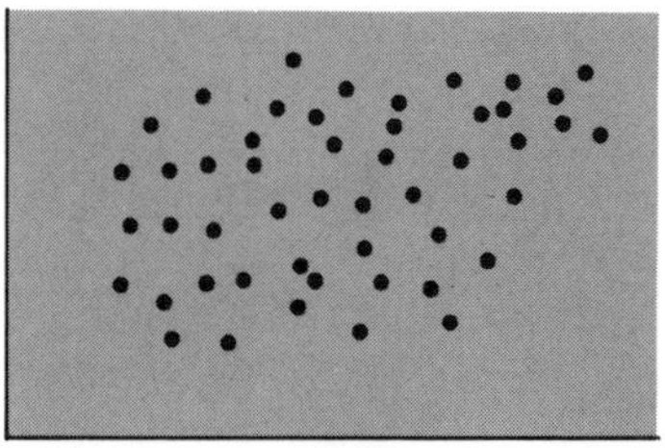

r = 0.34

(E)

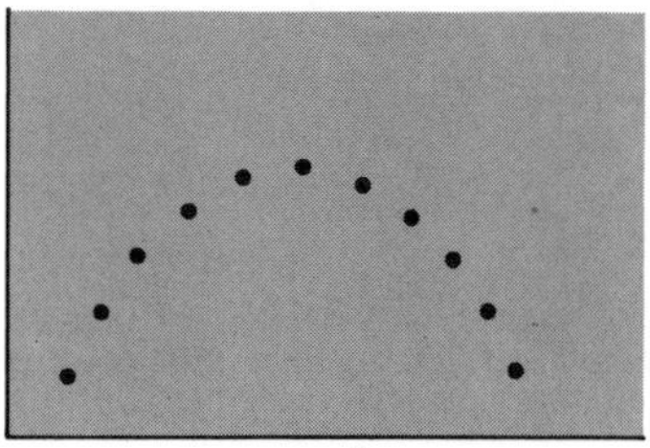

r = 0

(F)

In (F) there is certainly some association between the variables, but the coefficient of correlation measures only the *linear* association, the tendency of the points to lie on or near a straight line. In (F), the association is curvilinear.

EXAMPLE 1 The data that follows were obtained from student records. Calculate the correlation coefficient for the data.

Subject	*Grade-Point Average*	*Graduate Record Exam Score*
1	2.34	910
2	3.61	1,340
3	3.08	1,160
4	2.77	1,420
5	3.13	960
6	2.54	830
7	2.47	940
8	2.38	1,060
9	2.91	1,230
10	3.17	1,080
11	3.28	940
12	2.08	760
13	3.14	1,110
14	3.03	880
15	2.86	1,040
16	2.89	1,320
17	3.13	940
18	2.07	1,020
19	2.71	780
20	2.64	970

SOLUTION Using a calculator, we obtain

$$\sum x = 56.23, \quad \sum y = 20{,}690, \quad \sum x^2 = 161.2503,$$
$$\sum y^2 = 22{,}047{,}700, \quad \sum xy = 58{,}766.70.$$

Then

$$r = \frac{20(58{,}766.70) - (56.23)(20{,}690)}{\sqrt{[(20)(161.2503) - (56.23)^2][(20)(22{,}047{,}700) - (20{,}690)^2]}}$$
$$= \frac{11935.30}{\sqrt{813{,}794{,}422.5}}$$
$$\doteq 0.42$$

Data which are arranged in numerical order, usually from largest to smallest, and then numbered 1, 2, 3, ..., are said to be **in ranks**. These ranks are not normally distributed, but a coefficient of correlation developed by

Spearman can be used. This is called the **Spearman rank-difference correlation coefficient, *R***. In order to calculate R we arrange the data in ranks, computing the difference in rank, d, for each pair. The data of the weight-blood pressure example are used for the following:

Subject	*Weight* x	*Blood-Pressure* y	*Rank of* x	*Rank of* y	d	d^2
1	188	140	8	7	1	1
2	231	160	2	2.5	−.5	.25
3	176	130	10	9.5	.5	.25
4	194	130	6.5	9.5	−3	9
5	244	180	1	1	0	0
6	207	160	4	2.5	1.5	2.25
7	198	140	5	7	−2	4
8	217	150	3	4.5	−1.5	2.25
9	181	140	9	7	2	4
10	194	150	6.5	4.5	2	4
						27

If there are ties, the rank of *each* tied value is the mean of all positions the tied values occupy. In y, for instance, 160 occupies ranks 2 and 3, so each of these values has rank $(2 + 3)/2 = 2.5$; 140 occupies ranks 6, 7, and 8, so each of these values has rank $(6 + 7 + 8)/3 = 7$.

Then we apply the Spearman formula*

$$R = 1 - \frac{6(\sum d^2)}{n(n^2 - 1)}$$

Here $\sum d^2 = 27$, $n = 10$, so

$$R = 1 - \frac{6(27)}{10(99)} \doteq 0.84.$$

This value is slightly different than $r = 0.88$. The Spearman formula is not as sensitive to differences of degree; the difference between 244 and 217 is 27, while the difference between 188 and 176 is 12, but each pair differs in rank by 2.

There is, incidentally, a correction factor which can be applied if the number of ties is large, but it is rarely used unless the number of ties is extremely large.

* The same results will be obtained if the Pearson formula is used. The Spearman formula is the Pearson formula applied to a set of data given in ranks.

EXAMPLE 2 Calculate the rank-order correlation coefficient (R) for the data of example 1.

SOLUTION Using the procedure outlined already we have the following:

Subject	*x*	*y*	*Rank of x*	*Rank of y*	*d*	d^2
1	2.34	910	18	16	2	4
2	3.61	1,340	1	2	−1	1
3	3.08	1,160	7	5	2	4
4	2.77	1,420	12	1	11	121
5	3.13	960	5.5	12	−6.5	42.25
6	2.54	830	15	18	−3	9
7	2.47	940	16	14	2	4
8	2.38	1,060	17	8	9	81
9	2.91	1,230	9	4	5	25
10	3.17	1,080	3	7	−4	16
11	3.28	940	2	14	−12	144
12	2.08	760	19	20	−1	1
13	3.14	1,110	4	6	−2	4
14	3.03	880	8	17	−9	81
15	2.86	1,040	11	9	2	4
16	2.89	1,320	10	3	7	49
17	3.13	940	5.5	14	−8.5	72.25
18	2.07	1,020	20	10	−10	100
19	2.71	780	13	19	−6	36
20	2.64	970	14	11	3	9
						807.50

$$\text{so } R = 1 - \frac{6(807.50)}{20(399)} \doteq 0.39$$

Problems

1. A sociologist obtained some data about the incidence of absenteeism among city employees during the year and their relative "job-pride" as measured by a scale he had developed. Data on each of a dozen city employees is given here. Estimate r and R.

Days Absent	*8*	*1*	*3*	*11*	*6*	*7*	*3*	*8*	*10*	*24*	*6*	*2*
"Job-pride" index	63	82	59	73	84	67	81	72	63	94	81	90

ee with the highest "job-pride" index was absent the as due to unavoidable factors and recalculate r and R aining eleven employees.

ms that there is some relationship between the cost of a ow-Jones Stock Average. He investigates five random nonths and obtains the following data:

37	*41*	*29*	*33*	*19*
768.5	796.25	759.5	756.75	748.5

e data.

ers which may appear on a die are as follows:

	1	2	3	4	5	6
om	6	5	4	3	2	1

values. Are the results suprising?

5. A horse owner is investigating the relationship between weight carried and the finish position of several horses in his stable. Calculate r and R for the data given.

Weight Carried	*Position Finished*
110	2
113	6
120	3
115	4
110	6
115	5
117	4
123	2
106	1
108	4
110	1
110	3
120	5
105	7
115	1
110	2
115	4
103	1
118	3
115	2
110	7
115	6
105	2
110	3

6. A study was conducted to determine the relationship between Reading Index scores on the Florida 12th Grade Examination and cumulative high school grade-point average. A total of 898 students was included in the study. Unfortunately the coefficient of correlation was not calculated and the raw data has been misplaced (temporarily, one hopes). It is known that the mean score on the test was 46.8 with a standard deviation of 14.8, and the mean grade-point average was 2.53 with a standard deviation of 0.76. Fortunately the cross product was obtained prior to loss of the data and is known to be 110,988.12. This is the value of $\sum xy$. Using this information the correlation coefficient can be calculated by using the special formula

$$r = \frac{\sum xy - n\bar{x}\bar{y}}{(n-1)s_x \cdot s_y}$$

where s_x and s_y are the respective standard deviations. Calculate r for the above data.

13.2 SIGNIFICANCE OF THE CORRELATION COEFFICIENT

It is probably apparent that if $r = +1$ or -1 there is a high degree of association between two variables, and that if $r = 0$, there is little or no linear relationship between the two variables. When r takes on values other than $+1$, -1, or 0, however, it requires additional interpretation and explanation.

Suppose we have two related variables. If s_y is the standard deviation of the second variable, then

$$s_y = \sqrt{\frac{\sum(y - \bar{y})^2}{n-1}}$$

The value given by $\sum (y - \bar{y})^2$ is called the **total variation** for y. The amount of this variation which is accounted for by factors involved with the first variable, x, is called the **explained variation**. The remainder is called the **unexplained variation**. These terms assume some actual relationship between the variables, although this may not be the case. In any event, the ratio of explained variation to total variation is called the **coefficient of determination, $\mathbf{r^2}$**. As you might expect from the choice of symbols, r^2 is the square of the correlation coefficient.

Suppose that two variables have $r = 0.4$. Then $r^2 = 0.16$ so that only 16% of the total variation is explained by a relationship between the two variables. If $r = -0.8$, however, $r^2 = 0.64$ and 64% of the total variation is explained by some association between the variables. Thus, if we have two correlation coefficients, r_1 and r_2, the relative strength of the two coefficients is more accurately measured by the ratio between r_1^2 and r_2^2. In the previous example a correlation coefficient of -0.8 (or $+0.8$) indicates a relationship *four times* as strong as a coefficient of $+0.4$ (or -0.4).

The relationship between a sample correlation coefficient and the true population correlation coefficient is highly involved and will not be discussed here. The relationship between the two, however, is the same as the relationship between any sample statistic and its corresponding population parameter. Thus r can be considered an estimate for the true **population correlation coefficient, ρ(rho)**. Techniques exist to estimate ρ from r as is done with other parameters,* but they also will not be discussed here. For many purposes it is sufficient to determine whether or not there appears to be any linear relationship between the variables. If there is not, then the population correlation coefficient is actually zero, and we may test H_0: $\rho = 0$ against a suitable alternate hypothesis ($\rho \neq 0$, $\rho > 0$, $\rho < 0$). Since some variation in the sample is to be expected, even if H_0 is true, this hypothesis is tested with a sampling distribution based on the size of the sample, or the number of degrees of freedom, $n - 2$, since there are two variables. Critical values for a sample r are given in Table 7. The procedure for rejection of H_0 is the same with r as it was with t. H_0 is rejected if the sample r is greater than the critical value (or less, if negative) for the appropriate degrees of freedom.

EXAMPLE 1 A graduate student in psychology hypothesizes that success in graduate school as measured by grade-point average is correlated positively with scores on the Graduate Record Examination. To test his hypothesis, he takes a random sample of 27 graduate students and finds that the sample correlation coefficient is $r = 0.307$. Does this support his hypothesis at a 0.01 level of significance?

SOLUTION The research hypothesis is directional because he claims that $\rho > 0$; this means that high GRE scores and high GPA's go together and low GRE scores and low GPA's go together. Hence we have the following.

1. H_0: $\rho = 0$
   H_1: $\rho > 0$
2. $\alpha = 0.01$

* See for instance D. V. Huntsberger and P. Billingsley: *Elements of Statistical Inference*, Third Edition, Allyn & Bacon, Inc., Boston, Mass., 1973, sec. 11.7.

3. Criteria: Reject H_0 if $r > 0.487$ for 25 degrees of freedom.

4. Results: $r = 0.307$.

5. Conclusions: Fail to reject H_0. There may very well be some positive correlation, but there is not enough to reach a 0.05 level of significance. (That would require $r > 0.323$.) In this case it would probably be best to reserve judgment.

EXAMPLE 2 A medical experiment on laboratory animals was conducted to see if there was any correlation between dosage of a certain drug (in mg) and respiration rate. Test, at a significance level of 0.05, to see if the following data show any correlation.

Dosage (mg)	*0*	*3*	*2*	*3*	*1*	*2*	*0*	*2*	*1*	*3*	*0*	*1*
Respiration (min)	28	29	27	31	28	30	27	31	32	29	25	29

SOLUTION Using our experimental procedures we have:

1. $H_0: \rho = 0$
   $H_1: \rho \neq 0$

2. $\alpha = 0.05$

3. Criteria: Reject H_0 if $r > 0.576$ or if $r < -0.576$.

4. Results:

   $\sum x = 18, \sum y = 346, \sum xy = 532, \sum x^2 = 42, \sum y^2 = 10{,}020, n = 12,$

   so

$$r = \frac{12(532) - (18)(346)}{\sqrt{[12(42) - (18)^2][12(10{,}020) - (346)^2]}}$$

$$= \frac{156}{\sqrt{(180)(524)}}$$

$$= \frac{156}{\sqrt{94{,}320}}$$

$$\doteq 0.508$$

5. Conclusions: Since 0.508 is not greater than 0.576 nor less than −0.576, we fail to reject H_0 and conclude that there may be no apparent correlation.

CAUTION Avoid drawing conclusions about causal relationships on the basis of significant correlations. For instance, a well-known study showed a significantly high positive correlation between teacher's salaries and national consumption of alcoholic beverages over a period of time. Rather than showing that teachers spend their newly increased salaries on drink, this figure is a reflection of a general upward trend in the economy and increased population over the same period of time. Care should be taken to ascertain that no other variables have influence in the calculations. In example 1, for instance, an unobserved factor may be intelligence, of which the GRE and GPA scores may both be a function. Again, certain fields, such as English or social sciences, may have different amounts of correlation between these variables than, say, business, or education. In the second example, it is possible that the relationship may not be linear. Low amounts of the drug may increase respiration, but high amounts decrease it or increase it less than intermediate amounts. A rule that should always be followed in statistical inference is: *Look at the data.* Inappropriate tests should never be performed.

The Spearman R cannot be tested in the same manner as the Pearson r. It is necessary to perform a transformation using:

$$t = \frac{R}{\sqrt{1 - R^2}}\sqrt{n - 2}$$

and applying the results to the appropriate criteria using t with $n - 2$ degrees of freedom.

EXAMPLE 3 A husband and wife are each asked, independently, to rank eleven well-known public figures 1 through 11 in order of degree to which they admire them. Using the data given here, test the hypothesis that husband and wife tend to rate these figures in the same way. Use $\alpha = 0.01$.

Public Figure	*A*	*B*	*C*	*D*	*E*	*F*	*G*	*H*	*I*	*J*	*K*
Husband	6	11	4	5	1	7	2	9	8	3	10
Wife	4	10	3	2	1	8	6	5	11	7	9

SOLUTION Using appropriate procedures we have

1. $H_0: \rho = 0$

   $H_1: \rho > 0$

2. $\alpha = 0.01$

3. Criteria: Since the data are in ranks, the Spearman R must be used and converted to a t score. Then we shall reject H_0 if $t > 2.821$.

4. Results:

$$R = 1 - \frac{6(74)}{11(120)} = 1 - \frac{444}{1,320}$$
$$= 1 - 0.336 = 0.664$$

Then

$$t = \frac{0.644}{\sqrt{1 - (0.664)^2}}\sqrt{11 - 2}$$
$$= \frac{0.664}{0.748}(3)$$
$$= 2.663$$

5. Conclusions: Since $2.663 \not> 2.821$, we fail to reject H_0. Reserve judgment.

DISCUSSION A Pearson r of 0.664 would be significant at $\alpha = 0.05$, since $0.664 > 0.602$ (the critical value for $df = 9$). The use of R is less sensitive than the use of r, so R should generally be used only if the data are given in rank-order, or if the distribution of data is far from normal.

Problems

1. In a reforestation project, all the trees were planted at the same time and were watered by means of irrigation ditches. After ten years a random sample of the trees was examined to find their height and distance from the nearest irrigation ditch. Test the hypothesis that there is no correlation between these variables at a 0.05 level of significance.

Ht. (*ft*)	*Distance* (*ft*)	*Ht.* (*ft*)	*Dist.* (*ft*)	*Ht.* (*ft*)	*Dist.* (*ft*)
23	80	21	100	27	72
18	113	17	120	23	88
22	108	20	110	19	84
24	92	22	94	18	108
21	87	23	103	24	92
19	110	26	76	23	84
23	90	24	82	21	93

2. A medical researcher measures the blood sugar level of cross country runners immediately before and after the race. He wishes to discover whether there is any correlation between the differences (before and after) and success in the race. The race has eighteen entrants. All runners had a net decrease in blood sugar levels after the race. Estimate if there is any correlation (at a 0.01 level of significance) if the following are the net decreases of the eighteen entrants in the order of finish, first to last: 10.8; 11.7; 9.7; 9.4; 10.3; 11.2; 8.8; 10.2; 7.4; 8.1; 7.7; 9.2; 6.4; 8.3; 9.4; 7.3; 6.1; 8.3.
3. A statistician noted that he threw a pair of dice, one red, one green, five times and received the following pairs: 5-4, 3-3, 6-4, 5-5, 2-4. Calculate the correlation coefficient relating the two dice if they are listed red first. Is this result surprising? Is it significant?

13.3 LINEAR PREDICTION—REGRESSION

If two variables are significantly correlated, and if there is some theoretical or apparent basis for doing so, it is possible to predict values of one variable from those of the other. Of course, the actual pairs of values in the sample are known, but the results are generalized to the population from which the sample is drawn by means of a **regression equation**. This equation is obtained by a technique known as **least-squares** which assumes in this case that the relation between the variables can best be described by a straight line. If this is not so, other techniques, particularly those of numerical analysis, can be used.

Although the actual techniques used are rather involved, the results are simple. Suppose a sample of size n has two sets of measures, designated x and y. Predictions for values of y can be made from values of x by using the equation

REGRESSION EQUATION

$$y' = a + bx$$

where the coefficients, *a* and *b*, are given by

REGRESSION COEFFICIENTS

$$b = \frac{n(\sum xy) - (\sum x)(\sum y)}{n(\sum x^2) - (\sum x)^2}$$

$$a = \frac{\sum y - b \sum x}{n}$$

The symbol y' refers to the *predicted* value of y from a given value of x.

EXAMPLE 1 During the past three years 14 graduates of Atlantic Junior College entered the University of the East majoring in mathematics. Their grade-point averages at Atlantic and at the university are given here. Figure out a regression equation which can be used to predict the grade-point averages at the university for entering Atlantic graduates majoring in mathematics. Give the predicted averages for Atlantic graduates with averages of 2.68 and 3.24, respectively.

	Grade-Point Average	
Student	AJC	UE
1	2.08	1.84
2	3.04	2.66
3	3.82	3.92
4	2.44	2.52
5	2.61	1.88
6	2.94	2.42
7	3.03	2.56
8	2.31	2.88
9	2.86	2.62
10	2.56	2.44
11	2.84	2.06
12	3.16	2.96
13	3.40	3.30
14	2.12	2.00

SOLUTION Since we wish to predict UE grade-point averages from AJC grade-point averages, we designate the former y and the latter x and obtain the required sums as follows:

	x	y	x^2	xy
	2.08	1.84	4.3264	3.8272
	3.04	2.66	9.2416	8.0864
	3.82	3.92	14.5924	14.9744
	2.44	2.52	5.9536	6.1488
	2.61	1.88	6.8121	4.9068
	2.94	2.42	8.6436	7.1148
	3.03	2.56	9.1809	6.7468
	2.31	2.88	5.3361	6.6528
	2.86	2.62	8.1796	7.4932
	2.56	2.44	6.5536	6.2464
	2.84	2.06	8.0656	5.8504
	3.16	2.96	9.9856	9.3536
	3.40	3.30	11.5600	11.2200
	2.12	2.00	4.4944	4.2400
$\sum =$	39.21	36.06	112.9255	102.8616

Then

$$b = \frac{14(102.8616) - (39.21)(36.06)}{14(112.9255) - (39.21)^2}$$

$$= \frac{26.1498}{43.4329}$$

$$\doteq 0.60$$

$$a = \frac{36.06 - (0.60)(39.21)}{14}$$

$$\doteq \frac{12.53}{14}$$

$$\doteq 0.90$$

The regression equation, then, is $y' = 0.90 + 0.60x$.

An incoming math major from Atlantic Junior college with a grade-point average of 2.68 would be predicted to have a grade-point average at the university of y' where

$$\begin{aligned} y' &= 0.90 + 0.60(2.68) \\ &\doteq 0.90 + 1.61 \\ &\doteq 2.51 \end{aligned}$$

The predicted average of an incoming student with a 3.24 average would be calculated by

$$\begin{aligned} y' &= 0.90 + 0.60(3.24) \\ &\doteq 0.90 + 1.94 \\ &\doteq 2.84 \end{aligned}$$

DISCUSSION In this case, the correlation between the two sets of data is $r = 0.514$, just barely significant at a 0.05, one-tailed, level. Since $r^2 \doteq 0.26$, nearly three-quarters of the variation is unexplained. Thus, the actual predictive value of the regression equation is open to question. Nonetheless, a regression equation often yields valuable information, even if the correlation is not significant. In cases where r is close to $+1.00$ or -1.00, the regression equation is extremely valuable.

In cases where the data are not arranged sequentially, it is often desirable to predict each variable from the other. This can be done by exchanging the x and y designations for the variables. Thus two regression equations can be obtained. The second can be written

$$x' = a' + b'y$$

An examination of the two formulas for b and b' will show that

$$r = \sqrt{bb'}$$

and that correlation and regression are indeed intimately linked.

Problems

1. The duration of a particular illness is apparently related to the bacterial count of the infecting organism. Ten patients with the illness have a bacterial count taken and the duration of symptoms is observed. Derive a regression equation based on the data which will predict the duration of the illness based on the bacterial count.

Patient	*Count (1,000s)*	*Duration (Days)*
A	8	11
B	7	10
C	4	9
D	6	8
E	9	12
F	7	9
G	8	13
H	3	7
I	8	10
J	7	11

2. Using the data of problem 1, section 13.2, derive a regression equation to predict the probable height of a tree, given its distance from the nearest irrigation ditch.
3. Using the data of problem 1, section 13.2, derive a regression equation to predict the probable distance of a tree from the nearest irrigation ditch, given its height.
4. A corporation uses a screening test to aid in the discovery of potential sales ability of applicants for jobs as salesmen. Since situations vary, the regression equation is refigured monthly, based upon the latest available data for salesmen who have taken the test and been hired within the last six months. Using the data given here, figure out a suitable regression equation and use it to predict the probable adjusted gross sales of today's applicants who have scores of 83, 97, 112, 124, and 146.

Scores	*Adjusted Gross Sales (thousands of dollars)*
97	141
132	113
88	94
154	157
143	118
119	131
157	148
89	107
134	158
135	136
162	159
155	146
113	122
124	131
108	113
136	94
117	124
182	237
130	118
122	145

5. Scores made by students in a statistics class on the mid-term and final examinations are given here. Develop a regression equation which may be used to predict final examination scores from the mid-term score.

Student	*Mid-Term*	*Final*
1	98	90
2	68	82
3	100	97
4	74	78
5	88	77
6	98	93
7	45	82
8	85	77
9	64	80
10	87	99
11	91	98
12	94	77
13	96	95
14	80	95
15	89	92
16	70	80
17	64	75
18	75	65
19	99	88
20	67	78
21	75	63
22	96	93
23	49	53
24	100	90
25	76	53
26	71	88
27	77	84
28	73	58
29	55	88
30	65	63

13.4 SUMMARY

If two sets of data are obtained from the same sample, the degree of association between them can be measured by a correlation coefficient. If neither set of data is given in rank-order, the coefficient used is the

PEARSON PRODUCT-MOMENT CORRELATION COEFFICIENT	$$r = \frac{n \sum xy - \sum x \sum y}{\sqrt{[n \sum x^2 - (\sum x)^2][n \sum y^2 - (\sum y)^2]}}$$ for two sets of n measures

If either set is in rank-order, then the other set must be in rank-order also. If d is a difference between ranks for a pair of measures, we have the

SPEARMAN RANK-DIFFERENCE CORRELATION COEFFICIENT	$$R = 1 - \frac{6 \sum d^2}{n(n^2 - 1)}$$

The significance of a Pearson r is determined by testing the hypothesis $\rho = 0$ (where ρ is the population coefficient) against an appropriate alternate hypothesis. Values greater than critical values as given in Table 7 for appropriate significance level, or less than the negative value, where appropriate, are considered significant. A similar hypothesis for a Spearman R is tested by using a t-test on the statistic

$$t = \frac{R}{\sqrt{1 - R^2}} \sqrt{n - 2} \quad \text{for } n - 2 \; df$$

One can often obtain information about a population by trying to predict unknown values of one variable (y) from known values of another variable (x). If two sets of data are obtained on the same sample of some population, one may predict on the basis of this sample by means of a

REGRESSION EQUATION	$$y' = a + bx$$

where y' is the value *predicted* from x, and a and b are the

REGRESSION COEFFICIENTS

$$b = \frac{n \sum xy - \sum x \sum y}{n \sum x^2 - (\sum x)^2}$$

$$a = \frac{\sum y - b \sum x}{n}$$

The value of the regression equation is greatly increased if there is a significant correlation between the two variables.

Problems

1. Several farmers in the same county employed varying amounts of fertilizer and obtained varying yields of corn. Use the data given here to find a regression equation to predict yield from amount of fertilizer employed. Then calculate the correlation coefficient between the 2 variables, and determine if it is significant at a 0.05 level of significance. Data are given in hundred-weight for fertilizer and tons for corn.

Fertilizer	8.3	9.2	7.7	8.4	8.8	9.6	10.3	8.7	9.1	9.4
Yield	13.6	15.4	12.8	13.4	14.6	15.8	15.5	14.1	14.9	15.6

2. Two professors rated eleven students in terms of ability. Use the data given here to estimate if the correlation between the ratings is significant.

Student	*Professor X*	*Professor Y*
A	1	4
B	7	8
C	8	10
D	3	1
E	6	5
F	10	9
G	9	11
H	2	3
I	11	7
J	4	2
K	5	6

3. Using the weight-blood pressure data of section 13.1, figure out the 2 regression equations if x represents weight and y represents blood pressure, and verify that $r = \sqrt{bb'}$.

4. An experiment was conducted to study the effect of drug dosage on reduction of cholesterol level in males over a period of six weeks. Calculate the regression equation for predicting reduction of cholesterol level from weekly drug dosage (mg). Determine the correlation coefficient and test the hypothesis, at the 0.05 level that increase in dosage reduces the cholesterol level more rapidly.

Subject	*Weekly Dosage (mg)*	*Reduction in Cholesterol Level*
1	0.5	8
2	0.5	11
3	0.5	6
4	1.0	12
5	1.0	9
6	1.0	7
7	1.5	15
8	1.5	13
9	1.5	18
10	2.0	23
11	2.0	27
12	2.0	19
13	2.5	28
14	2.5	31
15	2.5	30
16	3.0	28
17	3.0	21
18	3.0	23

5. Considering the data in problem 4 as reduction of cholesterol level in each of six different groups, perform an analysis of variance testing the hypothesis that there are differences among the groups which may be due to differences in dosage. Discuss the advantages and disadvantages of the two procedures. Which makes better use of the data?

Appendix A

Mathematics Review

A certain amount of mathematical knowledge is required of students taking a course in elementary statistics. Although many students do not need this review, it is included because some areas seem to cause more difficulty than others.

DECIMALS

To add or subtract decimals, you should have the same number of digits following the decimal point in each decimal. Then it is best to write them vertically, aligning the decimal point.

EXAMPLES

612.312 + 83.4. Write as

$$\begin{array}{r} 612.312 \\ 83.400 \\ \hline 695.712 \end{array}$$

48.27 − 11.304. Write as

$$\begin{array}{r} 48.270 \\ -11.304 \\ \hline 36.966 \end{array}$$

To multiply two decimals, multiply in the normal manner, ignoring the decimals. Then count the total number of digits following the decimal point in the two decimals. The product will contain that many integers following the decimal point (begin counting from the right).

EXAMPLE 4.316 × 11.2 will have 3 + 1 or 4 digits following the decimal point in the product, thus:

```
 4.316
  11.2
------
  8632
 4316
4316
------
48.3392
```

To divide two decimals, arrange in usual fashion.

EXAMPLE To divide 11.3143 by 2.46 arrange as

$$2.46\overline{)11.3143}$$

Count the number of digits in the divisor (two, in this example). Point off that many digits to the right of the decimal point in the dividend, indicating the position with a caret, as follows:

$$2.46\overline{)11.31_{\wedge}43}$$

The decimal point in the quotient goes above the caret, as shown:

```
          4.5993
     ___________
2.46/11.31^43
       984
      ----
       1474
       1230
       ----
        2443
        2214
        ----
         2290
         2214
         ----
           760
```

Zeros may be added when needed, as in this example.

FRACTIONS

Addition and subtraction of fractions is accomplished by finding a common denominator. For instance, the fractions 2/3 and 3/5 have a common denominator 15, that is, 15 is a multiple of both 3 and 5. Since 2/3 = 10/15 and 3/5 = 9/15.

$$2/3 + 3/5 = 10/15 + 9/15 = 19/15 \quad \text{or} \quad 1\frac{4}{15}$$

Another way to obtain the sum is to note that

$$\frac{2}{3} + \frac{3}{5} = \frac{(2 \cdot 5) + (3 \cdot 3)}{15} = \frac{10 + 9}{15} = \frac{19}{15}$$

In fact, $(a/b) + (c/d) = (ad + bc)/bd$ regardless of the fractions. It is not absolutely necessary to find the *least* (or smallest) common denominator. For instance, 5/6 and 3/4 have 12 as the least common denominator, but 24 is also a common denominator (as are 36, 48, etc.). We can change 5/6 to 10/12, 3/4 to 9/12, to obtain

$$5/6 + 3/4 = 10/12 + 9/12 = 19/12 = 1\frac{7}{12}$$

but we can also use the rule stated as follows:

$$\frac{5}{6} + \frac{3}{4} = \frac{(5 \cdot 4) + (6 \cdot 3)}{24} = \frac{20 + 18}{24} = \frac{38}{24} = 1\frac{14}{24}$$

Since 14/24 = 7/12, the results are the same.

Reducing fractions is accomplished by factoring common terms from both numerator and denominator, then dividing them out (or noting that the fraction a/a equals one).

EXAMPLE

$$\frac{120}{165} = \frac{5 \cdot 24}{5 \cdot 33} = \frac{5}{5} \cdot \frac{24}{33} = 1 \cdot \frac{24}{33} = \frac{24}{33}$$

$$\frac{24}{33} = \frac{3 \cdot 8}{3 \cdot 11} = \frac{3}{3} \cdot \frac{8}{11} = 1 \cdot \frac{8}{11} = \frac{8}{11}$$

Thus,

$$\frac{120}{165} = \frac{3 \cdot 5 \cdot 8}{3 \cdot 5 \cdot 11} = \frac{8}{11}$$

This is especially useful in evaluating expressions such as

$$\frac{8 \cdot 7 \cdot 6 \cdot 5 \cdot 4}{5 \cdot 4 \cdot 3 \cdot 2 \cdot 1}$$

Since $5 \cdot 4$ appears in both numerator and denominator and since $3 \cdot 2 = 6$, we can cancel *like factors* and obtain

$$\frac{8 \cdot 7 \cdot \not{6} \cdot \not{5} \cdot \not{4}}{\not{5} \cdot \not{4} \cdot \not{3} \cdot \not{2} \cdot 1} = 56$$

In order to cancel numbers in a fraction in which the numerator or denominator consists of a sum, such as $(18 + 15)/12$, it is necessary to write the sum in factored form, and then bring out the common factor using the distributive principle. For example

$$\frac{18 + 15}{12} = \frac{(3 \cdot 6) + (3 \cdot 5)}{12} = \frac{3(6 + 5)}{12} = \frac{\not{3}(6 + 5)}{\not{3} \cdot 4} = \frac{6 + 5}{4}$$

Multiplying fractions is done by multiplying numerators and multiplying denominators, so that

$$\frac{3}{5} \cdot \frac{8}{11} = \frac{3 \cdot 8}{5 \cdot 11} = \frac{24}{55}$$

If common factors occur, you should cancel them before obtaining the answer,

$$\frac{5}{18} \cdot \frac{3}{10} = \frac{5 \cdot 3}{18 \cdot 10} = \frac{5 \cdot 3}{3 \cdot 6 \cdot 2 \cdot 5} = \frac{1}{6 \cdot 2} = \frac{1}{12}$$

Dividing fractions is best done simply by writing them as a complex fraction,

$$\frac{11}{18} \div \frac{13}{15} = \frac{\frac{11}{18}}{\frac{13}{15}}$$

then multiplying numerator and denominator of the complex fraction by both denominators of the two simple fractions, and simplifying

$$\frac{11}{18} \div \frac{13}{15} = \frac{\frac{11}{18}}{\frac{13}{15}} = \frac{\frac{11}{18} \cdot 18 \cdot 15}{\frac{13}{15} \cdot 18 \cdot 15} = \frac{\frac{11}{\not{18}} \cdot \not{18} \cdot 15}{\frac{13}{\not{15}} \cdot 18 \cdot \not{15}} = \frac{11 \cdot 15}{13 \cdot 18} = \frac{11 \cdot 5 \cdot 3}{13 \cdot 6 \cdot 3} = \frac{11 \cdot 5}{13 \cdot 6} = \frac{55}{78}$$

EXPONENTS

Exponents are those little numbers or symbols above and to the right of another number or symbol. The exponent tells how many times the other number, called the base, should be multiplied by itself. For instance, $2^3 = 2 \cdot 2 \cdot 2 = 8$. A symbol such as P^r indicates that P (whatever it represents) is used as a factor r times.

To combine numbers with exponents, it is generally necessary to perform the indicated operations. That is,

$$3^2 + 3^3 = (3 \cdot 3) + (3 \cdot 3 \cdot 3) = 9 + 27 = 36$$

and

$$3^2 \cdot 4^4 = 3 \cdot 3 \cdot 4 \cdot 4 \cdot 4 \cdot 4 = 576$$

If, however, two numbers with the same base are to be multiplied, the product is the base number with an exponent which is the sum of the exponents. Thus

$$3^2 \cdot 3^5 = 3^{2+5} = 3^7$$

and

$$P^a \cdot P^b = P^{a+b}$$

Similarly, division of two such numbers is accomplished by subtracting exponents

$$\frac{3^8}{3^5} = 3^{8-5} = 3^3 = 27$$

and

$$\frac{P^a}{P^b} = P^{a-b}$$

If $a = b$, we have $P^a/P^a = P^{a-a} = P^0$. Since $P^a/P^a = 1$, it follows that $P^0 = 1$. In addition, since $P^{-3} = P^{0-3} = P^0/P^3 = 1/P^3$, it follows that a negative exponent is a reciprocal fraction. For instance, $2^{-4} = 1/2^4 = \frac{1}{16}$.

SIGNED NUMBERS

Combining signed numbers is quite important. To do so, you should remember that if two numbers of different signs are added, the sum is their difference, with the sign of the numerically larger retained. Thus,

$$-3 + 5 = +2$$

$$-3 + (-5) = -8$$

$$3 + (-5) = -2$$

Subtraction of signed numbers is equivalent to the addition of signed numbers with the sign of the number subtracted changed, as in

$$-3 - 5 = -3 + (-5) = -8$$
$$-3 - (-5) = -3 + 5 = +2$$
$$3 - 5 = 3 + (-5) = -2$$
$$3 - (-5) = 3 + 5 = +8$$

Multiplication or division of signed numbers is done by multiplying or dividing in the ordinary way, and affixing a positive sign if the numbers are *both* positive or both negative, and a negative sign if there is one of each. Thus,

$$(-2)(-4) = +8 \quad \text{or} \quad (-2)(+4) = -8$$
$$\frac{-4}{-2} = +2 \quad \text{or} \quad \frac{-4}{+2} = -2$$

ORDER RELATIONS

If two numbers are not equal, one is greater than the other. This is indicated by use of the sign $>$ or $<$. Since 5 is greater than 3, we have $5 > 3$ or $3 < 5$. The latter is usually read "3 is less than 5." A double use of the signs can be used, for instance, to show that x is between a and b, as follows:

$$a < x < b$$

Technically, this says "a is less than x and x is less than b," but this is identical to saying that x is between a and b. To show the cases in which the end points may be included, the symbols $\geq$ and $\leq$ are employed. They are read "greater than or equal to" and "less than or equal to," respectively. Thus, if z can take a value of 1.96, or greater, this is indicated by $z \geq 1.96$. The relation $z > 1.96$ would rule out the possibility of z being equal to 1.96.

SQUARE ROOTS

Table 1 is provided to help you to take square roots. Numbers from 1.00 to 9.99 are listed in the first column in each page of the table. The second column gives the square root of the number listed, and the third column gives the square root of ten times the number listed. For instance, the square root of 1.62 is listed in the second column opposite 1.62 in the first column, and we have $\sqrt{1.62} = 1.273$. The square root of 14 is listed in the third column opposite 1.40, since $14 = 10 \times 1.40$. Thus $\sqrt{14} = 3.742$. To find the square root of any other number than those from 1.00 to 99.9, a few simple techniques may be used. First, if the number has more than three significant figures, you can estimate the square root by taking the numbers above and below it, or use another table, or desk calculator, or one of the techniques which are taught in the lower grades.

To calculate the square root of a number above 99.9, pair digits to the left of the decimal point, starting at the decimal point, thus

$$\sqrt{\overgroup{38}\,\overgroup{20}\,\overgroup{00}} \quad \text{or} \quad \sqrt{\overgroup{4}\,\overgroup{60}\,\overgroup{00}\,\overgroup{00}}$$

The numbers of pairs, or pairs plus a single digit, gives the number of digits to the left of the decimal point, in the square root. In the first example, there will be three; in the second, four. If all digits are paired, the root will be found in the $\sqrt{10n}$ column; if there is a single digit, it will be found in the $\sqrt{n}$ column. For $\sqrt{382{,}000}$, we note that there will be three digits left of the decimal point and that $\sqrt{38.2} = 6.181$. Then $\sqrt{382{,}000} = 618.1$. For $\sqrt{4{,}600{,}000}$, there will be four digits to the left of the decimal point and $\sqrt{4.60} = 2.145$. Then $\sqrt{4{,}600{,}000} = 2{,}145$.

To calculate the square root of a number less than 1.00, a similar technique applies. The pairing begins at the decimal point and stops when a pair contains a digit other than zero, thus,

$$\sqrt{0.\overgroup{00}\,\overgroup{04}\,\overgroup{32}} \quad \text{or} \quad \sqrt{0.\overgroup{00}\,\overgroup{00}\,\overgroup{00}\,\overgroup{74}\,3}$$

Each *pair* of zeros places a zero to the right of the decimal point in the square root. If the first pair with a non-zero digit has the first digit zero, the appropriate root is found in the $\sqrt{n}$ column; if the first digit is not zero, the root is found in the $\sqrt{10n}$ column. In the examples here, $\sqrt{4.32} = 2.078$, so $\sqrt{.000432} = .02078$; $\sqrt{74.3} = 8.620$, so $\sqrt{0.000000743} =$ 0.000862. This can be displayed easily by using the following format.

$$\begin{array}{l} 0.0\;\;0\;\;0\;\;6\;\;017 \\ \sqrt{0.\overgroup{00}\,\overgroup{00}\,\overgroup{00}\,\overgroup{36}\,2} \end{array}$$

LINEAR EQUATIONS

A linear equation is an equation containing one variable, usually x, y, or z, in which no expression containing the variable is of higher order than first degree. That is, the variable does not have exponents. Linear equations in two variables or more are not needed for this text and will not be discussed here. Usually a linear equation can be simplified by combining like terms on each side to obtain a form like

$$ax + b = cx + d$$

where the a, b, c, d represent some known constants. In combining terms, remember that multiplication and division should be performed before addition and subtraction. Terms within parentheses should be combined as far as possible before removing the parentheses. A negative sign on the outside of parentheses applies to all terms within the parentheses and removal of the parentheses necessitates changing the sign of all terms within the parentheses (see example that follows). Then proceed to get the x terms all on one side, usually the left, and all other terms on the right side (or other side) by adding the negative of the term you wish to transfer to both sides. Thus

$$ax + b + (-b) = cx + d + (-b)$$

$$ax = cx + (d - b)$$

$$ax + (-cx) = cx + (-cx) + (d - b)$$

$$(a - c)x = d - b$$

Finally, divide both sides by the coefficient of x, and

$$x = \frac{d - b}{a - c}$$

EXAMPLE Solve the equation:

$$8x + (4 - 2x) + 3 = 3x - 2(7 + 4x) + 11$$

for x, we combine terms as follows:

$$8x + 4 - 2x + 3 = 3x - (14 + 8x) + 11$$

$$6x + 7 = 3x - 14 - 8x + 11$$

$$6x + 7 = -5x - 3$$

Then we combine the terms so that the x terms are on the left side and the other terms are on the right side. Recall that the negative of $-5x$ is $+5x$.

$$6x + 7 + (-7) = -5x - 3 + (-7)$$

$$6x = -5x - 10$$

$$5x + 6x = -5x + 5x - 10$$

$$11x = -10$$

Finally, divide both sides by 11 to obtain

$$x = -\frac{10}{11}$$

If desired we can check our solution by substituting the values of x into the original equation.

$$\begin{aligned} 8\left(-\frac{10}{11}\right) + 4 - 2\left(-\frac{10}{11}\right) + 3 &= -\frac{80}{11} + \frac{44}{11} + \frac{20}{11} + \frac{33}{11} \\ &= -\frac{80}{11} + \frac{97}{11} \\ &= \frac{17}{11} \end{aligned}$$

$$\begin{aligned} 3\left(-\frac{10}{11}\right) - 2\left[7 + 4\left(-\frac{10}{11}\right)\right] + 11 &= -\frac{30}{11} - 2\left(\frac{77}{11} - \frac{40}{11}\right) + \frac{121}{11} \\ &= -\frac{30}{11} - 2\left(\frac{37}{11}\right) + \frac{121}{11} \\ &= -\frac{30}{11} - \frac{74}{11} + \frac{121}{11} \\ &= -\frac{104}{11} + \frac{121}{11} \\ &= \frac{17}{11} \end{aligned}$$

Since substitution of $-\frac{10}{11}$ for x yields the same value for each side, we conclude that this is the solution of the equation.

If the equation involves fractions two methods work equally well. Either the fractions can be carried along, combining where needed, or both sides of the equation can be multiplied by a common multiple of all the denominators (preferably the least common multiple). The resulting equation can be solved as before.

The following example will be solved both ways. The check is left to the student.

EXAMPLE Solve the equation:

$$\frac{2}{3}x - \frac{1}{3} = \frac{1}{2}x - \frac{1}{4} - \frac{1}{6}x$$

Solving by the first method, we have

$$\frac{2}{3}x - \frac{1}{3} = \frac{1}{2}x - \frac{1}{6}x - \frac{1}{4}$$

$$\frac{2}{3}x - \frac{1}{3} = \frac{3}{6}x - \frac{1}{6}x - \frac{1}{4}$$

$$\frac{2}{3}x - \frac{1}{3} = \frac{1}{3}x - \frac{1}{4} \qquad \left(\text{since } \frac{2}{6} = \frac{1}{3}\right)$$

$$\frac{2}{3}x - \frac{1}{3} + \frac{1}{3} = \frac{1}{3}x - \frac{1}{4} + \frac{1}{3}$$

$$\frac{2}{3}x = \frac{1}{3}x + \left(\frac{1}{3} - \frac{1}{4}\right)$$

$$\frac{2}{3}x = \frac{1}{3}x + \left(\frac{4}{12} - \frac{3}{12}\right)$$

$$\frac{2}{3}x = \frac{1}{3}x + \frac{1}{12}$$

$$\frac{2}{3}x - \frac{1}{3}x = \frac{1}{12}$$

$$\frac{1}{3}x = \frac{1}{12}$$

Then we multiply both sides by whatever is necessary to make the left side equal to x.

$$3 \cdot \frac{1}{3}x = 3 \cdot \frac{1}{12}$$

$$x = \frac{1}{4}$$

Now if we had had, say $\frac{3}{5}x = \frac{9}{2}$, we would have multiplied both sides by $\frac{5}{3}$ and had $\frac{5}{3} \cdot \frac{3}{5}x = \frac{5}{3} \cdot \frac{9}{2}$ or $x = \frac{15}{2}$.

Solving the equation in the second way, we multiply both sides by some common multiple of the denominators 3, 2, 4, 6. Obviously their product, 144, will work, but a little thought will reveal that 12 is the least common multiple so we have

$$12\left(\frac{2}{3}x - \frac{1}{3}\right) = 12\left(\frac{1}{2}x - \frac{1}{4} - \frac{1}{6}x\right)$$

$$8x - 4 = 6x - 3 - 2x$$

$$8x - 4 = 4x - 3$$

$$8x - 4 + 4 = 4x - 3 + 4$$

$$8x = 4x + 1$$

$$8x - 4x = 4x - 4x + 1$$

$$4x = 1$$

$$x = \frac{1}{4}$$

Appendix B

Nonparametric Tests

Most of the statistical tests described in this text require an important assumption to be met if they are to be correctly applied. This assumption is that the population of data from which a sample or samples are drawn is normally distributed. These statistical tests do not require that the data be drawn from populations which are absolutely normally distributed; considerable latitude is acceptable. If the distribution from which a sample of data is drawn is badly skewed (bunched at one end) or is otherwise grossly nonnormal, however, these statistical tests will not yield meaningful results.

Statisticians have devised alternate procedures which can be used to test hypotheses about nonnormally distributed data. Since these tests do not depend on the shape of the distribution, they are called **distribution-free tests**. Since they do not depend upon population parameters such as the mean and the standard deviation, they are also called **nonparametric tests**. We have already discussed two nonparametric techniques: the Chi-Square test for proportions and the Spearman R.

Most experimental situations yield data which are normally distributed. If the data comes from a distribution which is bounded on one end, however, there is a good chance that the distribution will not be normal. For example, income distributions are bounded at their lower end, at zero, while they are practically unlimited at their upper end. Distributions of incomes tend to bunch up around the lower, limited end. If you are working with data which comes from a bounded distribution such as this one, it is a good idea to construct a histogram from the sample data before conducting any statistical tests. If the histogram shows that the data is nonnormally distributed, nonparametric tests are called for.

Many nonparametric tests have been devised. Articles and books have been written about them.* Four convenient and widely used alternatives will be discussed here. These tests are not as sensitive as their parametric counterparts and will fail to reject H_0 as often in cases where either test is applicable.

THE SIGNS TEST

The most common research problem involves the testing of two samples to determine if there is a difference between them. This was discussed in section 10.4. If the samples are related such as in before-and-after tests or other matched pairs, the correlated test, t- or z-, can be used only if the normality assumption is met. A convenient and simple alternative is the **signs test** which is derived from the binomial distribution.

To use the signs test, you calculate the difference between the values of each pair and record the signs of these differences. If the groups are equivalent, we would expect approximately equal numbers of plus and minus signs. This would correspond to a null hypothesis that $P = 0.50$, where P is the probability that a particular difference is positive. If there is no difference in a pair, it is dropped from the calculations. If we observe that one sign occurs more frequently than the other, Table 9, page 359, with $p = 0.50$, will give us the probability of obtaining at least that many occurrences. If our test is one-tailed, this probability is the value of α. If the test is two-tailed we must double the obtained value to get α.

EXAMPLE 1 Twelve subjects are given a test on perceptive awareness and are then given an intensive course designed to improve perceptive awareness. This course is followed by a retest. Their scores are

* See for instance Sidney Siegel: *Non-parametric Statistics for the Behavioral Sciences*, McGraw-Hill Book Company, New York, 1956.

Subject	*Before*	*After*	*d*
A	81	82	+1
B	76	84	+8
C	53	50	−3
D	71	79	+8
E	66	78	+12
F	59	73	+14
G	88	84	−4
H	73	79	+6
I	80	97	+17
J	66	74	+8
K	58	51	−7
L	70	69	−1

Use the signs test to test the hypothesis that the course improves perceptive awareness.

SOLUTION Using the experimental procedure outlined before, we have:

1. H_0: The intensive course has no effect on perceptive awareness.

   H_1: The course improves perceptive awareness.

2. $\alpha = 0.05$ is a good choice.

3. Criteria: From Table 9 with $n = 12$, $p = 0.50$, we note that the probability of 10 or more positive values is 0.019, which is less than α, but for 9 or more the probability is 0.073, which is greater than α. Thus we will reject H_0 if N, the number of positive signs, is greater than or equal to 10.

4. Results: $N = 8$

5. Conclusions: Since $8 < 10$, we cannot reject H_0. Since the signs test is not very powerful, we may wish to use a stronger test, such as the Wilcoxon T-test, which is discussed in the next section. If we do obtain a significant result, this result is all the stronger because the test is not very powerful.

EXAMPLE 2 Forty retail outlets belonging to a department store chain are compared with each other with respect to certain characteristics. On the basis of this comparison, the forty outlets are divided into twenty matched pairs. Then the forty outlets are randomly assigned to one of two groups with one member of each matched pair in each group. The groups are called A and B. The managers of group A stores are given an intensive course in improved sales techniques. The managers of group B stores are promised bonuses for increased sales. At the end of six months, sales increases for the stores are compared in order to assess the most effective means of increasing sales. Use the signs test on the following data to determine if either method is more effective. Use the 0.05 level of significance.

	INCREASED SALES (THOUSANDS OF DOLLARS)		
PAIR	*A*	*B*	*Difference*
1	13	7	+6
2	2	4	−2
3	−1	3	−4
4	4	12	−8
5	11	14	−3
6	7	9	−2
7	12	6	+6
8	−3	2	−5
9	−4	−1	−3
10	7	3	+4
11	5	6	−1
12	2	5	−3
13	1	−1	+2
14	0	7	−7
15	5	8	−3
16	2	1	+1
17	3	9	−6
18	4	5	−1
19	−2	1	−3
20	3	5	−2

SOLUTION Using the experimental procedure, we have the following results:

1. H_0: There is no difference between the techniques in increasing sales.

   H_1: There is a difference between the techniques in increasing sales.

2. $\alpha = 0.05$

3. Criteria: Since our alternate hypothesis is two-sided, we look for the value of r for which 0.025 or less is in the tail. From Table 9, for $n = 20$, $p = 0.50$, we see that for $x \geq 15$, the probability is 0.021. Thus we can reject H_0 if the number of the more frequent sign, N, is 15 or more.

4. Results: $N = 15$

5. Conclusions: Since $N \geq 15$, we reject H_0 and conclude that a system of bonuses is more effective in increasing sales than an intensive training course.

For large samples, generally considered $n > 25$ for the signs test, the normal approximation to the binomial may be used, correcting for continuity. Since $p = 0.50$ for this, we have the mean equal to $\frac{1}{2}n$, the standard deviation equal to $\frac{1}{2}\sqrt{n}$. The actual value of z can be computed using the formula

$$z = \frac{x - \frac{1}{2}n}{\frac{1}{2}\sqrt{n}}$$

where x has been corrected for continuity. The value obtained can then be compared to the appropriate critical value for z.

NOTE As mentioned before, in the event of ties, all tie scores are dropped from the calculations and n is reduced accordingly.

Problems

1. Fifteen pairs of subjects were carefully matched by age, sex, intelligence, and family background, and were randomly divided into 2 groups, 1 of each pair in each group. One group was given a short course in hygienic awareness, the other group was not. At the end of 6 months, each group was given a test for hygienic awareness and the results recorded. High scores indicate more hygienic awareness. Use the signs test to determine if the course was effective. Use the 0.05 level of significance.

	SCORE	
PAIR NUMBER	*Instructed*	*Not Instructed*
1	13	10
2	19	19
3	16	14
4	17	19
5	10	8
6	22	17
7	16	12
8	27	14
9	15	11
10	13	17
11	18	11
12	21	13
13	15	15
14	20	17
15	12	15

2. A medical technologist noted a marked deterioration in a sample of plasma in the presence of certain radiations. She matched 10 pairs of plasma samples by type, RH factors, and antigen groups, and exposed 1 of each pair to the radiations. She then observed the amount of deterioration in each group. Test, at the 0.05 level, whether the radiation increased deterioration rate, using the signs test. Higher scores indicate a higher rate of deterioration.

PAIR NUMBER	DETERIORATION RATE *Exposed*	*Not Exposed*
1	0.32	0.19
2	0.18	0.21
3	0.28	0.21
4	0.34	0.17
5	0.22	0.24
6	0.17	0.12
7	0.41	0.24
8	0.23	0.28
9	0.30	0.18
10	0.26	0.17

THE WILCOXON T-TEST

In the event that the signs test is inconclusive, the **Wilcoxon T-test**, a somewhat more powerful nonparametric test, makes use of the magnitude of the differences. To perform this test, rank the absolute differences from smallest to largest, retaining the sign of the difference after ranking. The sum of the absolute values of the less frequent sign is designated T. This is then compared with Table 8. If the obtained value of T is less than the critical value, the probability is less than α (for a one-tailed test) that a type I error has been made. For a two-tailed test, as usual, the value of α is double that given in the table.

EXAMPLE 1 Test the perceptive awareness of example 1 of the previous section using the Wilcoxon T-test.

SOLUTION The table is reproduced below

Subject	*Before*	*After*	*d*	*Rank of d*	*Less Frequent Ranks*
A	81	82	1	1.5	
B	76	84	8	8	
C	53	50	−3	−3	3
D	71	79	8	8	
E	66	78	12	10	
F	59	73	14	11	
G	88	84	−4	−4	4
H	73	79	6	5	
I	80	97	17	12	
J	66	74	8	8	
K	58	51	−7	−6	6
L	70	69	−1	−1.5	1.5
Total					14.5

Differences are ranked from lowest to highest without regard to sign. The sign of the difference is affixed. Ties are given the mean rank of all the numbers. In this example, a difference of 8 occupies ranks 7, 8, 9, so each is assigned rank 8. Differences of zero are ignored.

Now there are fewer negative than positive differences, so the sum of the negative ranks is computed, and this is the value of T. In this case, then, $T = 14.5$. Using the experimental procedure we have:

1. H_0: The intensive course has no effect on perceptive awareness.

   H_1: The course improves intensive awareness.

2. $\alpha = 0.05$ is a good choice.

3. Criteria: The alternate hypothesis is one-sided, so this is a one-tailed test. From Table 8, we shall reject H_0 if $T < 17$.

4. Results: $T = 14.5$

5. Conclusions: Since $14.5 < 17$, we reject H_0 and conclude that the course is effective.

If the assumption of normality is valid, which it may be in this case, we could use a t-test. Here we would have $t = 2.228$ which is significant at a 0.05 level.

For samples greater than 25 a transformation may be used to link T with the normal distribution so that the obtained value of z may be compared with the usual critical values of z. This transformation is

$$z = \frac{T - \dfrac{N(N+1)}{4}}{\sqrt{\dfrac{N(N+1)(2N+1)}{24}}}$$

where T is the sum of the less frequent ranks and N is the sample size ($N > 25$).

Problems

1. Use the Wilcoxon T-test with the data of problem 1 of the preceding section.
2. Use the Wilcoxon T-test with the data of problem 2 of the preceding section.
3. A group of 30 individuals suffering from low blood glucose was given before and after blood serology tests to determine if a certain diet increased glucose levels in the blood. Given the following data, use the Wilcoxon T-test to determine, at the 0.05 level of significance, if the diet can be expected to raise glucose levels in the population from which the sample was drawn.

	BLOOD GLUCOSE LEVEL	
INDIVIDUAL	*Before Diet*	*After Diet*
1	68	74
2	81	80
3	76	79
4	66	74
5	79	84
6	63	60
7	81	74
8	74	82
9	71	79
10	67	73
11	70	76
12	64	73
13	71	65
14	78	83
15	67	79
16	75	83
17	63	66
18	74	69
19	80	73

INDIVIDUAL	BLOOD GLUCOSE LEVEL	
	Before Diet	*After Diet*
20	75	88
21	72	74
22	76	73
23	81	97
24	72	81
25	68	70
26	74	73
27	81	79
28	68	84
29	74	81
30	72	90

THE MANN-WHITNEY *U*-TEST

For independent samples, a nonparametric technique can be used if the normality assumption is violated. This alternative, the **Mann-Whitney *U*-Test**, also makes use of ranks. In this test the data in the two samples are ranked in order from highest to lowest as if they were one sample, and the sum of the ranks in each sample calculated. Then the U statistic is calculated from the formula

$$U = n_1 n_2 + \frac{n_1(n_1 + 1)}{2} - R_1$$

or

$$U = n_1 n_2 + \frac{n_2(n_2 + 1)}{2} - R_2$$

where n_1 and n_2 are the sizes of the samples and R_1 and R_2 are the rank-sums of the corresponding samples. For small samples, if both n_1 and n_2 are less than 10, special tables must be used,* and if the smaller of the two numbers is less than the corresponding critical value, H_0 can be rejected. If the samples are not small, the distribution of U can be related to the standard normal curve by the statistic

$$z = \frac{U - \dfrac{n_1 n_2}{2}}{\sqrt{\dfrac{n_1 n_2(n_1 + n_2 + 1)}{12}}}$$

* See for instance Sidney Siegel: *op cit*. pp. 274 ff.

EXAMPLE 1 Two groups of subjects are rated for social graces by three judges on a scale of 5, and each person assigned a "social grace" score which is the sum of the three numbers. The two groups come from different backgrounds, but the judges do not know which comes from which. The investigator wishes to test the hypothesis that there is no difference between the groups but has no hypothesis of direction of difference if there is one. His data is given here and he uses $\alpha = 0.05$. What conclusions can he draw?

BACKGROUND A		BACKGROUND B	
Score	*Rank*	*Score*	*Rank*
7	21	8	19
11	8.5	9	15.5
9	15.5	13	3.5
4	23	14	1.5
8	19	11	8.5
6	22	10	12
12	5.5	12	5.5
11	8.5	14	1.5
9	15.5	13	3.5
10	12	9	15.5
11	8.5	10	12
		8	19
	$R_1 = 159.0$		$R_2 = 117.0$

SOLUTION We have the following procedure:

1. H_0: There is no difference between the groups.

   H_1: There is a difference between the groups.

2. $\alpha = 0.05$

3. Criteria: Since n_1 and n_2 are both at least ten, we can use the normal approximation, so we will reject H_0 if $z > 1.96$ or if $z < -1.96$.

4. Results:

$$U = (11 \cdot 12) + \frac{11(11 + 1)}{2} - 159$$

$$= 198 - 159$$

$$= 39$$

Then

$$z = \frac{39 - \dfrac{11 \cdot 12}{2}}{\sqrt{\dfrac{(11)(12)(11 + 12 + 1)}{12}}} = \frac{-27}{\sqrt{264}} \doteq -1.66$$

5. Conclusion. We fail to reject H_0. If possible, take another sample.

COMMENT In relating U to the normal curve, one value of U will give a positive z, the other a negative z, but both will give the same absolute value. NOTE: Tied observations are again given the mean of the common ranks.

Problems

1. A biologist investigating the toxicity of a certain substance is also concerned with finding a chemical which will retard the substance's toxic properties. He puts equal amounts of the substance into two tanks filled with goldfish and then puts a predetermined amount of a chemical into tank A. He observes how many hours pass before the goldfish succumb to the effects of the toxic substance. The following gives the number of hours of survival of each goldfish in the two tanks after the tanks have been infected.

Tank A	55, 57, 61, 63, 65, 66, 66, 67, 68, 68, 69, 70, 70, 72, 76
Tank B	48, 52, 54, 55, 57, 60, 63, 65, 68, 69, 70, 70, 71, 73, 75

Can he assert, at $\alpha = 0.05$, that the chemical retards the toxic properties of the substance? Use the U-test.

2. A business manager wants to place an order for carbon paper and wishes to test two different brands, which cost the same, to determine if there is a difference in the number of copies each brand will make. She has a secretary randomly use twenty of each brand and record the number of copies made by each carbon before it becomes unusable. Using the data below, use the U-test and a 0.05 level of significance to determine whether it is reasonable to maintain that there is no difference in the number of copies for each brand.

Brand A: 16, 14, 11, 7, 13, 14, 18, 12, 14, 16, 14, 11, 12, 16, 14, 13, 11, 13, 16, 17

Brand B: 14, 17, 12, 14, 15, 18, 12, 11, 13, 8, 15, 19, 14, 17, 15, 16, 12, 17, 11, 13

THE KRUSKAL-WALLIS H-TEST

If several independent samples are involved, analysis of variance is the usual procedure. Lack of normality makes its value doubtful, so an alternative technique was developed called the **Kruskal-Wallis One-Way Analysis of Variance,** or the ***H*-test.**

As is done in the Mann-Whitney U-test, all data are ranked as if they were in one sample, then the rank-sums, $R_1, R_2, R_3, \ldots, R_k$ of each sample are calculated. The H-statistic is calculated from the formula

$$H = \frac{12}{N(N+1)}\left(\frac{R_1^2}{n_1} + \frac{R_2^2}{n_2} + \cdots + \frac{R_k^2}{n_k}\right) - 3(N+1)$$

where $n_1, n_2, \ldots, n_k$ are the number in each of k samples, $N = n_1 + n_2 + \cdots + n_k$, and $R_1, R_2, \ldots, R_k$ are the rank-sums of each sample. If there are ties, the usual procedure is followed, but since H is fairly sensitive to ties, if there are very many of them a correction must be made.* For small samples, H must be compared to critical values in a table,† but if there are at least 5 in each sample, H is distributed approximately as the Chi-Square with $k - 1$ degrees of freedom, and Table 4 can be used.

EXAMPLE 1 Suppose that the social graces groups were regrouped into 3 groups, the third group intermediate between the others, with the following results:

BACKGROUND X		BACKGROUND Y		BACKGROUND Z	
Score	*Rank*	*Score*	*Rank*	*Score*	*Rank*
7	21	8	19	11	8.5
9	15.5	13	3.5	12	5.5
4	23	14	1.5	9	15.5
8	19	11	8.5	9	15.5
6	22	12	5.5	10	12
11	8.5	14	1.5	10	12
10	12	13	3.5	8	19
11	8.5	9	15.5		
	$R_1 = 129.5$		$R_2 = 58.5$		$R_3 = 88.0$

Test the hypothesis at $\alpha = 0.05$.

* Siegel: *op cit.*, p. 188.
† Siegel: *op cit.*, pp. 282–283.

SOLUTION Applying the experimental procedure, we have:

1. H_0: There are no differences among the groups.
   H_1: There are differences among the groups.
2. $\alpha = 0.05$
3. Criteria: Since there are at least 5 in each sample, the sampling distribution for χ^2 can be used. Thus, reject H_0 if $H > 5.991$.
4. Results:

$$H = \frac{12}{23(24)}\left[\frac{(129.5)^2}{8} + \frac{(58.5)^2}{8} + \frac{(88)^2}{7}\right] - 3(24)$$

$$\doteq \frac{1}{46}(3{,}630.3482) - 72$$

$$\doteq 6.921$$

5. Conclusions: Since $6.921 > 5.991$, we reject H_0 and conclude that there are differences among the groups.

Note that re-classification of the subjects greatly increased the sensitivity of the testing procedure.

Problems

1. Residual levels of DDT, in parts per billion, were measured in the blood of fish, for samples taken from four estuaries of a certain bay. Use the H-test to determine, at the 0.05 level of significance, whether we can conclude that there is no difference among the groups in DDT blood level.

DDT LEVELS, PPB IN THE BLOOD

Estuary			
A	*B*	*C*	*D*
15	6	26	16
11	21	11	28
27	9	9	41
9	13	17	27
33	11	7	16
16	10	24	22
22	15	18	18
28	13	14	37
11	17	13	26
21	12	17	19
17	8	15	32
22	13	19	27

2. A businessman wishes to use the best sales technique to display and sell a certain product. He displays the product in 3 different displays in each of 3 comparable stores, changing the display weekly. He obtains 6 weekly sales reports on each display (2 at each store, staggering the weeks). Use the H-test to determine if there are differences among the groups at a 0.05 level of significance.

SALES, TOTAL UNITS

	Display	
1	*2*	*3*
86	77	81
79	80	75
83	69	73
81	74	84
75	71	76
79	72	85

SUMMARY

In the event that data are given in ranks, or that one cannot assume that they come from reasonably normally distributed populations, it is possible to use tests which do not make the assumption that the data are normally distributed. These are called **distribution-free** or **nonparametric tests**. Four alternative nonparametric tests were discussed in this appendix.

Instead of the paired or correlated t- or z-test, one may use the **signs test** or the **Wilcoxon *T*-test**. If two samples are independent, the **Mann-Whitney *U*-test** may be used instead of the t- or z-test. Finally, the **Kruskal-Wallis *H*-test** may be used in place of the one-way analysis of variance.

One of the advantages of these nonparametric tests is that they are relatively easy to use. Calculations may be done quickly, and, if significant results are obtained, no further work is necessary. It should be noted, however, that these tests are less powerful than their counterparts which depend on the normality assumption. Using these tests you are more likely to make a type II error. It is usually wise, therefore, to use the standard tests whenever possible, reserving the nonparametric tests for those cases in which the normality assumptions are not valid.

Problems

1. Use the signs test and the T-test for the data of example 1, section 10.4.
2. Use the U-test for the data of example 4, section 10.4. For $\alpha \leq 0.056$, U must be less than or equal to 3.

3. Use the signs test and the T-test for the data of problem 2, section 10.4.
4. Use the signs test and the T-test for the data of problem 7, section 10.4.
5. Use the appropriate nonparametric tests for problems 14 and 15 of section 10.6. For two samples of five, U must be less than or equal to 4 in order to be significant at the 0.05 level.
6. Use the H-test for the data of example 2, section 12.2.
7. Use the H-test for the data of problem 3, section 12.2.
8. Use the H-test for the data of example 1, section 12.3.
9. Use the H-test for the data of problem 1, section 12.3.
10. Use the H-test for the data of problem 3, section 12.3.
11. Use the H-test for the data of problem 1, section 12.4.
12. Use the H-test for the data of problem 2, section 12.4.

Tables

TABLE 1

Square Roots

n	$\sqrt{n}$	$\sqrt{10n}$	n	$\sqrt{n}$	$\sqrt{10n}$	n	$\sqrt{n}$	$\sqrt{10n}$
1.00	1.000	3.162	1.13	1.063	3.362	1.26	1.122	3.550
1.01	1.005	3.178	1.14	1.068	3.376	1.27	1.127	3.564
1.02	1.010	3.194	1.15	1.072	3.391	1.28	1.131	3.578
1.03	1.015	3.209	1.16	1.077	3.406	1.29	1.136	3.592
1.04	1.020	3.225	1.17	1.082	3.421	1.30	1.140	3.606
1.05	1.025	3.240	1.18	1.086	3.435	1.31	1.145	3.619
1.06	1.030	3.256	1.19	1.091	3.450	1.32	1.149	3.633
1.07	1.034	3.271	1.20	1.095	3.464	1.33	1.153	3.647
1.08	1.039	3.286	1.21	1.100	3.479	1.34	1.158	3.661
1.09	1.044	3.302	1.22	1.105	3.493	1.35	1.162	3.674
1.10	1.049	3.317	1.23	1.109	3.507	1.36	1.166	3.688
1.11	1.054	3.332	1.24	1.114	3.521	1.37	1.170	3.701
1.12	1.058	3.347	1.25	1.118	3.536	1.38	1.175	3.715

TABLE 1—Square Roots (*continued*)

n	$\sqrt{n}$	$\sqrt{10n}$	n	$\sqrt{n}$	$\sqrt{10n}$	n	$\sqrt{n}$	$\sqrt{10n}$
1.39	1.179	3.728	1.80	1.342	4.243	2.21	1.487	4.701
1.40	1.183	3.742	1.81	1.345	4.254	2.22	1.490	4.712
1.41	1.187	3.755	1.82	1.349	4.266	2.23	1.493	4.722
1.42	1.192	3.768	1.83	1.353	4.278	2.24	1.497	4.733
1.43	1.196	3.782	1.84	1.356	4.290	2.25	1.500	4.743
1.44	1.200	3.795	1.85	1.360	4.301	2.26	1.503	4.754
1.45	1.204	3.808	1.86	1.364	4.313	2.27	1.507	4.764
1.46	1.208	3.821	1.87	1.367	4.324	2.28	1.510	4.775
1.47	1.212	3.834	1.88	1.371	4.336	2.29	1.513	4.785
1.48	1.217	3.847	1.89	1.375	4.347	2.30	1.517	4.796
1.49	1.221	3.860	1.90	1.378	4.359	2.31	1.520	4.806
1.50	1.225	3.873	1.91	1.382	4.370	2.32	1.523	4.817
1.51	1.229	3.886	1.92	1.386	4.382	2.33	1.526	4.827
1.52	1.233	3.899	1.93	1.389	4.393	2.34	1.530	4.837
1.53	1.237	3.912	1.94	1.393	4.405	2.35	1.533	4.848
1.54	1.241	3.924	1.95	1.396	4.416	2.36	1.536	4.858
1.55	1.245	3.937	1.96	1.400	4.427	2.37	1.539	4.868
1.56	1.249	3.950	1.97	1.404	4.438	2.38	1.543	4.879
1.57	1.253	3.962	1.98	1.407	4.450	2.39	1.546	4.889
1.58	1.257	3.975	1.99	1.411	4.461	2.40	1.549	4.899
1.59	1.261	3.987	2.00	1.414	4.472	2.41	1.552	4.909
1.60	1.265	4.000	2.01	1.418	4.483	2.42	1.556	4.919
1.61	1.269	4.012	2.02	1.421	4.494	2.43	1.559	4.930
1.62	1.273	4.025	2.03	1.425	4.506	2.44	1.562	4.940
1.63	1.277	4.037	2.04	1.428	4.517	2.45	1.565	4.950
1.64	1.281	4.050	2.05	1.432	4.528	2.46	1.568	4.960
1.65	1.285	4.062	2.06	1.435	4.539	2.47	1.572	4.970
1.66	1.288	4.074	2.07	1.439	4.550	2.48	1.575	4.980
1.67	1.292	4.087	2.08	1.442	4.561	2.49	1.578	4.990
1.68	1.296	4.099	2.09	1.446	4.572	2.50	1.581	5.000
1.69	1.300	4.111	2.10	1.449	4.583	2.51	1.584	5.010
1.70	1.304	4.123	2.11	1.453	4.593	2.52	1.587	5.020
1.71	1.308	4.135	2.12	1.456	4.604	2.53	1.591	5.030
1.72	1.311	4.147	2.13	1.459	4.615	2.54	1.954	5.040
1.73	1.315	4.159	2.14	1.463	4.626	2.55	1.597	5.050
1.74	1.319	4.171	2.15	1.466	4.637	2.56	1.600	5.060
1.75	1.323	4.183	2.16	1.470	4.648	2.57	1.603	5.070
1.76	1.327	4.195	2.17	1.473	4.658	2.58	1.606	5.079
1.77	1.330	4.207	2.18	1.476	4.669	2.59	1.609	5.089
1.78	1.334	4.219	2.19	1.480	4.680	2.60	1.612	5.099
1.79	1.338	4.231	2.20	1.483	4.690	2.61	1.616	5.109

TABLE 1—Square Roots (*continued*)

n	$\sqrt{n}$	$\sqrt{10n}$	n	$\sqrt{n}$	$\sqrt{10n}$	n	$\sqrt{n}$	$\sqrt{10n}$
2.62	1.619	5.119	3.03	1.741	5.505	3.44	1.855	5.865
2.63	1.622	5.128	3.04	1.744	5.514	3.45	1.857	5.874
2.64	1.625	5.138	3.05	1.746	5.523	3.46	1.860	5.882
2.65	1.628	5.148	3.06	1.749	5.532	3.47	1.863	5.891
2.66	1.631	5.158	3.07	1.752	5.541	3.48	1.865	5.899
2.67	1.634	5.167	3.08	1.755	5.550	3.49	1.868	5.908
2.68	1.637	5.177	3.09	1.758	5.559	3.50	1.871	5.916
2.69	1.640	5.187	3.10	1.761	5.568	3.51	1.873	5.925
2.70	1.643	5.196	3.11	1.764	5.577	3.52	1.876	5.933
2.71	1.646	5.206	3.12	1.766	5.586	3.53	1.879	5.941
2.72	1.649	5.215	3.13	1.769	5.595	3.54	1.882	5.950
2.73	1.652	5.225	3.14	1.772	5.604	3.55	1.884	5.958
2.74	1.655	5.235	3.15	1.775	5.612	3.56	1.887	5.967
2.75	1.658	5.244	3.16	1.778	5.621	3.57	1.889	5.975
2.76	1.661	5.254	3.17	1.780	5.630	3.58	1.892	5.983
2.77	1.664	5.263	3.18	1.783	5.639	3.59	1.894	5.992
2.78	1.667	5.273	3.19	1.786	5.648	3.60	1.897	6.000
2.79	1.670	5.282	3.20	1.789	5.657	3.61	1.900	6.008
2.80	1.673	5.292	3.21	1.792	5.666	3.62	1.903	6.017
2.81	1.676	5.301	3.22	1.794	5.675	3.63	1.905	6.025
2.82	1.679	5.310	3.23	1.797	5.683	3.64	1.908	6.033
2.83	1.682	5.320	3.24	1.800	5.692	3.65	1.910	6.042
2.84	1.685	5.329	3.25	1.803	5.701	3.66	1.913	6.050
2.85	1.688	5.339	3.26	1.806	5.710	3.67	1.916	6.058
2.86	1.691	5.348	3.27	1.808	5.718	3.68	1.918	6.066
2.87	1.694	5.357	3.28	1.811	5.727	3.69	1.921	6.075
2.88	1.697	5.367	3.29	1.814	5.736	3.70	1.924	6.083
2.89	1.700	5.376	3.30	1.817	5.745	3.71	1.926	6.091
2.90	1.703	5.385	3.31	1.819	5.753	3.72	1.929	6.099
2.91	1.706	5.394	3.32	1.822	5.762	3.73	1.931	6.107
2.92	1.709	5.404	3.33	1.825	5.771	3.74	1.934	6.116
2.93	1.712	5.413	3.34	1.828	5.779	3.75	1.936	6.124
2.94	1.715	5.422	3.35	1.830	5.788	3.76	1.939	6.132
2.95	1.718	5.431	3.36	1.833	5.797	3.77	1.942	6.140
2.96	1.720	5.441	3.37	1.836	5.805	3.78	1.944	6.148
2.97	1.723	5.450	3.38	1.838	5.814	3.79	1.947	6.156
2.98	1.726	5.459	3.39	1.841	5.822	3.80	1.949	6.164
2.99	1.729	5.468	3.40	1.844	5.831	3.81	1.952	6.173
3.00	1.732	5.477	3.41	1.847	5.840	3.82	1.954	6.181
3.01	1.735	5.486	3.42	1.849	5.848	3.83	1.957	6.189
3.02	1.738	5.495	3.43	1.852	5.857	3.84	1.960	6.197

TABLE 1—Square Roots (*continued*)

n	$\sqrt{n}$	$\sqrt{10n}$	n	$\sqrt{n}$	$\sqrt{10n}$	n	$\sqrt{n}$	$\sqrt{10n}$
3.85	1.962	6.205	4.26	2.064	6.527	4.67	2.161	6.834
3.86	1.965	6.213	4.27	2.066	6.535	4.68	2.163	6.841
3.87	1.967	6.221	4.28	2.069	6.542	4.69	2.166	6.848
3.88	1.970	6.229	4.29	2.072	6.550	4.70	2.168	6.856
3.89	1.972	6.237	4.30	2.074	6.557	4.71	2.170	6.863
3.90	1.975	6.245	4.31	2.076	6.565	4.72	2.173	6.870
3.91	1.977	6.253	4.32	2.078	6.573	4.73	2.175	6.878
3.92	1.980	6.261	4.33	2.081	6.580	4.74	2.177	6.885
3.93	1.982	6.269	4.34	2.083	6.588	4.75	2.180	6.892
3.94	1.985	6.277	4.35	2.086	6.595	4.76	2.182	6.899
3.95	1.987	6.285	4.36	2.088	6.603	4.77	2.184	6.907
3.96	1.990	6.293	4.37	2.090	6.611	4.78	2.186	6.914
3.97	1.992	6.301	4.38	2.093	6.618	4.79	2.189	6.921
3.98	1.995	6.309	4.39	2.096	6.626	4.80	2.191	6.928
3.99	1.997	6.317	4.40	2.098	6.633	4.81	2.193	6.935
4.00	2.000	6.325	4.41	2.100	6.641	4.82	2.195	6.943
4.01	2.002	6.332	4.42	2.102	6.648	4.83	2.198	6.950
4.02	2.005	6.340	4.43	2.105	6.656	4.84	2.200	6.957
4.03	2.008	6.348	4.44	2.107	6.663	4.85	2.202	6.964
4.04	2.010	6.356	4.45	2.110	6.671	4.86	2.205	6.971
4.05	2.012	6.364	4.46	2.112	6.678	4.87	2.207	6.979
4.06	2.015	6.372	4.47	2.114	6.686	4.88	2.209	6.986
4.07	2.017	6.380	4.48	2.117	6.693	4.89	2.211	6.993
4.08	2.020	6.387	4.49	2.119	6.701	4.90	2.214	7.000
4.09	2.022	6.395	4.50	2.121	6.708	4.91	2.216	7.007
4.10	2.025	6.403	4.51	2.124	6.716	4.92	2.218	7.014
4.11	2.027	6.411	4.52	2.126	6.723	4.93	2.220	7.021
4.12	2.030	6.419	4.53	2.128	6.731	4.94	2.223	7.029
4.13	2.032	6.427	4.54	2.131	6.738	4.95	2.225	7.036
4.14	2.035	6.434	4.55	2.133	6.745	4.96	2.227	7.043
4.15	2.037	6.442	4.56	2.135	6.753	4.97	2.229	7.050
4.16	2.040	6.450	4.57	2.138	6.760	4.98	2.232	7.057
4.17	2.042	6.458	4.58	2.140	6.768	4.99	2.234	7.064
4.18	2.045	6.465	4.59	2.142	6.775	5.00	2.236	7.071
4.19	2.047	6.473	4.60	2.145	6.782	5.01	2.238	7.078
4.20	2.049	6.481	4.61	2.147	6.790	5.02	2.241	7.085
4.21	2.052	6.488	4.62	2.149	6.797	5.03	2.243	7.092
4.22	2.054	6.496	4.63	2.152	6.804	5.04	2.245	7.099
4.23	2.057	6.504	4.64	2.154	6.812	5.05	2.247	7.106
4.24	2.059	6.512	4.65	2.156	6.819	5.06	2.249	7.113
4.25	2.062	6.519	4.66	2.159	6.826	5.07	2.252	7.120

TABLE 1—Square Roots (*continued*)

n	$\sqrt{n}$	$\sqrt{10n}$	n	$\sqrt{n}$	$\sqrt{10n}$	n	$\sqrt{n}$	$\sqrt{10n}$
5.08	2.254	7.127	5.49	2.343	7.409	5.90	2.429	7.681
5.09	2.256	7.134	5.50	2.345	7.416	5.91	2.431	7.688
5.10	2.258	7.141	5.51	2.347	7.423	5.92	2.433	7.694
5.11	2.261	7.148	5.52	2.349	7.430	5.93	2.435	7.701
5.12	2.263	7.155	5.53	2.352	7.436	5.94	2.437	7.707
5.13	2.265	7.162	5.54	2.354	7.443	5.95	2.439	7.714
5.14	2.267	7.169	5.55	2.356	7.450	5.96	2.441	7.720
5.15	2.269	7.176	5.56	2.358	7.457	5.97	2.443	7.727
5.16	2.272	7.183	5.57	2.360	7.463	5.98	2.445	7.733
5.17	2.274	7.190	5.58	2.362	7.470	5.99	2.447	7.740
5.18	2.276	7.197	5.59	2.364	7.477	6.00	2.449	7.746
5.19	2.278	7.204	5.60	2.366	7.483	6.01	2.452	7.752
5.20	2.280	7.211	5.61	2.369	7.490	6.02	2.454	7.759
5.21	2.283	7.218	5.62	2.371	7.596	6.03	2.456	7.765
5.22	2.285	7.225	5.63	2.373	7.503	6.04	2.458	7.772
5.23	2.287	7.232	5.64	2.375	7.510	6.05	2.460	7.778
5.24	2.289	7.239	5.65	2.377	7.517	6.06	2.462	7.785
5.25	2.291	7.246	5.66	2.379	7.523	6.07	2.464	7.791
5.26	2.293	7.253	5.67	2.381	7.530	6.08	2.466	7.797
5.27	2.296	7.259	5.68	2.383	7.537	6.09	2.468	7.804
5.28	2.298	7.266	5.69	2.385	7.543	6.10	2.470	7.810
5.29	2.300	7.273	5.70	2.387	7.550	6.11	2.472	7.817
5.30	2.302	7.280	5.71	2.390	7.556	6.12	2.474	7.823
5.31	2.304	7.287	5.72	2.392	7.563	6.13	2.476	7.829
5.32	2.307	7.294	5.73	2.394	7.570	6.14	2.478	7.836
5.33	2.309	7.301	5.74	2.396	7.576	6.15	2.480	7.842
5.34	2.311	7.308	5.75	2.398	7.583	6.16	2.482	7.849
5.35	2.313	7.314	5.76	2.400	7.589	6.17	2.484	7.855
5.36	2.315	7.321	5.77	2.402	7.596	6.18	2.486	7.861
5.37	2.317	7.328	5.78	2.404	7.603	6.19	2.488	7.868
5.38	2.319	7.335	5.79	2.406	7.609	6.20	2.490	7.874
5.39	2.322	7.342	5.80	2.408	7.616	6.21	2.492	7.880
5.40	2.324	7.348	5.81	2.410	7.622	6.22	2.494	7.887
5.41	2.326	7.355	5.82	2.412	7.629	6.23	2.496	7.893
5.42	2.328	7.362	5.83	2.415	7.635	6.24	2.498	7.899
5.43	2.330	7.369	5.84	2.417	7.642	6.25	2.500	7.906
5.44	2.332	7.376	5.85	2.419	7.649	6.26	2.502	7.912
5.45	2.335	7.382	5.86	2.421	7.655	6.27	2.504	7.918
5.46	2.337	7.389	5.87	2.423	7.662	6.28	2.506	7.925
5.47	2.339	7.396	5.88	2.425	7.668	6.29	2.508	7.931
5.48	2.341	7.403	5.89	2.427	7.675	6.30	2.510	7.937

TABLE 1—Square Roots (*continued*)

n	$\sqrt{n}$	$\sqrt{10n}$	n	$\sqrt{n}$	$\sqrt{10n}$	n	$\sqrt{n}$	$\sqrt{10n}$
6.31	2.512	7.943	6.72	2.592	8.198	7.13	2.670	8.444
6.32	2.514	7.950	6.73	2.594	8.204	7.14	2.672	8.450
6.33	2.516	7.956	6.74	2.596	8.210	7.15	2.674	8.456
6.34	2.518	7.962	6.75	2.598	8.216	7.16	2.676	8.462
6.35	2.520	7.969	6.76	2.600	8.222	7.17	2.678	8.468
6.36	2.522	7.975	6.77	2.602	8.228	7.18	2.680	8.473
6.37	2.524	7.981	6.78	2.604	8.234	7.19	2.681	8.479
6.38	2.526	7.987	6.79	2.606	8.240	7.20	2.683	8.485
6.39	2.528	7.994	6.80	2.608	8.246	7.21	2.685	8.491
6.40	2.530	8.000	6.81	2.610	8.252	7.22	2.687	8.497
6.41	2.532	8.006	6.82	2.612	8.258	7.23	2.689	8.503
6.42	2.534	8.012	6.83	2.613	8.264	7.24	2.691	8.509
6.43	2.536	8.019	6.84	2.615	8.270	7.25	2.693	8.515
6.44	2.538	8.025	6.85	2.617	8.276	7.26	2.694	8.521
6.45	2.540	8.031	6.86	2.619	8.283	7.27	2.696	8.526
6.46	2.542	8.037	6.87	2.621	8.289	7.28	2.698	8.532
6.47	2.544	8.044	6.88	2.623	8.295	7.29	2.700	8.538
6.48	2.546	8.050	6.89	2.625	8.301	7.30	2.702	8.544
6.49	2.548	8.056	6.90	2.627	8.307	7.31	2.704	8.550
6.50	2.550	8.062	6.91	2.629	8.313	7.32	2.706	8.556
6.51	2.551	8.068	6.92	2.631	8.319	7.33	2.707	8.562
6.52	2.553	8.075	6.93	2.632	8.325	7.34	2.709	8.567
6.53	2.555	8.081	6.94	2.634	8.331	7.35	2.711	8.573
6.54	2.557	8.087	6.95	2.636	8.337	7.36	2.713	8.579
6.55	2.559	8.093	6.96	2.638	8.343	7.37	2.715	8.585
6.56	2.561	8.099	6.97	2.640	8.349	7.38	2.717	8.591
6.57	2.563	8.106	6.98	2.642	8.355	7.39	2.718	8.597
6.58	2.565	8.112	6.99	2.644	8.361	7.40	2.720	8.602
6.59	2.567	8.118	7.00	2.646	8.367	7.41	2.722	8.608
6.60	2.569	8.124	7.01	2.648	8.373	7.42	2.724	8.614
6.61	2.571	8.130	7.02	2.650	8.379	7.43	2.726	8.620
6.62	2.573	8.136	7.03	2.651	8.385	7.44	2.728	8.626
6.63	2.575	8.142	7.04	2.653	8.390	7.45	2.729	8.631
6.64	2.577	8.149	7.05	2.655	8.396	7.46	2.731	8.637
6.65	2.579	8.155	7.06	2.657	8.402	7.47	2.733	8.643
6.66	2.581	8.161	7.07	2.659	8.408	7.48	2.735	8.649
6.67	2.583	8.167	7.08	2.661	8.414	7.49	2.737	8.654
6.68	2.585	8.173	7.09	2.663	8.420	7.50	2.739	8.660
6.69	2.587	8.179	7.10	2.665	8.426	7.51	2.740	8.666
6.70	2.588	8.185	7.11	2.666	8.432	7.52	2.742	8.672
6.71	2.590	8.191	7.12	2.668	8.438	7.53	2.744	8.678

TABLE 1—Square Roots (*continued*)

n	$\sqrt{n}$	$\sqrt{10n}$	n	$\sqrt{n}$	$\sqrt{10n}$	n	$\sqrt{n}$	$\sqrt{10n}$
7.54	2.746	8.683	7.95	2.820	8.916	8.36	2.891	9.143
7.55	2.748	8.689	7.96	2.821	8.922	8.37	2.893	9.149
7.56	2.750	8.695	7.97	2.823	8.927	8.38	2.895	9.154
7.57	2.751	8.701	7.98	2.825	8.933	8.39	2.897	9.160
7.58	2.753	8.706	7.99	2.827	8.939	8.40	2.898	9.165
7.59	2.755	8.712	8.00	2.828	8.944	8.41	2.900	9.171
7.60	2.757	8.718	8.01	2.830	8.950	8.42	2.902	9.176
7.61	2.759	8.724	8.02	2.832	8.955	8.43	2.903	9.182
7.62	2.760	8.729	8.03	2.834	8.961	8.44	2.905	9.187
7.63	2.762	8.735	8.04	2.835	8.967	8.45	2.907	9.192
7.64	2.764	8.741	8.05	2.837	8.972	8.46	2.909	9.198
7.65	2.766	8.746	8.06	2.839	8.978	8.47	2.910	9.203
7.66	2.768	8.752	8.07	2.841	8.983	8.48	2.912	9.209
7.67	2.769	8.758	8.08	2.843	8.989	8.49	2.914	9.214
7.68	2.771	8.764	8.09	2.844	8.994	8.50	2.915	9.220
7.69	2.773	8.769	8.10	2.846	9.000	8.51	2.917	9.225
7.70	2.775	8.775	8.11	2.848	9.006	8.52	2.919	9.230
7.71	2.777	8.781	8.12	2.850	9.011	8.53	2.921	9.236
7.72	2.778	8.786	8.13	2.851	9.017	8.54	2.922	9.241
7.73	2.780	8.792	8.14	2.853	9.022	8.55	2.924	9.247
7.74	2.782	8.798	8.15	2.855	9.028	8.56	2.926	9.252
7.75	2.784	8.803	8.16	2.857	9.033	8.57	2.927	9.257
7.76	2.786	8.809	8.17	2.858	9.039	8.58	2.929	9.263
7.77	2.787	8.815	8.18	2.860	9.044	8.59	2.931	9.268
7.78	2.789	8.820	8.19	2.862	9.050	8.60	2.933	9.274
7.79	2.791	8.826	8.20	2.864	9.055	8.61	2.934	9.279
7.80	2.793	8.832	8.21	2.865	9.061	8.62	2.936	9.284
7.81	2.795	8.837	8.22	2.867	9.066	8.63	2.938	9.290
7.82	2.796	8.843	8.23	2.869	9.072	8.64	2.939	9.295
7.83	2.798	8.849	8.24	2.871	9.077	8.65	2.941	9.301
7.84	2.800	8.854	8.25	2.872	9.083	8.66	2.943	9.306
7.85	2.802	8.860	8.26	2.874	9.088	8.67	2.944	9.311
7.86	2.804	8.866	8.27	2.876	9.094	8.68	2.946	9.317
7.87	2.805	8.871	8.28	2.877	9.099	8.69	2.948	9.322
7.88	2.807	8.877	8.29	2.879	9.105	8.70	2.950	9.327
7.89	2.809	8.883	8.30	2.881	9.110	8.71	2.951	9.333
7.90	2.811	8.888	8.31	2.883	9.116	8.72	2.953	9.338
7.91	2.812	8.894	8.32	2.884	9.121	8.73	2.955	9.343
7.92	2.814	8.899	8.33	2.886	9.127	8.74	2.956	9.349
7.93	2.816	8.905	8.34	2.888	9.132	8.75	2.958	9.354
7.94	2.818	8.911	8.35	2.890	9.138	8.76	2.960	9.359

TABLE 1—Square Roots (*continued*)

n	$\sqrt{n}$	$\sqrt{10n}$	n	$\sqrt{n}$	$\sqrt{10n}$	n	$\sqrt{n}$	$\sqrt{10n}$
8.77	2.961	9.365	9.18	3.030	9.581	9.59	3.097	9.793
8.78	2.963	9.370	9.19	3.032	9.586	9.60	3.099	9.798
8.79	2.965	9.376	9.20	3.033	9.592	9.61	3.100	9.803
8.80	2.966	9.381	9.21	3.035	9.597	9.62	3.102	9.808
8.81	2.968	9.386	9.22	3.036	9.602	9.63	3.103	9.813
8.82	2.970	9.391	9.23	3.038	9.607	9.64	3.105	9.818
8.83	2.972	9.397	9.24	3.040	9.612	9.65	3.106	9.823
8.84	2.973	9.402	9.25	3.041	9.618	9.66	3.108	9.829
8.85	2.975	9.407	9.26	3.043	9.623	9.67	3.110	9.834
8.86	2.977	9.413	9.27	3.045	9.628	9.68	3.111	9.839
8.87	2.978	9.418	9.28	3.046	9.633	9.69	3.113	9.844
8.88	2.980	9.423	9.29	3.048	9.638	9.70	3.114	9.849
8.89	2.982	9.429	9.30	3.050	9.644	9.71	3.116	9.854
8.90	2.983	9.434	9.31	3.051	9.649	9.72	3.118	9.859
8.91	2.985	9.439	9.32	3.053	9.654	9.73	3.119	9.864
8.92	2.987	9.445	9.33	3.055	9.659	9.74	3.121	9.869
8.93	2.988	9.450	9.34	3.056	9.664	9.75	3.122	9.874
8.94	2.990	9.455	9.35	3.058	9.670	9.76	3.124	9.879
8.95	2.992	9.460	9.36	3.059	9.675	9.77	3.126	9.884
8.96	2.993	9.466	9.37	3.061	9.680	9.78	3.127	9.889
8.97	2.995	9.471	9.38	3.063	9.685	9.79	3.129	9.894
8.98	2.997	9.476	9.39	3.064	9.690	9.80	3.130	9.899
8.99	2.998	9.482	9.40	3.066	9.695	9.81	3.132	9.905
9.00	3.000	9.487	9.41	3.068	9.701	9.82	3.134	9.910
9.01	3.002	9.492	9.42	3.069	9.706	9.83	3.135	9.915
9.02	3.003	9.497	9.43	3.071	9.711	9.84	3.137	9.920
9.03	3.005	9.503	9.44	3.072	9.716	9.85	3.138	9.925
9.04	3.007	9.508	9.45	3.074	9.721	9.86	3.140	9.930
9.05	3.008	9.513	9.46	3.076	9.726	9.87	3.142	9.935
9.06	3.010	9.518	9.47	3.077	9.731	9.88	3.143	9.940
9.07	3.012	9.524	9.48	3.079	9.737	9.89	3.145	9.945
9.08	3.013	9.529	9.49	3.081	9.742	9.90	3.146	9.950
9.09	3.015	9.534	9.50	3.082	9.747	9.91	3.148	9.955
9.10	3.017	9.539	9.51	3.084	9.752	9.92	3.150	9.960
9.11	3.018	9.545	9.52	3.085	9.757	9.93	3.151	9.965
9.12	3.020	9.550	9.53	3.087	9.762	9.94	3.153	9.970
9.13	3.022	9.555	9.54	3.089	9.767	9.95	3.154	9.975
9.14	3.023	9.560	9.55	3.090	9.772	9.96	3.156	9.980
9.15	3.025	9.566	9.56	3.092	9.778	9.97	3.157	9.985
9.16	3.027	9.571	9.57	3.094	9.783	9.98	3.159	9.990
9.17	3.028	9.576	9.58	3.095	9.788	9.99	3.161	9.995

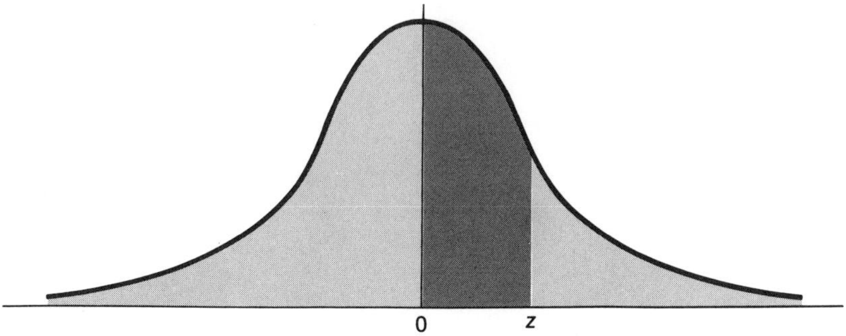

The entries in Table 2 give the area under the standard normal curve between 0 and z (shaded area shown). This is equivalent to the probability that a random variable with a standard normal distribution has a value between 0 and z.

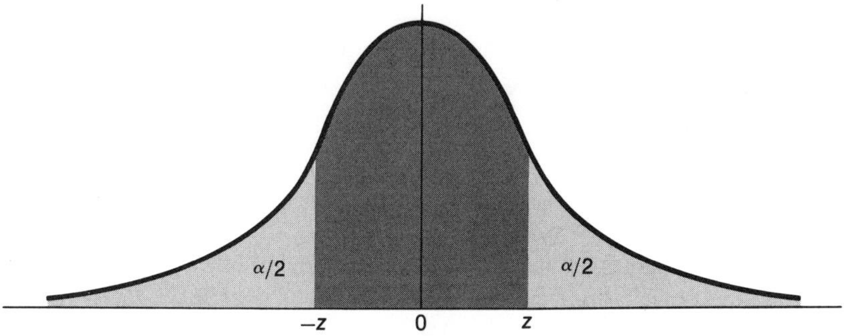

If a random variable has a standard normal distribution, a $1 - \alpha$ confidence interval can be obtained using the value of z for which the area under the standard normal curve between z and $-z$ (shaded area shown) is $1 - \alpha$.

TABLE 2

Areas Under the Normal Curve

z	.00	.01	.02	.03	.04	.05	.06	.07	.08	.09
0.0	.0000	.0040	.0080	.0120	.0160	.0199	.0239	.0279	.0319	.0359
0.1	.0398	.0438	.0478	.0517	.0557	.0596	.0636	.0675	.0714	.0753
0.2	.0793	.0832	.0871	.0910	.0948	.0987	.1026	.1064	.1103	.1141
0.3	.1179	.1217	.1255	.1293	.1331	.1368	.1406	.1443	.1480	.1517
0.4	.1554	.1591	.1628	.1664	.1700	.1736	.1772	.1808	.1844	.1879
0.5	.1915	.1950	.1985	.2019	.2054	.2088	.2123	.2157	.2190	.2224
0.6	.2257	.2291	.2324	.2357	.2389	.2422	.2454	.2486	.2517	.2549
0.7	.2580	.2611	.2642	.2673	.2704	.2734	.2764	.2794	.2823	.2852
0.8	.2881	.2910	.2939	.2967	.2995	.3023	.3051	.3078	.3106	.3133
0.9	.3159	.3186	.3212	.3238	.3264	.3289	.3315	.3340	.3365	.3389
1.0	.3413	.3438	.3461	.3485	.3508	.3531	.3554	.3577	.3599	.3621
1.1	.3643	.3665	.3686	.3708	.3729	.3749	.3770	.3790	.3810	.3830
1.2	.3849	.3869	.3888	.3907	.3925	.3944	.3962	.3980	.3997	.4015
1.3	.4032	.4049	.4066	.4082	.4099	.4115	.4131	.4147	.4162	.4177
1.4	.4192	.4207	.4222	.4236	.4251	.4265	.4279	.4292	.4306	.4319
1.5	.4332	.4345	.4357	.4370	.4382	.4394	.4406	.4418	.4429	.4441
1.6	.4452	.4463	.4474	.4484	.4495	.4505	.4515	.4525	.4535	.4545
1.7	.4554	.4564	.4573	.4582	.4591	.4599	.4608	.4616	.4625	.4633
1.8	.4641	.4649	.4656	.4664	.4671	.4678	.4686	.4693	.4699	.4706
1.9	.4713	.4719	.4726	.4732	.4738	.4744	.4750	.4756	.4761	.4767
2.0	.4772	.4778	.4783	.4788	.4793	.4798	.4803	.4808	.4812	.4817
2.1	.4821	.4826	.4830	.4834	.4838	.4842	.4846	.4850	.4854	.4857
2.2	.4861	.4864	.4868	.4871	.4875	.4878	.4881	.4884	.4887	.4890
2.3	.4893	.4896	.4898	.4901	.4904	.4906	.4909	.4911	.4913	.4916
2.4	.4918	.4920	.4922	.4925	.4927	.4929	.4931	.4932	.4934	.4936
2.5	.4938	.4940	.4941	.4943	.4945	.4946	.4948	.4949	.4951	.4952
2.6	.4953	.4955	.4956	.4957	.4959	.4960	.4961	.4962	.4963	.4964
2.7	.4965	.4966	.4967	.4968	.4969	.4970	.4971	.4972	.4973	.4974
2.8	.4974	.4975	.4976	.4977	.4977	.4978	.4979	.4979	.4980	.4981
2.9	.4981	.4982	.4982	.4983	.4984	.4984	.4985	.4985	.4986	.4987
3.0	.4987	.4987	.4987	.4988	.4988	.4989	.4989	.4989	.4990	.4990
3.1	.4990	.4991	.4991	.4991	.4992	.4992	.4992	.4992	.4993	.4993
3.2	.4993	.4993	.4994	.4994	.4994	.4994	.4994	.4995	.4995	.4995
3.3	.4995	.4995	.4996	.4996	.4996	.4996	.4996	.4996	.4996	.4997
3.4	.4997	.4997	.4997	.4997	.4997	.4997	.4997	.4997	.4998	.4998
3.5	.4998	.4998	.4998	.4998	.4998	.4998	.4998	.4998	.4998	.4998

Table 2 is taken from Table B.2 of *Introduction to Statistical Inference* by E. S. Keeping, Van Nostrand Reinhold.

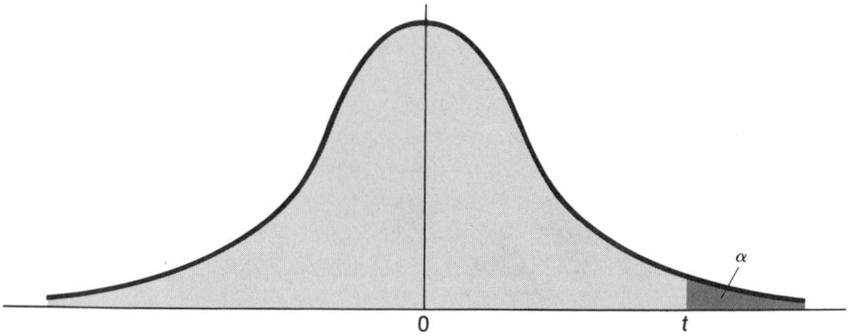

For a one-tailed t-test on a random variable with a t distribution, the probability is α, the proportion in the tail, that t exceeds the critical value given for appropriate degrees of freedom.

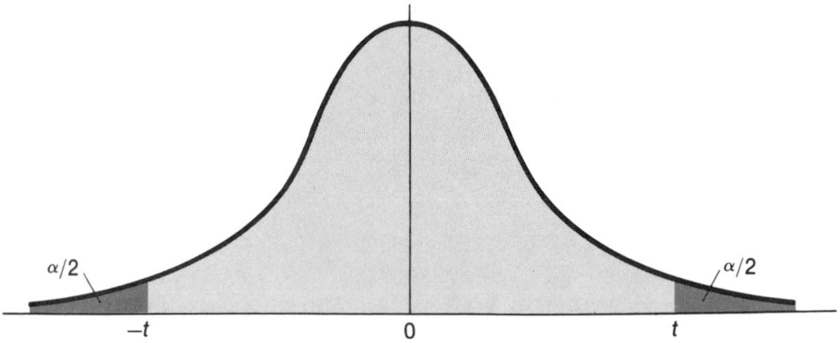

For a two-tailed t test on a random variable with a t distribution, the probability is α, twice the proportion in the tail, that either t exceeds the critical value given or is less than negative of the critical value given for appropriate degrees of freedom.

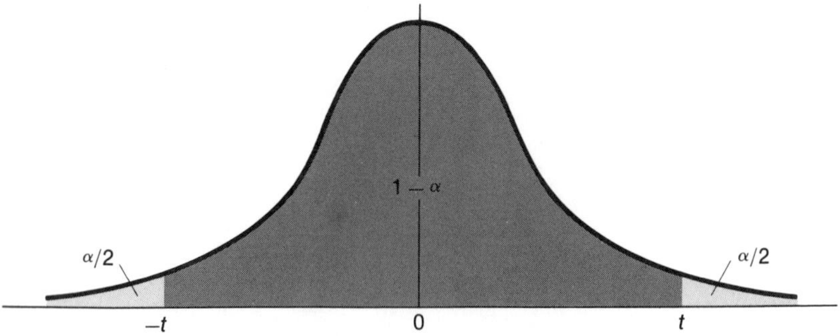

If a random variable has a t distribution, a $1 - \alpha$ confidence interval can be obtained using the value of t for which the area under the curve between t and $-t$ (shaded area shown) is $1 - \alpha$.

TABLE 3

Critical Values of *t*

Degrees of Freedom	*Proportion in the Tail* 0.050	0.025	0.010	0.005
1	6.314	12.706	31.821	63.657
2	2.920	4.303	6.945	9.925
3	2.353	3.182	4.541	5.841
4	2.132	2.776	3.747	4.604
5	2.015	2.571	3.365	4.032
6	1.943	2.447	3.143	3.707
7	1.895	2.365	2.998	3.499
8	1.860	2.306	2.896	3.355
9	1.833	2.262	2.821	3.250
10	1.812	2.228	2.764	3.169
11	1.796	2.201	2.718	3.106
12	1.782	2.179	2.681	3.055
13	1.771	2.160	2.650	3.012
14	1.761	2.145	2.624	2.977
15	1.753	2.131	2.602	2.947
16	1.746	2.120	2.583	2.921
17	1.740	2.110	2.567	2.898
18	1.734	2.101	2.552	2.878
19	1.729	2.093	2.539	2.861
20	1.725	2.086	2.528	2.845
21	1.721	2.080	2.518	2.831
22	1.717	2.074	2.508	2.819
23	1.714	2.069	2.500	2.807
24	1.711	2.064	2.492	2.797
25	1.708	2.060	2.485	2.787
26	1.706	2.056	2.479	2.779
27	1.703	2.052	2.473	2.771
28	1.701	2.048	2.467	2.763
29	1.699	2.045	2.462	2.756
∞	1.645	1.960	2.326	2.576

Table 3 is taken from Table 12 of the *Biometrika Tables for Statisticians*, Volume 1, Third Edition, by Pearson and Hartley.

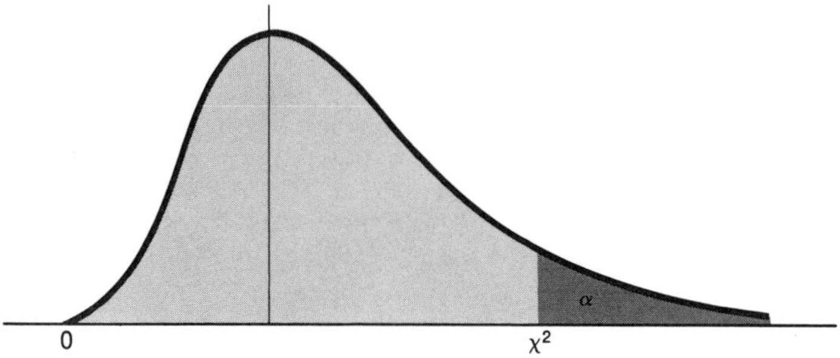

The entries in Table 4 give the values of Chi-Square for which the area to the right of the critical value listed for given degrees of freedom is α (shaded area shown). If a random variable has a Chi-Square distribution, α is the probability that a value of the variable exceeds the critical value.

TABLE 4

Critical Values of Chi-Square

Degrees of Freedom	$\alpha = 0.05$	$\alpha = 0.01$
1	3.841	6.635
2	5.991	9.210
3	7.815	11.345
4	9.488	13.277
5	11.070	15.086
6	12.592	16.812
7	14.067	18.475
8	15.507	20.090
9	16.919	21.666
10	18.307	23.209
11	19.675	24.725
12	21.026	26.217
13	22.362	27.688
14	23.685	29.141
15	24.996	30.578
16	26.296	32.000
17	27.587	33.409
18	28.869	34.805
19	30.144	36.191
20	31.410	37.566
21	32.671	38.932
22	33.924	40.289
23	35.172	41.638
24	36.415	42.980
25	37.652	44.314
26	38.885	45.642
27	40.113	46.963
28	41.337	48.278
29	42.557	49.588
30	43.773	50.892

Table 4 is taken from Table 8 of the *Biometrika Tables for Statisticians*, Volume 1, Third Edition, by Pearson and Hartley.

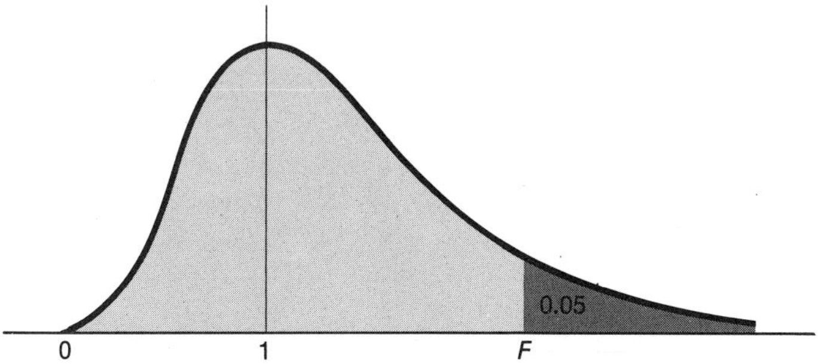

The entries in Table 5 give the values of F for which the area to the right of the critical value listed for given degrees of freedom is 0.05 (shaded area shown). If a random variable has an F distribution, 0.05 is the probability that a value of the variable exceeds the critical value.

TABLE 5

Critical Values of F for $\alpha = 0.05$

Degrees of Freedom Within Groups	*Degrees of Freedom Between Groups*									
	1	*2*	*3*	*4*	*5*	*6*	*7*	*8*	*9*	*10*
1	161	200	216	225	230	234	237	239	241	242
2	18.5	19.0	19.2	19.2	19.3	19.3	19.4	19.4	19.4	19.4
3	10.1	9.55	9.28	9.12	9.01	8.94	8.89	8.85	8.81	8.79
4	7.71	6.94	6.59	6.39	6.26	6.16	6.09	6.04	6.00	5.96
5	6.61	5.79	5.41	5.19	5.05	4.95	4.88	4.82	4.77	4.74
6	5.99	5.14	4.76	4.53	4.39	4.28	4.21	4.15	4.10	4.06
7	5.59	4.74	4.35	4.12	3.97	3.87	3.79	3.73	3.68	3.64
8	5.32	4.46	4.07	3.84	3.69	3.58	3.50	3.44	3.39	3.35
9	5.12	4.26	3.86	3.63	3.48	3.37	3.29	3.23	3.18	3.14
10	4.96	4.10	3.71	3.48	3.33	3.22	3.14	3.07	3.02	2.98
11	4.84	3.98	3.59	3.36	3.20	3.09	3.01	2.95	2.90	2.85
12	4.75	3.89	3.49	3.26	3.11	3.00	2.91	2.85	2.80	2.75
13	4.67	3.81	3.41	3.18	3.03	2.92	2.83	2.77	2.71	2.67
14	4.60	3.74	3.34	3.11	2.96	2.85	2.76	2.70	2.65	2.60
15	4.54	3.68	3.29	3.06	2.90	2.79	2.71	2.64	2.59	2.54
16	4.49	3.63	3.24	3.01	2.85	2.74	2.66	2.59	2.54	2.49
17	4.45	3.59	3.20	2.96	2.81	2.70	2.61	2.55	2.49	2.45
18	4.41	3.55	3.16	2.93	2.77	2.66	2.58	2.51	2.46	2.41
19	4.38	3.52	3.13	2.90	2.74	2.63	2.54	2.48	2.42	2.38
20	4.35	3.49	3.10	2.87	2.71	2.60	2.51	2.45	2.39	2.35
21	4.32	3.47	3.07	2.84	2.68	2.57	2.49	2.42	2.37	2.32
22	4.30	3.44	3.05	2.82	2.66	2.55	2.46	2.40	2.34	2.30
23	4.28	3.42	3.03	2.80	2.64	2.53	2.44	2.37	2.32	2.27
24	4.26	3.40	3.01	2.78	2.62	2.51	2.42	2.36	2.30	2.25
25	4.24	3.39	2.99	2.76	2.60	2.49	2.40	2.34	2.28	2.24
26	4.22	3.37	2.98	2.74	2.59	2.47	2.39	2.32	2.27	2.22
27	4.21	3.35	2.96	2.73	2.57	2.46	2.37	2.30	2.25	2.20
28	4.20	3.34	2.95	2.71	2.56	2.44	2.36	2.29	2.24	2.19
29	4.18	3.33	2.93	2.70	2.54	2.43	2.35	2.28	2.22	2.18
30	4.17	3.32	2.92	2.69	2.53	2.42	2.34	2.27	2.21	2.16
40	4.08	3.23	2.84	2.61	2.45	2.34	2.25	2.18	2.12	2.07
50	4.03	3.18	2.79	2.56	2.40	2.29	2.20	2.13	2.07	2.02
60	4.00	3.15	2.76	2.52	2.37	2.25	2.17	2.10	2.04	1.99
70	3.98	3.13	2.74	2.50	2.35	2.23	2.14	2.07	2.01	1.97
80	3.96	3.11	2.72	2.48	2.33	2.21	2.12	2.05	1.99	1.95
100	3.94	3.09	2.70	2.46	2.30	2.19	2.10	2.03	1.97	1.92
120	3.92	3.07	2.68	2.45	2.29	2.18	2.09	2.02	1.96	1.91
∞	3.84	2.99	2.60	2.37	2.21	2.09	2.01	1.94	1.88	1.83

Table 5 is taken from Table 18 of the *Biometrika Tables for Statisticians*, Volume 1, Third Edition, by Pearson and Hartley.

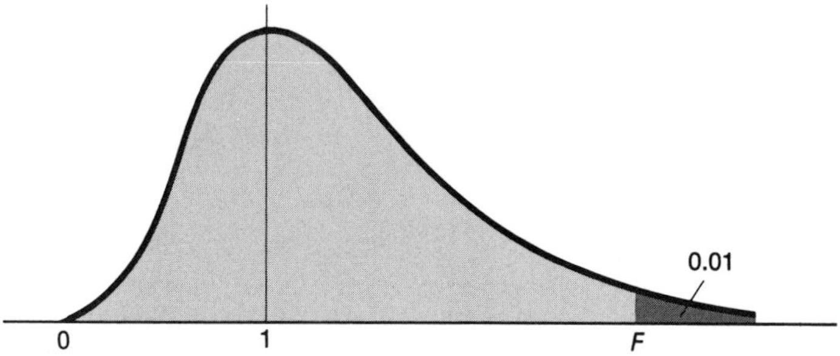

The entries in Table 6 give the values of F for which the area to the right of the critical value listed for given degrees of freedom is 0.01 (shaded area shown). If a random variable has an F distribution, 0.01 is the probability that a value of the variable exceeds the critical value.

TABLE 6

Critical Values of F for $\alpha = 0.01$

Degrees of Freedom Within Groups	Degrees of Freedom Between Groups 1	2	3	4	5	6	7	8	9	10
1	4,052	5,000	5,403	5,625	5,764	5,859	5,928	5,982	6,023	6,056
2	98.5	99.0	99.2	99.2	99.3	99.3	99.4	99.4	99.4	99.4
3	34.1	30.8	29.5	28.7	28.2	27.9	27.7	27.5	27.3	27.2
4	21.2	18.0	16.7	16.0	15.5	15.2	15.0	14.8	14.7	14.5
5	16.3	13.3	12.1	11.4	11.0	10.7	10.5	10.3	10.2	10.1
6	13.7	10.9	9.78	9.15	8.75	8.47	8.26	8.10	7.98	7.87
7	12.2	9.55	8.45	7.85	7.46	7.19	6.99	6.84	6.72	6.62
8	11.3	8.65	7.59	7.01	6.63	6.37	6.18	6.03	5.91	5.81
9	10.6	8.02	6.99	6.42	6.06	5.80	5.61	5.47	5.35	5.26
10	10.0	7.56	6.55	5.99	5.64	5.39	5.20	5.06	4.94	4.85
11	9.65	7.21	6.22	5.67	5.32	5.07	4.89	4.74	4.63	4.54
12	9.33	6.93	5.95	5.41	5.06	4.82	4.64	4.50	4.39	4.30
13	9.07	6.70	5.74	5.21	4.86	4.62	4.44	4.30	4.19	4.10
14	8.86	6.51	5.56	5.04	4.70	4.46	4.28	4.14	4.03	3.94
15	8.68	6.36	5.42	4.89	4.56	4.32	4.14	4.00	3.89	3.80
16	8.53	6.23	5.29	4.77	4.44	4.20	4.03	3.89	3.78	3.69
17	8.40	6.11	5.19	4.67	4.34	4.10	3.93	3.79	3.68	3.59
18	8.29	6.01	5.09	4.58	4.25	4.01	3.84	3.71	3.60	3.51
19	8.19	5.93	5.01	4.50	4.17	3.94	3.77	3.63	3.52	3.43
20	8.10	5.85	4.94	4.43	4.10	3.87	3.70	3.56	3.46	3.37
21	8.02	5.78	4.87	4.37	4.04	3.81	3.64	3.51	3.40	3.31
22	7.95	5.72	4.82	4.31	3.99	3.76	3.59	3.45	3.35	3.26
23	7.88	5.66	4.76	4.26	3.94	3.71	3.54	3.41	3.30	3.21
24	7.82	5.61	4.72	4.22	3.90	3.67	3.50	3.36	3.26	3.17
25	7.77	5.57	4.68	4.18	3.86	3.63	3.46	3.32	3.22	3.13
26	7.72	5.53	4.64	4.14	3.82	3.59	3.42	3.29	3.17	3.09
27	7.68	5.49	4.60	4.11	3.79	3.56	3.39	3.26	3.14	3.06
28	7.64	5.45	4.57	4.07	3.76	3.53	3.36	3.23	3.11	3.03
29	7.60	5.42	4.54	4.04	3.73	3.50	3.33	3.20	3.08	3.00
30	7.56	5.39	4.51	4.02	3.70	3.47	3.30	3.17	3.06	2.98
40	7.31	5.18	4.31	3.83	3.51	3.29	3.12	2.99	2.89	2.80
50	7.17	5.06	4.20	3.72	3.41	3.18	3.02	2.88	2.78	2.70
60	7.08	4.98	4.13	3.65	3.34	3.12	2.95	2.82	2.72	2.63
70	7.01	4.92	4.08	3.60	3.29	3.07	2.91	2.77	2.67	2.59
80	6.96	4.88	4.04	3.56	3.25	3.04	2.87	2.74	2.64	2.55
100	6.90	4.82	3.98	3.51	3.20	2.99	2.82	2.69	2.59	2.51
120	6.85	4.79	3.95	3.48	3.17	2.96	2.79	2.66	2.56	2.47
∞	6.63	4.61	3.78	3.32	3.02	2.80	2.64	2.51	2.41	2.32

Table 6 is taken from Table 18 of the *Biometrika Tables for Statisticians*, Volume 1, Third Edition, by Pearson and Hartley.

If the population correlation coefficient (ρ) is equal to zero, the probability is α, the proportion in the tail, that a sample r is greater than the critical value given for a one-tailed test with appropriate degrees of freedom.

If the population correlation coefficient (ρ) is equal to zero, the probability is α, twice the proportion in the tail, that a sample r is either greater than the critical value or less than the negative of the critical value given for a two-tailed test with appropriate degrees of freedom.

TABLE 7

Critical Values of *r*

Degrees of Freedom	Proportion in the Tail 0.050	0.025	0.010	0.005
1	.9877	.9969	.9995	.9999
2	.9000	.9500	.9800	.9900
3	.8054	.8783	.9343	.9587
4	.7293	.8114	.8822	.9172
5	.6694	.7545	.8329	.8745
6	.6215	.7067	.7887	.8343
7	.5822	.6664	.7498	.7977
8	.5494	.6319	.7155	.7646
9	.5214	.6021	.6851	.7348
10	.4973	.5760	.6581	.7079
11	.4762	.5529	.6339	.6835
12	.4575	.5324	.6120	.6614
13	.4409	.5139	.5923	.6411
14	.4259	.4973	.5742	.6226
15	.4124	.4821	.5577	.6055
16	.4000	.4683	.5425	.5897
17	.3887	.4555	.5285	.5751
18	.3783	.4438	.5155	.5614
19	.3687	.4329	.5034	.5487
20	.3598	.4227	.4921	.5368
25	.3233	.3809	.4451	.4869
30	.2960	.3494	.4093	.4487
35	.2746	.3246	.3810	.4182
40	.2573	.3044	.3578	.3932
45	.2428	.2875	.3384	.3721
50	.2306	.2732	.3218	.3541
60	.2108	.2500	.2948	.3248
70	.1954	.2319	.2737	.3017
80	.1829	.2172	.2565	.2830
90	.1726	.2050	.2422	.2673
100	.1638	.1946	.2301	.2540

Table 7 is taken from Table VI of Fisher and Yates: *Statistical Tables for Biological, Agricultural, and Medical Research*, published by Oliver & Boyd, Edinburgh, and by permission of the authors and publishers.

TABLE 8

Critical Values of *T*

Sample Size *n*	*Probability that T is less than the Critical Value* 0.050	0.025	0.010	0.005
5	1			
6	2	1		
7	4	2		
8	6	4	2	
9	8	6	3	2
10	11	8	5	3
11	14	11	7	5
12	17	14	10	7
13	21	17	13	10
14	26	21	16	13
15	30	25	20	16
16	36	30	24	19
17	41	35	28	23
18	47	40	33	28
19	54	46	38	32
20	60	52	43	37
21	68	59	49	43
22	75	66	56	49
23	83	73	62	55
24	92	81	69	61
25	101	90	77	68

Table 8 is taken from *Some Rapid Approximate Statistical Procedures* (1964) p. 28 by F. Wilcoxon and R. A. Wilcox. It is reproduced with the kind permission of R. A. Wilcox and Lederle Laboratories.

If there is no difference in the population or populations from which the sample or samples were obtained, the probability is the probability shown that a sample value of T will be *less than* the critical value given for a one-tailed test with a sample size n and a one-sided alternative.

If there is no difference in the population or populations from which the sample or samples were obtained, the probability is twice the probability shown that a sample value of T will be *less than* the critical value given for a two-tailed test with a sample size n and a two-sided alternative.

TABLE 9—Cumulative Binomial Probabilities

		p													
n	*r*	.01	.05	.10	.20	.30	.40	.50	.60	.70	.80	.90	.95	.99	*r*
2	0	1	1	1	1	1	1	1	1	1	1	1	1	1	0
	1	020	098	190	360	510	640	750	840	910	960	990	998	1−	1
	2	0+	002	010	040	090	160	250	360	490	640	810	902	980	2
3	0	1	1	1	1	1	1	1	1	1	1	1	1	1	0
	1	030	143	271	488	657	784	875	936	973	992	999	1−	1−	1
	2	0+	007	028	104	216	352	500	648	784	896	972	993	1−	2
	3	0+	0+	001	008	027	064	125	216	343	512	729	857	970	3
4	0	1	1	1	1	1	1	1	1	1	1	1	1	1	0
	1	039	185	344	590	760	870	938	974	992	998	1−	1−	1−	1
	2	001	014	052	181	348	525	688	821	916	973	996	1−	1−	2
	3	0+	0+	004	027	084	179	312	475	652	819	948	986	999	3
	4	0+	0+	0+	002	008	026	062	130	240	410	656	815	961	4
5	0	1	1	1	1	1	1	1	1	1	1	1	1	1	0
	1	049	226	410	672	832	922	969	990	998	1−	1−	1−	1−	1
	2	001	023	081	263	472	663	812	913	969	993	1−	1−	1−	2
	3	0+	001	009	058	163	317	500	683	837	942	991	999	1−	3
	4	0+	0+	0+	007	031	087	188	337	528	737	919	977	999	4
	5	0+	0+	0+	0+	002	010	031	078	168	328	590	774	951	5
6	0	1	1	1	1	1	1	1	1	1	1	1	1	1	0
	1	059	265	469	738	882	953	984	996	999	1−	1−	1−	1−	1
	2	001	033	114	345	580	767	891	959	989	998	1−	1−	1−	2
	3	0+	002	016	099	256	456	656	821	930	983	999	1−	1−	3
	4	0+	0+	001	017	070	179	344	544	744	901	984	998	1−	4
	5	0+	0+	0+	002	011	041	109	233	420	655	886	967	999	5
	6	0+	0+	0+	0+	001	004	016	047	118	262	531	735	941	6

TABLE 9—Cumulative Binomial Probabilities (*continued*)

		p													
n	*r*	.01	.05	.10	.20	.30	.40	.50	.60	.70	.80	.90	.95	.99	*r*
7	0	1	1	1	1	1	1	1	1	1	1	1	1	1	0
	1	068	302	522	790	918	972	992	998	1−	1−	1−	1−	1−	1
	2	002	044	150	423	671	841	938	981	996	1−	1−	1−	1−	2
	3	0+	004	026	148	353	580	773	904	971	995	1−	1−	1−	3
	4	0+	0+	003	033	126	290	500	710	874	967	997	1−	1−	4
	5	0+	0+	0+	005	029	096	227	420	647	852	974	996	1−	5
	6	0+	0+	0+	0+	004	019	062	159	329	577	850	956	998	6
	7	0+	0+	0+	0+	0+	002	008	028	082	210	478	698	932	7
8	0	1	1	1	1	1	1	1	1	1	1	1	1	1	0
	1	077	337	570	832	942	983	996	999	1−	1−	1−	1−	1−	1
	2	003	057	187	497	745	894	965	991	999	1−	1−	1−	1−	2
	3	0+	006	038	203	448	685	855	950	989	999	1−	1−	1−	3
	4	0+	0+	005	056	194	406	637	826	942	990	1−	1−	1−	4
	5	0+	0+	0+	010	058	174	363	594	806	944	995	1−	1−	5
	6	0+	0+	0+	001	011	050	145	315	552	797	962	994	1−	6
	7	0+	0+	0+	0+	001	009	035	106	255	503	813	943	997	7
	8	0+	0+	0+	0+	0+	001	004	017	058	168	430	663	923	8
9	0	1	1	1	1	1	1	1	1	1	1	1	1	1	0
	1	086	370	613	866	960	990	998	1−	1−	1−	1−	1−	1−	1
	2	003	071	225	564	804	929	980	996	1−	1−	1−	1−	1−	2
	3	0+	008	053	262	537	768	910	975	996	1−	1−	1−	1−	3
	4	0+	001	008	086	270	517	746	901	975	997	1−	1−	1−	4

	5	0+	0+	001	020	099	267	500	733	901	980	999	1−	1−	5
	6	0+	0+	0+	003	025	099	254	483	730	914	992	999	1−	6
	7	0+	0+	0+	0+	004	025	090	232	463	738	947	992	1−	7
	8	0+	0+	0+	0+	0+	004	020	071	196	436	775	929	997	8
	9	0+	0+	0+	0+	0+	0+	002	010	040	134	387	630	914	9
10	0	1	1	1	1	1	1	1	1	1	1	1	1	1	0
	1	096	401	651	893	972	994	999	1−	1−	1−	1−	1−	1−	1
	2	004	086	264	624	851	954	989	998	1−	1−	1−	1−	1−	2
	3	0+	012	070	322	617	833	945	988	998	1−	1−	1−	1−	3
	4	0+	001	013	121	350	618	828	945	989	999	1−	1−	1−	4
	5	0+	0+	002	033	150	367	623	834	953	994	1−	1−	1−	5
	6	0+	0+	0+	006	047	166	377	633	850	967	998	1−	1−	6
	7	0+	0+	0+	001	011	055	172	382	650	879	987	999	1−	7
	8	0+	0+	0+	0+	002	012	055	167	383	678	930	988	1−	8
	9	0+	0+	0+	0+	0+	002	011	046	149	376	736	914	996	9
	10	0+	0+	0+	0+	0+	0+	001	006	028	107	349	599	904	10
11	0	1	1	1	1	1	1	1	1	1	1	1	1	1	0
	1	105	431	686	914	980	996	1−	1−	1−	1−	1−	1−	1−	1
	2	005	102	303	678	887	970	994	999	1−	1−	1−	1−	1−	2
	3	0+	015	090	383	687	881	967	994	999	1−	1−	1−	1−	3
	4	0+	002	019	161	430	704	887	971	996	1−	1−	1−	1−	4
	5	0+	0+	003	050	210	467	726	901	978	998	1−	1−	1−	5
	6	0+	0+	0+	012	078	247	500	753	922	988	1−	1−	1−	6
	7	0+	0+	0+	002	022	099	274	533	790	950	997	1−	1−	7
	8	0+	0+	0+	0+	004	029	113	296	570	839	981	998	1−	8
	9	0+	0+	0+	0+	001	006	033	119	313	617	910	985	1−	9
	10	0+	0+	0+	0+	0+	001	006	030	113	322	697	898	995	10
	11	0+	0+	0+	0+	0+	0+	0+	004	020	086	314	569	895	11

TABLE 9—Cumulative Binomial Probabilities (*continued*)

n	r	p: .01	.05	.10	.20	.30	.40	.50	.60	.70	.80	.90	.95	.99	r
12	0	1	1	1	1	1	1	1	1	1	1	1	1	1	0
	1	114	460	718	931	986	998	1−	1−	1−	1−	1−	1−	1−	1
	2	006	118	341	725	915	980	997	1−	1−	1−	1−	1−	1−	2
	3	0+	020	111	442	747	917	981	997	1−	1−	1−	1−	1−	3
	4	0+	002	026	205	507	775	927	985	998	1−	1−	1−	1−	4
	5	0+	0+	004	073	276	562	806	943	991	999	1−	1−	1−	5
	6	0+	0+	001	019	118	335	613	842	961	996	1−	1−	1−	6
	7	0+	0+	0+	004	039	158	387	665	882	981	999	1−	1−	7
	8	0+	0+	0+	001	009	057	194	438	724	927	996	1−	1−	8
	9	0+	0+	0+	0+	002	015	073	225	493	795	974	998	1−	9
	10	0+	0+	0+	0+	0+	003	019	083	253	558	889	980	1−	10
	11	0+	0+	0+	0+	0+	0+	003	020	085	275	659	882	994	11
	12	0+	0+	0+	0+	0+	0+	0+	002	014	069	282	540	886	12
13	0	1	1	1	1	1	1	1	1	1	1	1	1	1	0
	1	122	487	746	945	990	999	1−	1−	1−	1−	1−	1−	1−	1
	2	007	135	379	766	936	987	998	1−	1−	1−	1−	1−	1−	2
	3	0+	025	134	498	798	942	989	999	1−	1−	1−	1−	1−	3
	4	0+	003	034	253	579	831	954	992	999	1−	1−	1−	1−	4
	5	0+	0+	006	099	346	647	867	968	996	1−	1−	1−	1−	5
	6	0+	0+	001	030	165	426	709	902	982	999	1−	1−	1−	6
	7	0+	0+	0+	007	062	229	500	771	938	993	1−	1−	1−	7
	8	0+	0+	0+	001	018	098	291	574	835	970	999	1−	1−	8
	9	0+	0+	0+	0+	004	032	133	353	654	901	994	1−	1−	9

	10	0+	0+	0+	0+	001	008	046	169	421	747	966	997	1−	10
	11	0+	0+	0+	0+	0+	001	011	058	202	502	866	975	1−	11
	12	0+	0+	0+	0+	0+	0+	002	013	064	234	621	865	993	12
	13	0+	0+	0+	0+	0+	0+	0+	001	010	055	254	513	878	13
14	0	1	1	1	1	1	1	1	1	1	1	1	1	1	0
	1	131	512	771	956	993	999	1−	1−	1−	1−	1−	1−	1−	1
	2	008	153	415	802	953	992	999	1−	1−	1−	1−	1−	1−	2
	3	0+	030	158	552	839	960	994	999	1−	1−	1−	1−	1−	3
	4	0+	004	044	302	645	876	971	996	1−	1−	1−	1−	1−	4
	5	0+	0+	009	130	416	721	910	982	998	1−	1−	1−	1−	5
	6	0+	0+	001	044	219	514	788	942	992	1−	1−	1−	1−	6
	7	0+	0+	0+	012	093	308	605	850	969	998	1−	1−	1−	7
	8	0+	0+	0+	002	031	150	395	692	907	988	1−	1−	1−	8
	9	0+	0+	0+	0+	008	058	212	486	781	956	999	1−	1−	9
	10	0+	0+	0+	0+	002	018	090	279	584	870	991	1−	1−	10
	11	0+	0+	0+	0+	0+	004	029	124	355	698	956	996	1−	11
	12	0+	0+	0+	0+	0+	001	006	040	161	448	842	970	1−	12
	13	0+	0+	0+	0+	0+	0+	001	008	047	198	585	847	992	13
	14	0+	0+	0+	0+	0+	0+	0+	001	007	044	229	488	869	14
15	0	1	1	1	1	1	1	1	1	1	1	1	1	1	0
	1	140	537	794	965	995	1−	1−	1−	1−	1−	1−	1−	1−	1
	2	010	171	451	833	965	995	1−	1−	1−	1−	1−	1−	1−	2
	3	0+	036	184	602	873	973	996	1−	1−	1−	1−	1−	1−	3
	4	0+	005	056	352	703	909	982	998	1−	1−	1−	1−	1−	4
	5	0+	001	013	164	485	783	941	991	999	1−	1−	1−	1−	5
	6	0+	0+	002	061	278	597	849	966	996	1−	1−	1−	1−	6
	7	0+	0+	0+	018	131	390	696	905	985	999	1−	1−	1−	7
	8	0+	0+	0+	004	050	213	500	787	950	996	1−	1−	1−	8
	9	0+	0+	0+	001	015	095	304	610	869	982	1−	1−	1−	9

TABLE 9—Cumulative Binomial Probabilities (*continued*)

		p													
n	*r*	.01	.05	.10	.20	.30	.40	.50	.60	.70	.80	.90	.95	.99	*r*
15	10	0+	0+	0+	0+	004	034	151	403	722	939	998	1−	1−	10
	11	0+	0+	0+	0+	001	009	059	217	515	836	987	999	1−	11
	12	0+	0+	0+	0+	0+	002	018	091	297	648	944	995	1−	12
	13	0+	0+	0+	0+	0+	0+	004	027	127	398	816	964	1−	13
	14	0+	0+	0+	0+	0+	0+	0+	005	035	167	549	829	990	14
	15	0+	0+	0+	0+	0+	0+	0+	0+	005	035	206	463	860	15
16	0	1	1	1	1	1	1	1	1	1	1	1	1	1	0
	1	149	560	815	972	997	1−	1−	1−	1−	1−	1−	1−	1−	1
	2	011	189	485	859	974	997	1−	1−	1−	1−	1−	1−	1−	2
	3	001	043	211	648	901	982	998	1−	1−	1−	1−	1−	1−	3
	4	0+	007	068	402	754	935	989	999	1−	1−	1−	1−	1−	4
	5	0+	001	017	202	550	833	962	995	1−	1−	1−	1−	1−	5
	6	0+	0+	003	082	340	671	895	981	998	1−	1−	1−	1−	6
	7	0+	0+	001	027	175	473	773	942	993	1−	1−	1−	1−	7
	8	0+	0+	0+	007	074	284	598	858	974	999	1−	1−	1−	8
	9	0+	0+	0+	001	026	142	402	716	926	993	1−	1−	1−	9
	10	0+	0+	0+	0+	007	058	227	527	825	973	999	1−	1−	10
	11	0+	0+	0+	0+	002	019	105	329	660	918	997	1−	1−	11
	12	0+	0+	0+	0+	0+	005	038	167	450	798	983	999	1−	12
	13	0+	0+	0+	0+	0+	001	011	065	246	598	932	993	1−	13
	14	0+	0+	0+	0+	0+	0+	002	018	099	352	789	957	999	14
	15	0+	0+	0+	0+	0+	0+	0+	003	026	141	515	811	989	15
	16	0+	0+	0+	0+	0+	0+	0+	0+	003	028	185	440	851	16

17	0	1	1	1	1	1	1	1	1	1	1	1	1	1	0
	1	157	582	833	977	998	1−	1−	1−	1−	1−	1−	1−	1−	1
	2	012	208	518	882	981	998	1−	1−	1−	1−	1−	1−	1−	2
	3	001	050	238	690	923	988	999	1−	1−	1−	1−	1−	1−	3
	4	0+	009	083	451	798	954	994	1−	1−	1−	1−	1−	1−	4
	5	0+	001	022	242	611	874	975	997	1−	1−	1−	1−	1−	5
	6	0+	0+	005	106	403	736	928	989	999	1−	1−	1−	1−	6
	7	0+	0+	001	038	225	552	834	965	997	1−	1−	1−	1−	7
	8	0+	0+	0+	011	105	359	685	908	987	1−	1−	1−	1−	8
	9	0+	0+	0+	003	040	199	500	801	960	997	1−	1−	1−	9
	10	0+	0+	0+	0+	013	092	315	641	895	989	1−	1−	1−	10
	11	0+	0+	0+	0+	003	035	166	448	775	962	999	1−	1−	11
	12	0+	0+	0+	0+	001	011	072	264	597	894	995	1−	1−	12
	13	0+	0+	0+	0+	0+	003	025	126	389	758	978	999	1−	13
	14	0+	0+	0+	0+	0+	0+	006	046	202	549	917	991	1−	4
	15	0+	0+	0+	0+	0+	0+	001	012	077	310	762	950	999	15
	16	0+	0+	0+	0+	0+	0+	0+	002	019	118	482	792	988	16
	17	0+	0+	0+	0+	0+	0+	0+	0+	002	023	167	418	843	17
18	0	1	1	1	1	1	1	1	1	1	1	1	1	1	0
	1	165	603	850	982	998	1−	1−	1−	1−	1−	1−	1−	1−	1
	2	014	226	550	901	986	999	1−	1−	1−	1−	1−	1−	1−	2
	3	001	058	266	729	940	992	999	1−	1−	1−	1−	1−	1−	3
	4	0+	011	098	499	835	967	996	1−	1−	1−	1−	1−	1−	4
	5	0+	002	028	284	667	906	985	999	1−	1−	1−	1−	1−	5
	6	0+	0+	006	133	466	791	952	994	1−	1−	1−	1−	1−	6
	7	0+	0+	001	051	278	626	881	980	999	1−	1−	1−	1−	7
	8	0+	0+	0+	016	141	437	760	942	994	1−	1−	1−	1−	8
	9	0+	0+	0+	004	060	263	593	865	979	999	1−	1−	1−	9

TABLE 9—Cumulative Binomial Probabilities (*continued*)

		p													
n	r	.01	.05	.10	.20	.30	.40	.50	.60	.70	.80	.90	.95	.99	r
18	10	0+	0+	0+	001	021	135	407	737	940	996	1−	1−	1−	10
	11	0+	0+	0+	0+	006	058	240	563	859	984	1−	1−	1−	11
	12	0+	0+	0+	0+	001	020	119	374	722	949	999	1−	1−	12
	13	0+	0+	0+	0+	0+	006	048	209	534	867	994	1−	1−	13
	14	0+	0+	0+	0+	0+	001	015	094	333	716	972	998	1−	14
	15	0+	0+	0+	0+	0+	0+	004	033	165	501	902	989	1−	15
	16	0+	0+	0+	0+	0+	0+	001	008	060	271	734	942	999	16
	17	0+	0+	0+	0+	0+	0+	0+	001	014	099	450	774	986	17
	18	0+	0+	0+	0+	0+	0+	0+	0+	002	018	150	397	835	18
19	0	1	1	1	1	1	1	1	1	1	1	1	1	1	0
	1	174	623	865	986	999	1−	1−	1−	1−	1−	1−	1−	1−	1
	2	015	245	580	917	990	999	1−	1−	1−	1−	1−	1−	1−	2
	3	001	067	295	763	954	995	1−	1−	1−	1−	1−	1−	1−	3
	4	0+	013	115	545	867	977	998	1−	1−	1−	1−	1−	1−	4
	5	0+	002	035	327	718	930	990	999	1−	1−	1−	1−	1−	5
	6	0+	0+	009	163	526	837	968	997	1−	1−	1−	1−	1−	6
	7	0+	0+	002	068	334	692	916	988	999	1−	1−	1−	1−	7
	8	0+	0+	0+	023	182	512	820	965	997	1−	1−	1−	1−	8
	9	0+	0+	0+	007	084	333	676	912	989	1−	1−	1−	1−	9
	10	0+	0+	0+	002	033	186	500	814	967	998	1−	1−	1−	10
	11	0+	0+	0+	0+	011	088	324	667	916	993	1−	1−	1−	11
	12	0+	0+	0+	0+	003	035	180	488	818	977	1−	1−	1−	12
	13	0+	0+	0+	0+	001	012	084	308	666	932	998	1−	1−	13
	14	0+	0+	0+	0+	0+	003	032	163	474	837	991	1−	1−	14

	15	0+	0+	0+	0+	0+	001	010	070	282	673	965	998	1−	15
	16	0+	0+	0+	0+	0+	0+	002	023	133	455	885	987	1−	16
	17	0+	0+	0+	0+	0+	0+	0+	005	046	237	705	933	999	17
	18	0+	0+	0+	0+	0+	0+	0+	001	010	083	420	755	985	18
	19	0+	0+	0+	0+	0+	0+	0+	0+	001	014	135	377	826	19
20	0	1	1	1	1	1	1	1	1	1	1	1	1	1	0
	1	182	642	878	988	999	1−	1−	1−	1−	1−	1−	1−	1−	1
	2	017	264	608	931	992	999	1−	1−	1−	1−	1−	1−	1−	2
	3	001	075	323	794	965	996	1−	1−	1−	1−	1−	1−	1−	3
	4	0+	016	133	589	893	984	999	1−	1−	1−	1−	1−	1−	4
	5	0+	003	043	370	762	949	994	1−	1−	1−	1−	1−	1−	5
	6	0+	0+	011	196	584	874	979	998	1−	1−	1−	1−	1−	6
	7	0+	0+	002	087	392	750	942	994	1−	1−	1−	1−	1−	7
	8	0+	0+	0+	032	228	584	868	979	999	1−	1−	1−	1−	8
	9	0+	0+	0+	010	113	404	748	943	995	1−	1−	1−	1−	9
	10	0+	0+	0+	003	048	245	588	872	983	999	1−	1−	1−	10
	11	0+	0+	0+	001	017	128	412	755	952	997	1−	1−	1−	11
	12	0+	0+	0+	0+	005	057	252	596	887	990	1−	1−	1−	12
	13	0+	0+	0+	0+	001	021	132	416	772	968	1−	1−	1−	13
	14	0+	0+	0+	0+	0+	006	058	250	608	913	998	1−	1−	14
	15	0+	0+	0+	0+	0+	002	021	126	416	804	989	1−	1−	15
	16	0+	0+	0+	0+	0+	0+	006	051	238	630	957	997	1−	16
	17	0+	0+	0+	0+	0+	0+	001	016	107	411	867	984	1−	17
	18	0+	0+	0+	0+	0+	0+	0+	004	035	206	677	925	999	18
	19	0+	0+	0+	0+	0+	0+	0+	001	008	069	392	736	983	19
	20	0+	0+	0+	0+	0+	0+	0+	0+	001	012	122	358	818	20

TABLE 9—Cumulative Binomial Probabilities (*continued*)

		p													
n	*r*	.01	.05	.10	.20	.30	.40	.50	.60	.70	.80	.90	.95	.99	*r*
21	0	1	1	1	1	1	1	1	1	1	1	1	1	1	0
	1	190	659	891	991	999	1−	1−	1−	1−	1−	1−	1−	1−	1
	2	019	283	635	942	994	1−	1−	1−	1−	1−	1−	1−	1−	2
	3	001	085	352	821	973	998	1−	1−	1−	1−	1−	1−	1−	3
	4	0+	019	152	630	914	989	999	1−	1−	1−	1−	1−	1−	4
	5	0+	003	052	414	802	963	996	1−	1−	1−	1−	1−	1−	5
	6	0+	0+	014	231	637	904	987	999	1−	1−	1−	1−	1−	6
	7	0+	0+	003	109	449	800	961	996	1−	1−	1−	1−	1−	7
	8	0+	0+	001	043	277	650	905	988	999	1−	1−	1−	1−	8
	9	0+	0+	0+	014	148	476	808	965	998	1−	1−	1−	1−	9
	10	0+	0+	0+	004	068	309	668	915	991	1−	1−	1−	1−	10
	11	0+	0+	0+	001	026	174	500	826	974	999	1−	1−	1−	11
	12	0+	0+	0+	0+	009	085	332	691	932	996	1−	1−	1−	12
	13	0+	0+	0+	0+	002	035	192	524	852	986	1−	1−	1−	13
	14	0+	0+	0+	0+	001	012	095	350	723	957	999	1−	1−	14
	15	0+	0+	0+	0+	0+	004	039	200	551	891	997	1−	1−	15
	16	0+	0+	0+	0+	0+	001	013	096	363	769	986	1−	1−	16
	17	0+	0+	0+	0+	0+	0+	004	037	198	586	948	997	1−	17
	18	0+	0+	0+	0+	0+	0+	001	011	086	370	848	981	1−	18
	19	0+	0+	0+	0+	0+	0+	0+	002	027	179	648	915	999	19
	20	0+	0+	0+	0+	0+	0+	0+	0+	006	058	365	717	981	20
	21	0+	0+	0+	0+	0+	0+	0+	0+	001	009	109	341	810	21

22	0	1	1	1	1	1	1	1	1	1	1	1	1	1	0
	1	198	676	902	993	1−	1−	1−	1−	1−	1−	1−	1−	1−	1
	2	020	302	661	952	996	1−	1−	1−	1−	1−	1−	1−	1−	2
	3	001	095	380	846	979	998	1−	1−	1−	1−	1−	1−	1−	3
	4	0+	022	172	668	932	992	1−	1−	1−	1−	1−	1−	1−	4
	5	0+	004	062	457	835	973	998	1−	1−	1−	1−	1−	1−	5
	6	0+	001	018	267	687	928	992	1−	1−	1−	1−	1−	1−	6
	7	0+	0+	004	133	506	842	974	998	1−	1−	1−	1−	1−	7
	8	0+	0+	001	056	329	710	933	993	1−	1−	1−	1−	1−	8
	9	0+	0+	0+	020	186	546	857	979	999	1−	1−	1−	1−	9
	10	0+	0+	0+	006	092	376	738	945	996	1−	1−	1−	1−	10
	11	0+	0+	0+	002	039	228	584	879	986	1−	1−	1−	1−	11
	12	0+	0+	0+	0+	014	121	416	772	961	998	1−	1−	1−	12
	13	0+	0+	0+	0+	004	055	262	624	908	994	1−	1−	1−	13
	14	0+	0+	0+	0+	001	021	143	454	814	980	1−	1−	1−	14
	15	0+	0+	0+	0+	0+	007	067	290	671	944	999	1−	1−	15
	16	0+	0+	0+	0+	0+	002	026	158	494	867	996	1−	1−	16
	17	0+	0+	0+	0+	0+	0+	008	072	313	733	982	999	1−	17
	18	0+	0+	0+	0+	0+	0+	002	027	165	543	938	996	1−	18
	19	0+	0+	0+	0+	0+	0+	0+	008	068	332	828	978	1−	19
	20	0+	0+	0+	0+	0+	0+	0+	002	021	154	620	905	999	20
	21	0+	0+	0+	0+	0+	0+	0+	0+	004	048	339	698	980	21
	22	0+	0+	0+	0+	0+	0+	0+	0+	0+	007	098	324	802	22
23	0	1	1	1	1	1	1	1	1	1	1	1	1	1	0
	1	206	693	911	994	1−	1−	1−	1−	1−	1−	1−	1−	1−	1
	2	022	321	685	960	997	1−	1−	1−	1−	1−	1−	1−	1−	2
	3	002	105	408	867	984	999	1−	1−	1−	1−	1−	1−	1−	3
	4	0+	026	193	703	946	995	1−	1−	1−	1−	1−	1−	1−	4

TABLE 9—Cumulative Binomial Probabilities (*continued*)

		p													
n	*r*	.01	.05	.10	.20	.30	.40	.50	.60	.70	.80	.90	.95	.99	*r*
23	5	0+	005	073	499	864	981	999	1−	1−	1−	1−	1−	1−	5
	6	0+	001	023	305	731	946	995	1−	1−	1−	1−	1−	1−	6
	7	0+	0+	006	160	560	876	983	999	1−	1−	1−	1−	1−	7
	8	0+	0+	001	072	382	763	953	996	1−	1−	1−	1−	1−	8
	9	0+	0+	0+	027	229	612	895	987	999	1−	1−	1−	1−	9
	10	0+	0+	0+	009	120	444	798	965	998	1−	1−	1−	1−	10
	11	0+	0+	0+	003	055	287	661	919	993	1−	1−	1−	1−	11
	12	0+	0+	0+	001	021	164	500	836	979	999	1−	1−	1−	12
	13	0+	0+	0+	0+	007	081	339	713	945	997	1−	1−	1−	13
	14	0+	0+	0+	0+	002	035	202	556	880	991	1−	1−	1−	14
	15	0+	0+	0+	0+	001	013	105	388	771	973	1−	1−	1−	15
	16	0+	0+	0+	0+	0+	004	047	237	618	928	999	1−	1−	16
	17	0+	0+	0+	0+	0+	001	017	124	440	840	994	1−	1−	17
	18	0+	0+	0+	0+	0+	0+	005	054	269	695	977	999	1−	18
	19	0+	0+	0+	0+	0+	0+	001	019	136	501	927	995	1−	19
	20	0+	0+	0+	0+	0+	0+	0+	005	054	297	807	974	1−	20
	21	0+	0+	0+	0+	0+	0+	0+	001	016	133	592	895	998	21
	22	0+	0+	0+	0+	0+	0+	0+	0+	003	040	315	679	978	22
	23	0+	0+	0+	0+	0+	0+	0+	0+	0+	006	089	307	794	23
24	0	1	1	1	1	1	1	1	1	1	1	1	1	1	0
	1	214	708	920	995	1−	1−	1−	1−	1−	1−	1−	1−	1−	1
	2	024	339	708	967	998	1−	1−	1−	1−	1−	1−	1−	1−	2
	3	002	116	436	885	988	999	1−	1−	1−	1−	1−	1−	1−	3
	4	0+	030	214	736	958	996	1−	1−	1−	1−	1−	1−	1−	4

	5	0+	006	085	540	889	987	999	1−	1−	1−	1−	1−	1−	5
	6	0+	001	028	344	771	960	997	1−	1−	1−	1−	1−	1−	6
	7	0+	0+	007	189	611	904	989	999	1−	1−	1−	1−	1−	7
	8	0+	0+	002	089	435	808	968	998	1−	1−	1−	1−	1−	8
	9	0+	0+	0+	036	275	672	924	992	1−	1−	1−	1−	1−	9
	10	0+	0+	0+	013	153	511	846	978	999	1−	1−	1−	1−	10
	11	0+	0+	0+	004	074	350	729	947	996	1−	1−	1−	1−	11
	12	0+	0+	0+	001	031	213	581	886	988	1−	1−	1−	1−	12
	13	0+	0+	0+	0+	012	114	419	787	969	999	1−	1−	1−	13
	14	0+	0+	0+	0+	004	053	271	650	926	996	1−	1−	1−	14
	15	0+	0+	0+	0+	001	022	154	489	847	987	1−	1−	1−	15
	16	0+	0+	0+	0+	0+	008	076	328	725	964	1−	1−	1−	16
	17	0+	0+	0+	0+	0+	002	032	192	565	911	998	1−	1−	17
	18	0+	0+	0+	0+	0+	001	011	096	389	811	993	1−	1−	18
	19	0+	0+	0+	0+	0+	0+	003	040	229	656	972	999	1−	19
	20	0+	0+	0+	0+	0+	0+	001	013	111	460	915	994	1−	20
	21	0+	0+	0+	0+	0+	0+	0+	004	042	264	786	970	1−	21
	22	0+	0+	0+	0+	0+	0+	0+	001	012	115	564	884	998	22
	23	0+	0+	0+	0+	0+	0+	0+	0+	002	033	292	661	976	23
	24	0+	0+	0+	0+	0+	0+	0+	0+	0+	005	080	292	786	24
25	0	1	1	1	1	1	1	1	1	1	1	1	1	1	0
	1	222	723	928	996	1−	1−	1−	1−	1−	1−	1−	1−	1−	1
	2	026	358	729	973	998	1−	1−	1−	1−	1−	1−	1−	1−	2
	3	002	127	463	902	991	1−	1−	1−	1−	1−	1−	1−	1−	3
	4	0+	034	236	766	967	998	1−	1−	1−	1−	1−	1−	1−	4
	5	0+	007	098	579	910	991	1−	1−	1−	1−	1−	1−	1−	5
	6	0+	001	033	383	807	971	998	1−	1−	1−	1−	1−	1−	6
	7	0+	0+	009	220	659	926	993	1−	1−	1−	1−	1−	1−	7
	8	0+	0+	002	109	488	846	978	999	1−	1−	1−	1−	1−	8
	9	0+	0+	0+	047	323	726	946	996	1−	1−	1−	1−	1−	9

TABLE 9—Cumulative Binomial Probabilities (*continued*)

		p													
n	*r*	.01	.05	.10	.20	.30	.40	.50	.60	.70	.80	.90	.95	.99	*r*
25	10	0+	0+	0+	017	189	575	885	987	1−	1−	1−	1−	1−	10
	11	0+	0+	0+	006	098	414	788	966	998	1−	1−	1−	1−	11
	12	0+	0+	0+	002	044	268	655	922	994	1−	1−	1−	1−	12
	13	0+	0+	0+	0+	017	154	500	846	983	1−	1−	1−	1−	13
	14	0+	0+	0+	0+	006	078	345	732	956	998	1−	1−	1−	14
	15	0+	0+	0+	0+	002	034	212	586	902	994	1−	1−	1−	15
	16	0+	0+	0+	0+	0+	013	115	425	811	983	1−	1−	1−	16
	17	0+	0+	0+	0+	0+	004	054	274	677	953	1−	1−	1−	17
	18	0+	0+	0+	0+	0+	001	022	154	512	891	998	1−	1−	18
	19	0+	0+	0+	0+	0+	0+	007	074	341	780	991	1−	1−	19
	20	0+	0+	0+	0+	0+	0+	002	029	193	617	967	999	1−	20
	21	0+	0+	0+	0+	0+	0+	0+	009	090	421	902	993	1−	21
	22	0+	0+	0+	0+	0+	0+	0+	002	033	234	764	966	1−	22
	23	0+	0+	0+	0+	0+	0+	0+	0+	009	098	537	873	998	23
	24	0+	0+	0+	0+	0+	0+	0+	0+	002	027	271	642	974	24
	25	0+	0+	0+	0+	0+	0+	0+	0+	0+	004	072	277	778	25

Table 9 is taken from Table IV, Part B, of *Probability: A First Course*, Second Edition, 1970, by Mosteller, Rourke, and Thomas, Addison-Wesley.

TABLE 10

Poisson Probabilities

x \ μ	0.5	1	2	3	4	5	6	7	8	9	10
0	0.607	0.368	0.135	0.050	0.018	0.007	0.002	0.001			
1	0.303	0.368	0.271	0.149	0.073	0.034	0.015	0.006	0.003	0.001	
2	0.076	0.184	0.271	0.224	0.147	0.084	0.045	0.022	0.010	0.005	0.002
3	0.013	0.061	0.180	0.224	0.195	0.140	0.089	0.052	0.029	0.015	0.008
4	0.002	0.015	0.090	0.168	0.195	0.175	0.134	0.091	0.057	0.034	0.019
5		0.003	0.036	0.101	0.156	0.175	0.161	0.128	0.092	0.061	0.038
6		0.001	0.012	0.050	0.104	0.146	0.161	0.149	0.122	0.091	0.063
7			0.003	0.022	0.059	0.104	0.138	0.149	0.140	0.117	0.090
8			0.001	0.008	0.030	0.065	0.103	0.130	0.140	0.132	0.113
9				0.003	0.013	0.036	0.069	0.101	0.124	0.132	0.125
10				0.001	0.005	0.018	0.041	0.071	0.099	0.119	0.125
11					0.002	0.008	0.023	0.045	0.072	0.097	0.114
12					0.001	0.003	0.011	0.026	0.048	0.073	0.095
13						0.001	0.005	0.014	0.030	0.050	0.073
14							0.002	0.007	0.017	0.032	0.052
15							0.001	0.003	0.009	0.019	0.035
16								0.001	0.005	0.011	0.022
17								0.001	0.002	0.006	0.013
18									0.001	0.003	0.007
19										0.001	0.004
20										0.001	0.002
21											0.001

TABLE 11

Random Numbers

	1	2	3	4	5	6	7	8	9	10	
1	48461	14952	72619	73689	52059	37086	60050	86192	67049	64739	1
2	76534	38149	49692	31366	52093	15422	20498	33901	10319	43397	2
3	70437	25861	38504	14752	23757	59660	67844	78815	23758	86814	3
4	59584	03370	42806	11393	71722	93804	09095	07856	55589	46020	4
5	04285	58554	16085	51555	27501	73883	33427	33343	45507	50063	5
6	77340	10412	69189	85171	29082	44785	83638	02583	96483	76553	6
7	59183	62687	91778	80354	23512	97219	65921	02035	59847	91403	7
8	91800	04281	39979	03927	82564	28777	59049	97532	54540	79472	8
9	12066	24817	81099	48940	69554	55925	48379	12866	51232	21580	9
10	69907	91751	53512	23748	65906	91385	84983	27915	48491	91068	10
11	80467	04873	54053	25955	48518	13815	37707	68687	15570	08890	11
12	78057	67835	28302	45048	56761	97725	58438	91528	24645	18544	12
13	05648	39387	78191	88415	60269	94880	58812	42931	71898	61534	13
14	22304	39246	01350	99451	61862	78688	30339	60222	74052	25740	14
15	61346	50269	67005	40442	33100	16742	61640	21046	31909	72641	15
16	66793	37696	27965	30459	91011	51426	31006	77468	61029	57108	16
17	86411	48809	36698	42453	83061	43769	39948	87031	30767	13953	17
18	62098	12825	81744	28882	27369	88183	65846	92545	09065	22655	18
19	68775	06261	54265	16203	23340	84750	16317	88686	86842	00879	19
20	52679	19595	13687	74872	89181	01939	18447	10787	76246	80072	20
21	84096	87152	20719	25215	04349	54434	72344	93008	83282	31670	21
22	63964	55937	21417	49944	38356	98404	14850	17994	17161	98981	22
23	31191	75131	72386	11689	95727	05414	88727	45583	22568	77700	23
24	30545	68523	29850	67833	05622	89975	79042	27142	99257	32349	24
25	52573	91001	52315	26430	54175	30122	31796	98842	37600	26025	25

26	16586	81842	01076	99414	31574	94719	34656	80018	86988	79234	26
27	81841	88481	61191	25013	30272	23388	22463	65774	10029	58376	27
28	43563	66829	72838	08074	57080	15446	11034	98143	74989	26885	28
29	19945	84193	57581	77252	85604	45412	43556	27518	90572	00563	29
30	79374	23796	16919	99691	80276	32818	62953	78831	54395	30705	30
31	48503	26615	43980	09810	38289	66679	73799	48418	12647	40044	31
32	32049	65541	37937	41105	70106	89706	40829	40789	59547	00783	32
33	18547	71562	95493	34112	76895	46766	96395	31718	48302	45893	33
34	03180	96742	61486	43305	34183	99605	67803	13491	09243	29557	34
35	94822	24738	67749	83748	59799	25210	31093	62925	72061	69991	35
36	34330	60599	85828	19152	68499	27977	35611	96240	62747	89529	36
37	43770	81537	59527	95674	76692	86420	69930	10020	72881	12532	37
38	56908	77192	50623	41215	14311	42834	80651	93750	59957	31211	38
39	32787	07189	80539	75927	75475	73965	11796	72140	48944	74156	39
40	52441	78392	11733	57703	29133	71164	55355	31006	25526	55790	40
41	22377	54723	18227	28449	04570	18882	00023	67101	06895	08915	41
42	18376	73460	88841	39602	34049	20589	05701	08249	74213	25220	42
43	53201	28610	87957	21497	64729	64983	71551	99016	87903	63875	43
44	34919	78901	59710	27396	02593	05665	11964	44134	00273	76358	44
45	33617	92159	21971	16901	57383	34262	41744	60891	57624	06962	45
46	70010	40964	98780	72418	52571	18415	64362	90636	38034	04909	46
47	19282	68447	35665	31530	59832	49181	21914	65742	89815	39231	47
48	91429	73328	13266	54898	68795	40948	80808	63887	89939	47938	48
49	97637	78393	33021	05867	86520	45363	43066	00988	64040	09803	49
50	95150	07625	05255	83254	93943	52325	93230	62668	79529	65964	50

Answers with Solutions to Selected Problems

CHAPTER 1

SECTION 1.1

1. (a) $168 - 72 = 96$
   (b) $96 \div 12 = 8$
   (c)

161–168	161–170
153–160	151–160
145–152	141–150
137–144	131–140
129–136	121–130
121–128	111–120
113–120	101–110
105–112	91–100
97–104	81–90
89–96	71–80
81–88	
73–80	
65–72	

2.

Sales (in Dollars)	*No. of Outlets*	*Rel. Freq.*	*Cum. Freq.*
10–20	15	15/80	15
21–31	20	20/80	35
32–42	14	14/80	49
43–53	6	6/80	55
54–64	10	10/80	65
65–75	8	8/80	73
76–86	5	5/80	78
87–97	2	2/80	80

3. (a) The range is 43 − 29 or 14, so if we let the class interval be equal to two we will have the required number of intervals.

(b)

Length (cm)	*No. of Mullet*	*Rel. Freq.*	*Cum. Freq.*
29–30	2	2/72	2
31–32	4	4/72	6
33–34	10	10/72	16
35–36	20	20/72	36
37–38	24	24/72	60
39–40	7	7/72	67
41–42	4	4/72	71
43–44	1	1/72	72

Note: We could start at 28 and end at 43. The table would be somewhat different.

4.

Pollution Indices	*No. of Days*	*Rel. Freq.*	*Cum. Freq.*
26–30	2	2/120	2
31–35	3	3/120	5
36–40	5	5/120	10
41–45	16	16/120	26
46–50	15	15/120	41
51–55	15	15/120	56
56–60	20	20/120	76
61–65	13	13/120	89
66–70	15	15/120	104
71–75	7	7/120	111
76–80	2	2/120	113
81–85	4	4/120	117
86–90	3	3/120	120

SECTION 1.2

1.

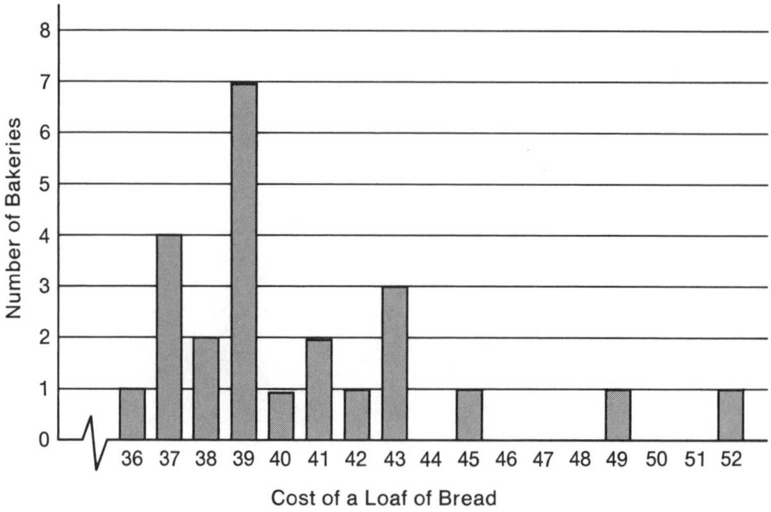

2.

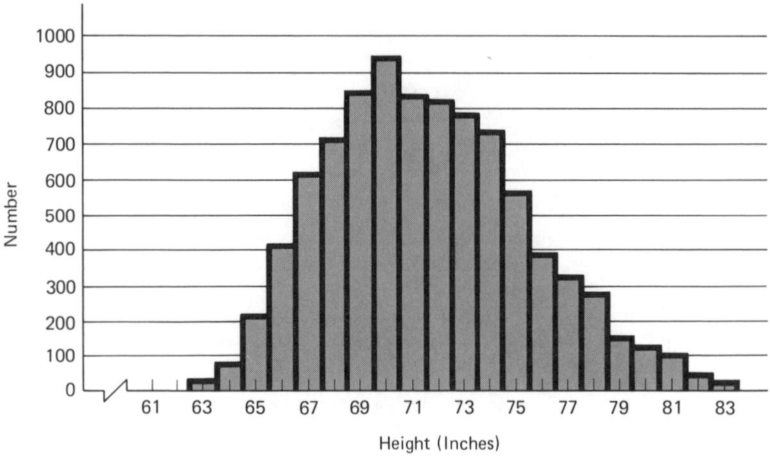

3.

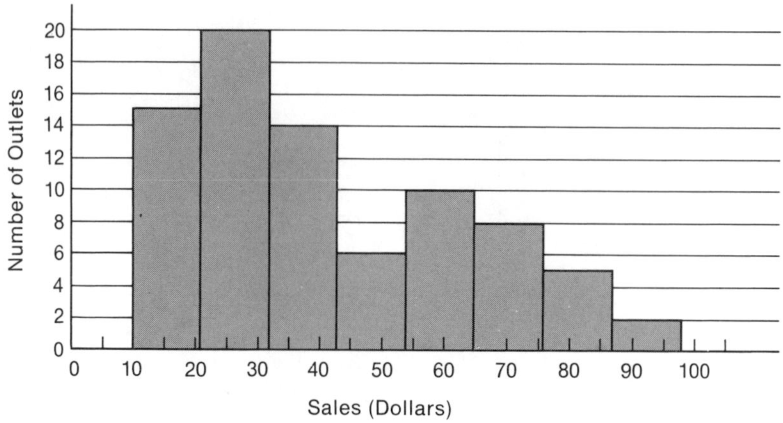
Number of Outlets
20
18
16
14
12
10
8
6
4
2
0
0 10 20 30 40 50 60 70 80 90 100
Sales (Dollars)

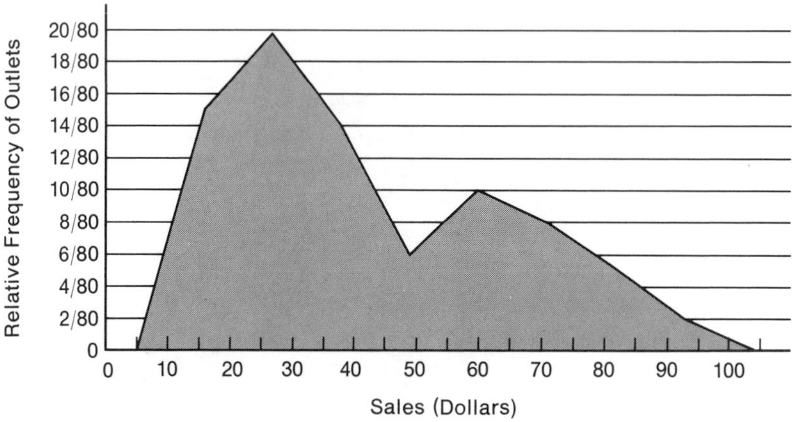
Relative Frequency of Outlets
20/80
18/80
16/80
14/80
12/80
10/80
8/80
6/80
4/80
2/80
0
0 10 20 30 40 50 60 70 80 90 100
Sales (Dollars)

4.

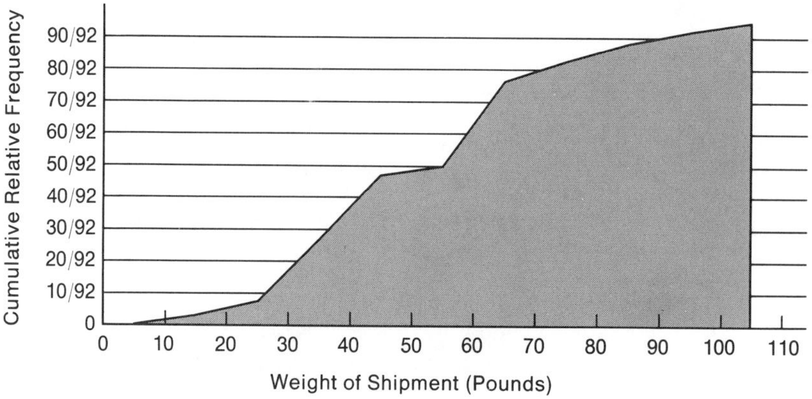
Cumulative Relative Frequency
90/92
80/92
70/92
60/92
50/92
40/92
30/92
20/92
10/92
0
0 10 20 30 40 50 60 70 80 90 100 110
Weight of Shipment (Pounds)

5.

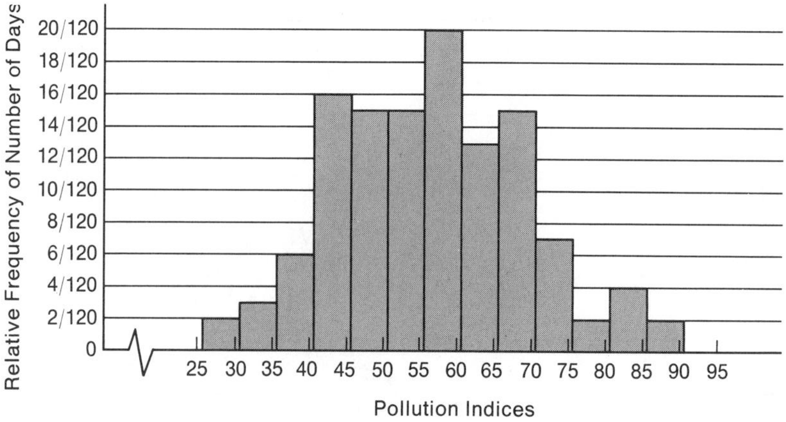
Relative Frequency of Number of Days
20/120
18/120
16/120
14/120
12/120
10/120
8/120
6/120
4/120
2/120
0
25 30 35 40 45 50 55 60 65 70 75 80 85 90 95
Pollution Indices

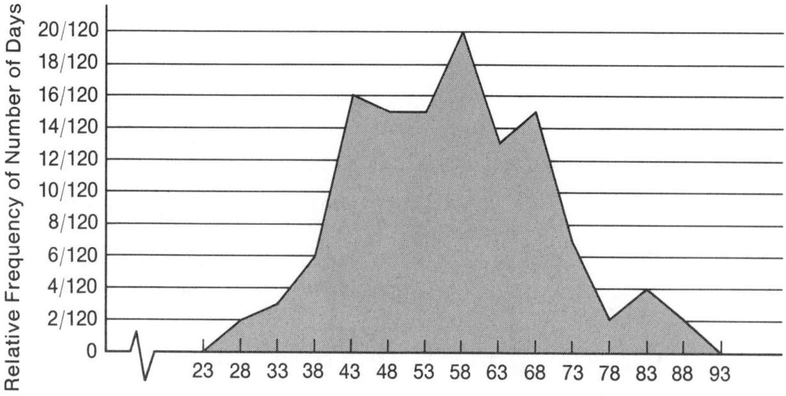
Relative Frequency of Number of Days
20/120
18/120
16/120
14/120
12/120
10/120
8/120
6/120
4/120
2/120
0
23 28 33 38 43 48 53 58 63 68 73 78 83 88 93

6.

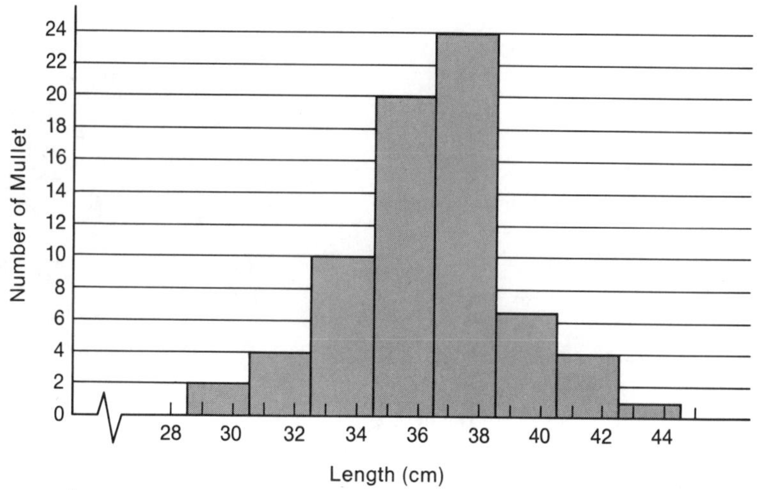
Number of Mullet
24
22
20
18
16
14
12
10
8
6
4
2
0
28
30
32
34
36
38
40
42
44
Length (cm)

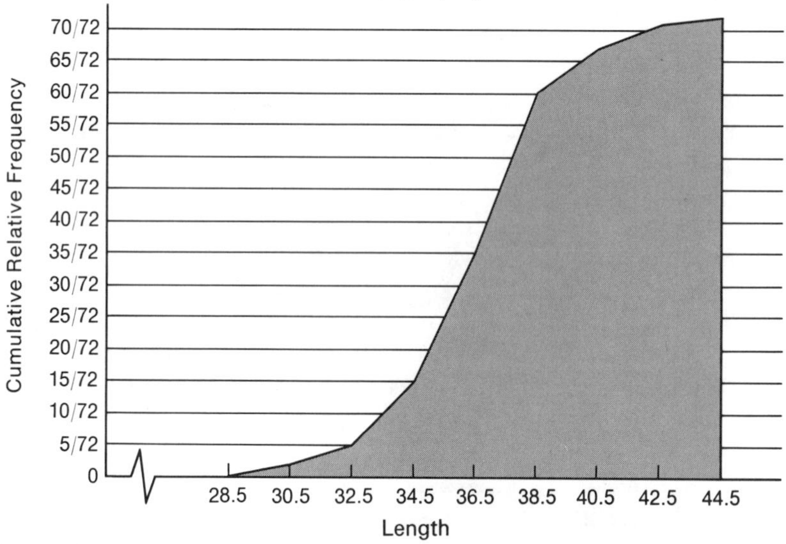
Cumulative Relative Frequency
70/72
65/72
60/72
55/72
50/72
45/72
40/72
35/72
30/72
25/72
20/72
15/72
10/72
5/72
0
28.5
30.5
32.5
34.5
36.5
38.5
40.5
42.5
44.5
Length

7.

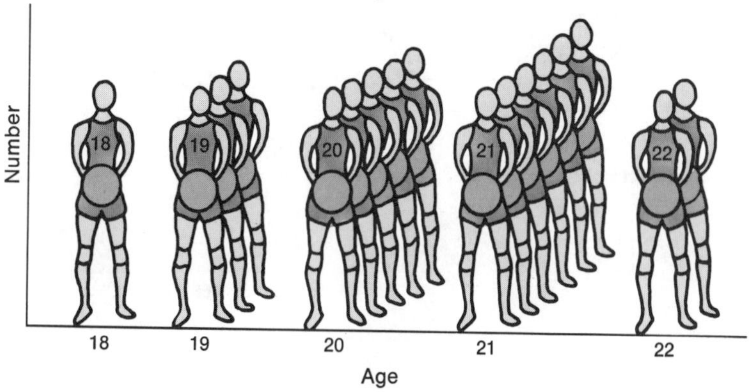
Number
18
19
20
21
22
18
19
20
21
22
Age

8.

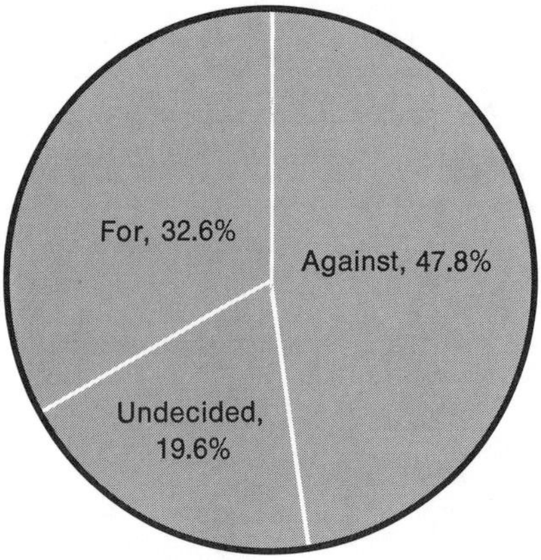
For, 32.6%
Against, 47.8%
Undecided, 19.6%

9. (a)

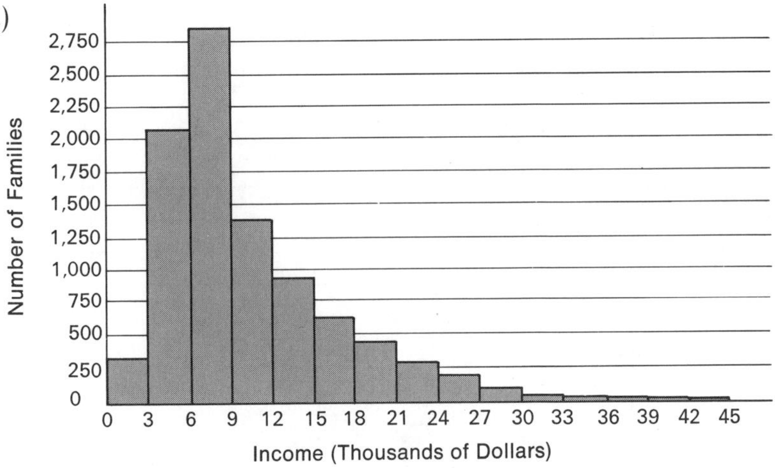
Number of Families
2,750
2,500
2,250
2,000
1,750
1,500
1,250
1,000
750
500
250
0
0 3 6 9 12 15 18 21 24 27 30 33 36 39 42 45
Income (Thousands of Dollars)

(b)

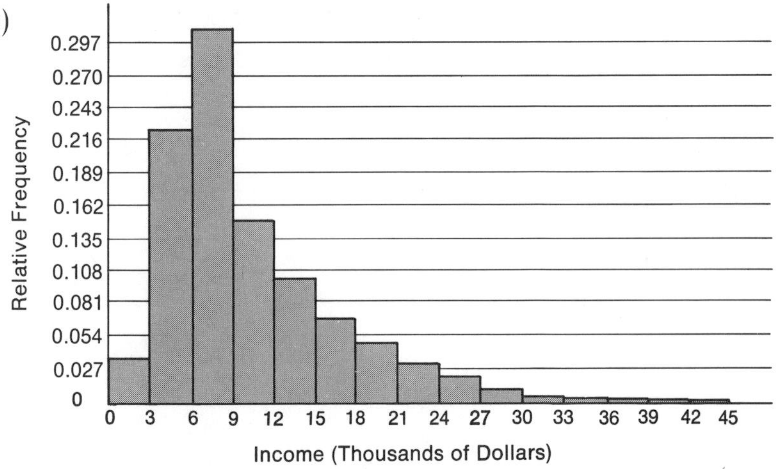
Relative Frequency
0.297
0.270
0.243
0.216
0.189
0.162
0.135
0.108
0.081
0.054
0.027
0
0 3 6 9 12 15 18 21 24 27 30 33 36 39 42 45
Income (Thousands of Dollars)

(c)

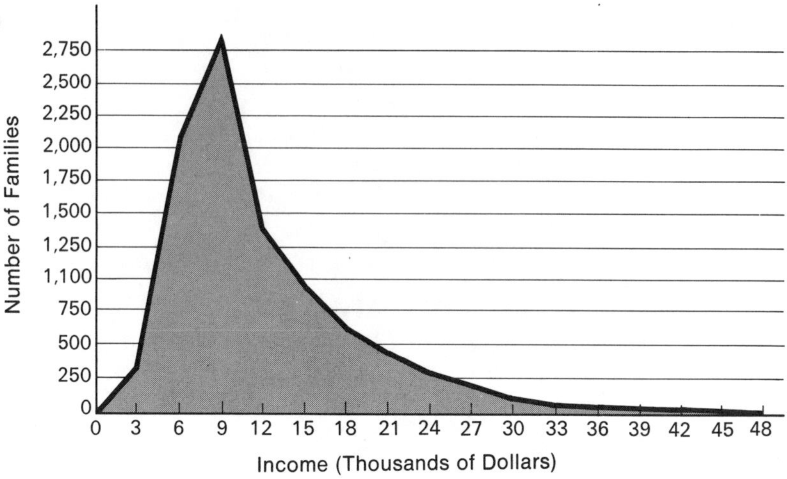
Number of Families
2,750
2,500
2,250
2,000
1,750
1,500
1,250
1,100
750
500
250
0
0 3 6 9 12 15 18 21 24 27 30 33 36 39 42 45 48
Income (Thousands of Dollars)

(d)

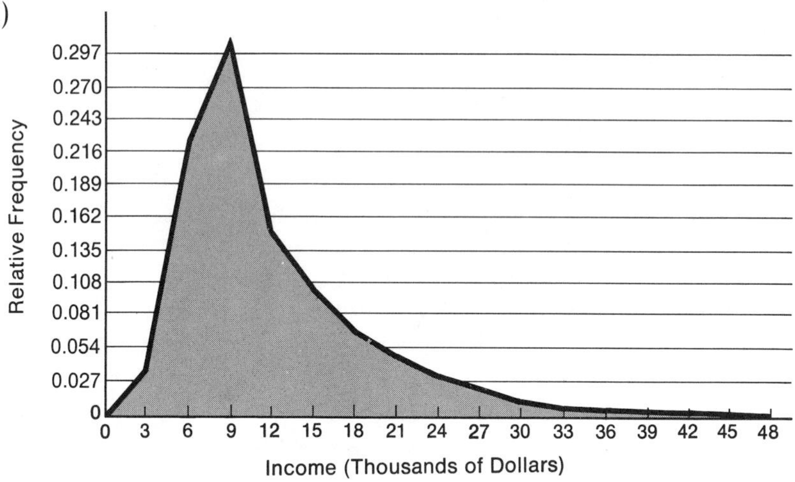
Relative Frequency
0.297
0.270
0.243
0.216
0.189
0.162
0.135
0.108
0.081
0.054
0.027
0
0 3 6 9 12 15 18 21 24 27 30 33 36 39 42 45 48
Income (Thousands of Dollars)

(e)

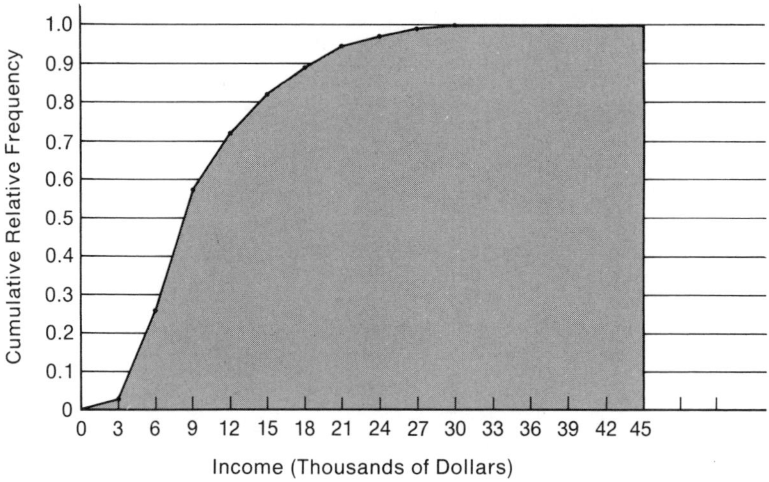

SECTION 1.3

1. 363 − 107 = 256; 256 ÷ 11 ≐ 23; 25 × 11 = 275, so we can conveniently begin at, say, 100.5 and go to 375.5. Thus we might have the following (other answers are possible):

Class	*Class mark*	*The class may also be written as*
100.5–125.5	113	101–125
125.5–150.5	138	126–150
150.5–175.5	163	151–175
175.5–200.5	188	176–200
200.5–225.5	213	201–225
225.5–250.5	238	226–250
250.5–275.5	263	251–275
275.5–300.5	288	276–300
300.5–325.5	313	301–325
325.5–350.5	338	326–350
350.5–375.5	363	351–375

2. (a) yes (b) no (c) yes (d) no (e) no (f) no

3. The class interval is 50, so each class extends 25 units above and below the class mark, so the class limits are 100.5, 150.5, 200.5, 250.5, 300.5, 350.5, 400.5, 450.5, 500.5, 550.5, 600.5.

4. (a) 16.995 and 18.995 (b) 13.995 (c) 2.000 (d) 10.94.

5.

Sales	*Number of Days*
48	1
47	1
46	1
45	0
44	0
43	2
42	3
41	6
40	2
39	0
38	0
37	4
36	1
35	1
34	7
33	2
32	5
31	2
30	2
29	3
28	4
27	2
26	2
25	1
24	4
23	1
22	3

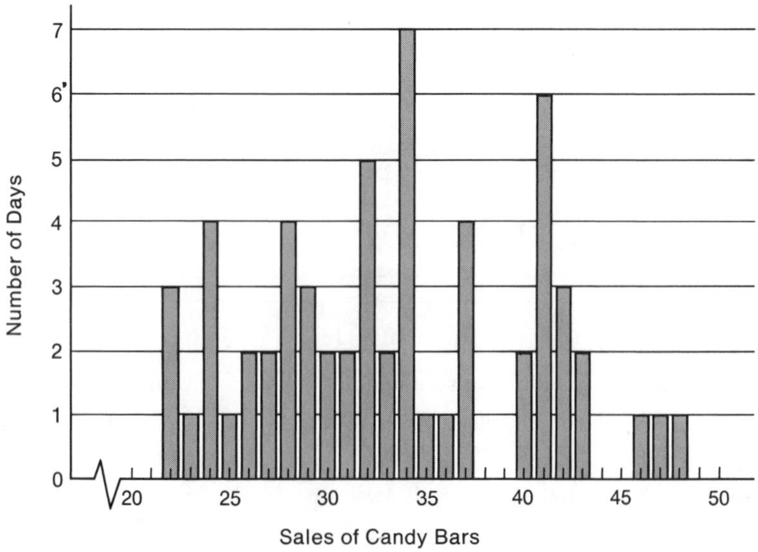

6. (a)

Age	*No. of Patients*	*Cum. Freq.*
1–10	29	29
11–20	10	39
21–30	15	54
31–40	12	66
41–50	10	76
51–60	21	97
61–70	8	105
71–80	8	113
81–90	7	120

(b)

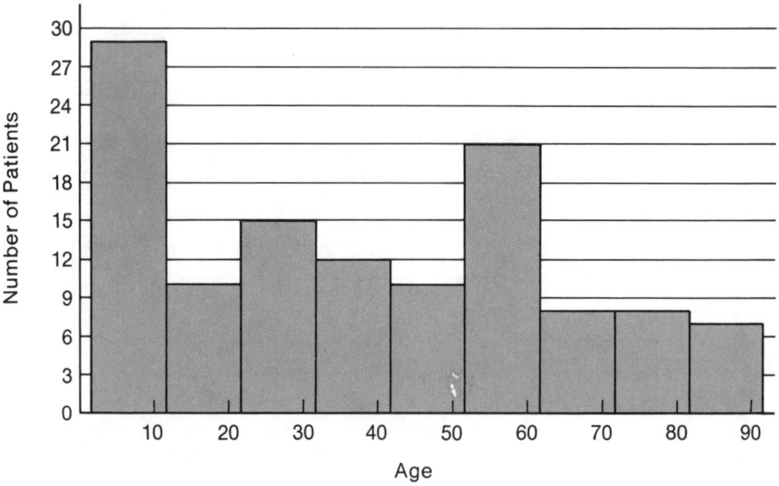

(c)

Age	*Relative Frequency*
1–10	29/120
11–20	10/120
21–30	15/120
31–40	12/120
41–50	10/120
51–60	21/120
61–70	8/120
71–80	8/120
81–90	7/120

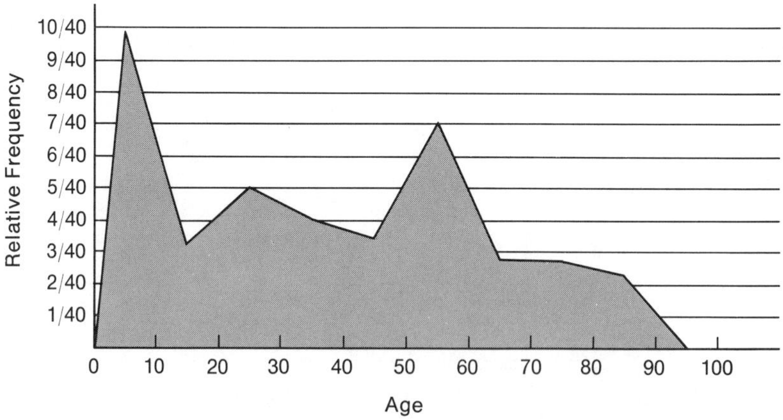

(d)

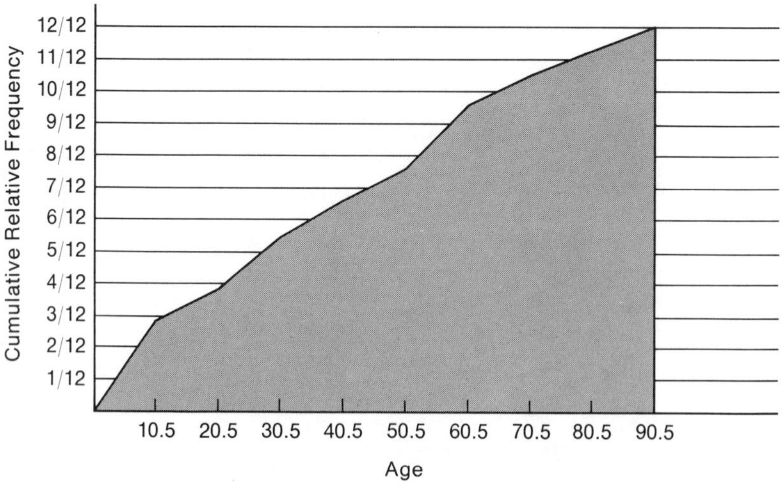

7. The total expenses are \$800, so each expenditure must be represented as a fraction of \$800. If we wish, we may calculate the number of degrees for each one as well. Thus housing is 0.475 (171°), food is 0.1375 (49.5°), utilities, 0.075 (27°), transportation, 0.16875 (60.75°), medical expenses, 0.01875 (6.75°), clothing, 0.03125 (11.25°), savings, 0.03125 (11.25°) and miscellaneous, 0.0625 (22.5°). The chart below is one representation.

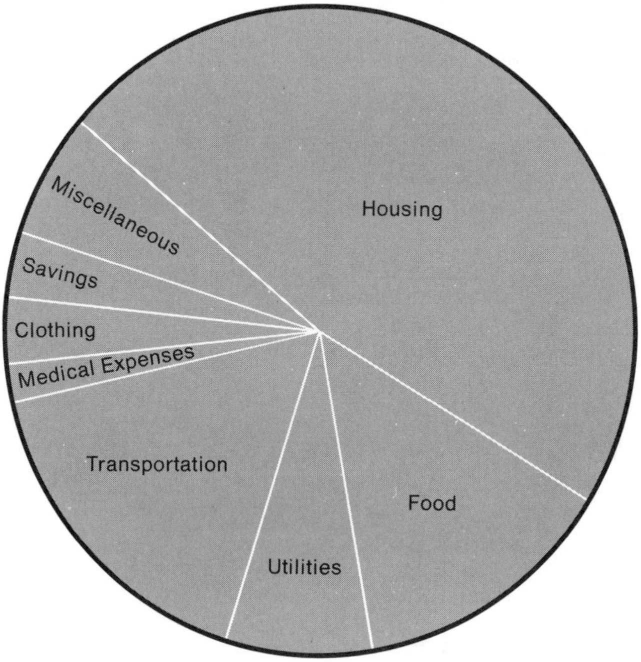

CHAPTER 2

SECTION 2.1

1. $\bar{x} = 25$; $md = (23 + 26)/2 = 24.5$; there is no mode.
2. Using 25 as an assumed mean, we have $\sum (x - 25) = 3$, so $\bar{x} = 25 + \frac{3}{12} = 25.25$; $md = 25$; mode $= 25$.
3. $\bar{x} = 4.31$; $md = 3.02$; no mode.
4. $\bar{x} = 172.51$; $md = 171$; mode $= 170$ and 173.

5. $\bar{x} = 40.5$; $md = 39$; mode $= 39$.

6. $$\bar{x} = \frac{\begin{matrix} 2(11.995) + 7(13.995) + 13(15.995) + 24(17.995) \\ + 11(19.995) + 4(21.995) \end{matrix}}{61}$$

$$= \frac{1{,}069.695}{61} \doteq 17.536 \doteq 17.54;$$

Using the integers, we have

$$\bar{x} = \frac{2(12) + 7(14) + 13(16) + 24(18) + 11(20) + 4(22)}{61}$$

$$= \frac{1{,}070}{61} \doteq 17.541 \doteq 17.54;$$

$md \doteq 17.715$ (or about 17.7); mode $= 17.995 \doteq 18$.

7. From raw scores, $\bar{x} \doteq 71.6$; $md = 71$; mode $= 70$. From the frequency table, $\bar{x} = 71.6$; $md = 71.3$; mode $= 70$.

8. $\bar{x} = 41.0$; $md = 36$; mode $= 37$.

9. $\bar{x} = 36.22$; $md = 36.5$; mode $= 37$.

10. $\bar{x} = 56.7$; $md = 57$; mode $= 54$.

11. $\bar{x} \doteq 33.3$; $md = 33.0$; mode $= 34$.

12. $\bar{x} \doteq 37.2$; $md = 37$; mode $= 7$.

13. $\bar{x} \doteq \$10{,}055.00$; $md \doteq \$8{,}361.00$; mode $= \$7{,}500.00$. (See answer to problem 14, section 2.2.).

SECTION 2.2

1. $s^2 = 16$; $s = 4$.

2. $s^2 \doteq 19.2955$; $s \doteq 4.39$.

3. $s^2 \doteq 9.86$; $s \doteq 3.14$.

4. Q_1 is at measure number 4.25; Q_3 is at 12.75.

Thus $Q_1 = 1.89 + 0.25(2.03 - 1.89) = 1.925 \doteq 1.92$;
$Q_3 = 5.57 + 0.75(6.83 - 5.57) = 6.515 \doteq 6.52$.

Other answers are possible, depending upon the convention used.

5. $s \doteq 14.7$; $Q_1 = 169$, $Q_3 = 174$; $Q = 2.5$.

6. $s \doteq 3.87$.

7. $s \doteq 2.35$.

8. Using raw data, $s \doteq 3.88$; using the frequency table, $s \doteq 3.96$.

9. $s \doteq 22.47$.

10. $s \doteq 2.73$.

11. $s \doteq 13.05$.

12. $s \doteq 6.88$.

13. $s \doteq 25.00$.

14. One solution would be to use coding. Since the mode occurs in the 6,000–8,999 interval, it would be reasonable to use either that interval or the one above it to contain the assumed mean. Probably anything from 7,500 to 11,000 would be reasonable, but since the distribution is skewed to the right, it would tend to be higher. In addition, the class marks will end in 500, so it would be convenient to use the $\hat{x}$ ending in 500, particularly a class mark. The most convenient class mark, then, would be 10,500. Finally, to simplify calculations, we should use the incomes in thousands, so use $\hat{x} = 10.5$. The calculations follow:

x	f	$x - 10.5$	$(x - 10.5) \cdot f$	$(x - 10.5)^2$	$(x - 10.5)^2 \cdot f$
43.5	3	33	99	1,089	3,267
40.5	7	30	210	900	6,300
37.5	12	27	324	729	8,748
34.5	17	24	408	576	9,792
31.5	28	21	588	441	12,348
28.5	60	18	1,080	324	19,440
25.5	216	15	3,240	225	48,600
22.5	268	12	3,216	144	38,592
19.5	448	9	4,032	81	36,288
16.5	621	6	3,726	36	11,556
13.5	949	3	2,847	9	8,541
10.5	1,421	0	0	0	0
7.5	2,846	−3	−8,532	9	22,596
4.5	2,123	−6	−738	36	76,428
1.5	299	−9	−2,646	81	23,814
Total	9,311		−4,146		340,110

Then $\bar{x} = 10.5 + -4{,}146/9{,}311 = 10.05472$ thousands, or about \$10,055,

$$\text{and } s^2 = \frac{340{,}110 - (9{,}311) \cdot (-4{,}146/9{,}311)^2}{9{,}310}$$

$$\doteq \frac{340{,}110 - 1{,}846.13}{9{,}310} = \frac{338{,}253.87}{9{,}310} \doteq 36.3334$$

Thus $s \doteq 6.028$ thousands, or about \$6,028.00. Using the class marks and the raw data formula, we have

$$s = \sqrt{\frac{1{,}279{,}581{,}750{,}000 - (93{,}619{,}500)^2/9{,}311}{9{,}310}} \doteq 6{,}028$$

For Q, we must first determine Q_1 and Q_3. There are 9,312 data points in the set, and $\frac{1}{4}(9{,}312) = 2{,}328$ and $\frac{3}{4}(9{,}312) = 6{,}984$, so we have the location of Q_1 and Q_3. Now Q_1 is in the interval 3,000 to 5,999, above 2,034 of the data points in the interval so

$$Q_1 = \$2{,}999.50 + \frac{2{,}034}{2{,}124}(\$3{,}000)$$

$$\doteq \$2{,}999.50 + \$2{,}872.88$$

$$\doteq \$5{,}872.38$$

or, to the nearest dollar, $Q_1 \doteq \$5{,}872.00$.

Since 6,682 data points lie below \$12,000, Q_3 lies in the interval from 12,000 to 14,999, above 302 of the data points so

$$Q_3 = \$11{,}999.50 + \frac{302}{950}(\$3{,}000.00)$$

$$\doteq \$12{,}953.00$$

Then $Q = \frac{1}{2}(\$12{,}953.00 - \$5{,}872.00) \doteq \$3{,}540.00$.

SECTION 2.3

1. -2.5; -1.92; -1.17; -0.25; 0.5; 1.67; 2.83.

2. p_{10} (the tenth percentile) is 161; $p_{22} = 169$; $p_{43} = 171$; $p_{71} = 174$; $p_{93} = 186$; $p_{96} = 201$.

3. 100.58; 141.04; 181.84; 236.92 (or 101; 141; 182; 237).

4. 25; 50; 62; 88.

5. A: $z = 0.45$; B: $z = 0.40$; C: $z = 0.25$. Best on test A, worst on test C.

6. p_5 is at 18,573 from the bottom, so $p_5 = \$2{,}999.50 + (6{,}809/84{,}914) \times (\$3{,}000.00)$ or about \$3,240.00. p_{27} is at 100,292, so $p_{27} = \$5{,}999.50 + (3{,}615/113{,}753)(\$3{,}000.00) \doteq \$6{,}095.00$. p_{44} is at 163,439, so $p_{44} = \$5{,}999.50 + (66{,}762/113{,}753)(\$3{,}000.00) \doteq \$7{,}760.00$. p_{63} is at 234,015, so $p_{63} = \$8{,}999.50 + (23{,}586/56{,}855)(\$3{,}000.00) \doteq \$10{,}244.00$. p_{82} is at 304,591, so $p_{82} = \$11{,}999.50 + (37{,}308/37{,}957)(\$3{,}000.00) \doteq \$14{,}948.00$. p_{94} is at 349,165, so $p_{94} = \$20{,}999.50 + (1{,}132/10{,}717)(\$3{,}000.00) \doteq \$21{,}316.00$.

SECTION 2.4

1. (a) $\bar{x} = 76.68$; $md = 77$; mode $= 85$.
   (b) $s \doteq 6.40$.
   (c) $Q_1 = 70.5$; $Q_3 = 82$; $Q = 5.75$.
   (d) For 87, $z = 1.61$; for 66, $z = -1.67$.
   (e) $p_{11} = 68$; $p_{40} = 76$; $p_{59} = 78$.

2. (a) $\bar{x} = 48.5$; $md = 47.5$.
   (b) $s = 15.70$.
   (c) $Q_1 = 43$; $Q_3 = 61$; $Q = 9$.
   (d) For 76, $z = 1.75$; for 18, $z = -1.94$.
   (e) $p_{11} = 32$; $p_{40} = 44$; $p_{59} = 52$.

3. (a) $\bar{x} = 12.19$; $md = 11$; mode $= 11$.
   (b) $s \doteq 5.79$.
   (c) $Q_1 = 8$; $Q_3 = 17$; $Q = 4.5$.
   (d) $p_5 = 3$; $p_{28} = 8$; $p_{57} = 13$; $p_{78} = 17$.

4. (a) $\bar{x} = 45.5$; $md = 44.5$; mode $= 44$.
   (b) $s \doteq 8.18$.
   (c) $Q_1 = 39$; $Q_3 = 51$; $Q = 6$.
   (d) $p_{15} = 37$; $p_{44} = 44$; $p_{58} = 47$; $p_{83} = 54$.

5. (1) $V = 16$ (2) $V = 17.39$ (3) $V = 72.85$ (4) $V = 8.53$. Thus distribution (3) is the most variable, distribution (4) the least.

6. (A) $\bar{x} = \$170$, $s \doteq \$20.92$ (B) $\bar{x} = \$215$, $s \doteq \$26.32$

7. For city A, a fee of \$200 was associated with a z of 1.43, while in city B, a fee of \$250 had a z of 1.33; thus the highest fee in city A was farther from

the mean of that city in terms of standard deviations than the corresponding fee in city *B*, hence was most expensive relative to that of the colleagues.

8. For Mr. Jones, $z = 1.2$; for Mr. Adams, $z = 1.0$. Thus Mr. Jones' blood pressure is higher than Mr. Adams', compared to his group.

CHAPTER 3

SECTION 3.1

3. Theoretically the 13 hearts divide the 39 remaining cards into 14 parts which, on the average, will be equal. Therefore, each part will contain 39/14 or about 2.79 cards, on the average. The part before the first heart will average this many and, adding the first heart, we obtain 3.79.

4. 75.

5. 96; 1/8.

6. 2/3; 1/12. The fact that he is a man has no bearing on the outcome. If it had been stated "What is the probability that Fred Jones will be selected if it is known that a man has been chosen?" then the answer would have been 1/8.

7. 4/9; 2/3.

8. 32.

9. 6.

SECTION 3.2

1. (a) $8^4 = 4{,}096$ (b) $8 \cdot 7 \cdot 6 \cdot 5 = 1{,}680$.

2. 2,598,960.

3. $1/\binom{52}{13}$ times the number of perfect hands (8). The probability can then be written as $8/\binom{52}{13}$, or $(8 \cdot 13!\,39!)/52!$ which is about $1.26(10)^{-11}$ (i.e., 0.0000000000126).

4. The order of finish by stable was Jones, Foster, Jones, Foster, Jones. Since there are 3! or 6 possible orders for the Jones horses to finish among themselves and 2 possible orders for the Foster horses to finish between them, the total number of ways is $6 \cdot 2$ or 12.

5. (a) 30 (b) 45 (c) 55 (d) 46 (e) 56.

6. 150.

7. (a) 120 (b) 151,200 (c) 20,160 (d) 10,080 (e) 34,650.

8. 1/455.

9. 302,400.

10. $6! = 720$; for the second part, one approach is to add 4 "spaces" to the 6 friends. Then we have 10 "things" to place, 6 different, 4 alike, so the solution is $10!/4!$ or 151,200.

SECTION 3.3

1. None since $P(A) + P(B) > 1$ and the sum of the probabilities of all the mutually exclusive points in a sample space is exactly 1.

2. $P(A \text{ or } B) = P(A) + P(B) = 0.9$.

3. Since all the possible outcomes are stated and they are mutually exclusive, the sum must be 1. But $0.44 + 0.29 + 0.17 = 0.90$, which is not possible. He is mistaken.

4. $2/15 + 0.9 \doteq 1.033$ (or 31/30) which is greater than 1. Since these are mutually exclusive, the statement is false. However, $2/15 + 0.8 = 14/15$. Since there are other possibilities (tie, cancellation), the second fan's statement is not inconsistent with the rules of probability. This does not, however, increase his likelihood of being correct.

5. 0.7.

6. (a) 0.4 (b) 0.7 (c) 0.7 (d) 0 (*not* $\varnothing$).

7. Four are neither, so
   (a) 0.1 (b) 0.3 (c) 0.6 (d) 0.4 (e) 0.6 (f) 0.9 (g) 0.7 (h) 0.7.

8. (a) 0.38 (b) 0.28 (c) $1 - (0.21 + 0.11) = 0.68$ (d) $1 - 0.49 = 0.51$.

9. (a) $P(R \text{ and } K) = 0.06$ (b) $P(T \text{ and } K) = 0.08$ (c) $P(R' \text{ and } K) = 0.14$ (d) $P(R' \text{ and } T') = 0.3$. (Note that R and T are *not* independent, but mutually exclusive.) (e) $P(K' \text{ or } T) = 0.88$ (f) $P(R' \text{ and } K') = 0.56$.

10. (a) 0.60 (b) 0.44 (c) 0.48 (d) 0.44.

11. (a) 0.46 (b) 0.67.

12. 82.

13. 0.08.

14. 0.05.

SECTION 3.4

1. (a) 1/50 (b) 1/15 (c) 2/15 (d) 2/15.
2. Since 10 take neither, 90 take one or the other or both. Since $80 + 60 = 140$, and $140 - 90 = 50$, then 50 take both. Thus $P(E \text{ and } M) = 0.50$. Now $P(M) = 0.60$, but $P(M \mid E) = 0.50/0.80 = 0.625$. Since they are not equal, the events are not independent. Other approaches are possible.
3. If A and B are mutually exclusive, $P(A \text{ or } B) = 0$; if either even occurs, the other cannot, so $P(A \mid B) = P(B \mid A) = 0$.
4. 1/32.
5. (a) 0.36 (b) 0.34.
6. 2.
7. (a) 0.56 (b) 0.73 (c) 0.48 (d) 0.62 (e) 0.40 (f) 0.30 (g) 0.60 (h) 0.70 (i) 0.27 (j) 0.38.
8. (a) 1/216 (b) 215/216 (c) 1/36 (d) 5/9 (e) 5/108 or about 0.046.
9. 3/20.
10. 0.00000016; about 0.92.
11. 1/12.
12. 0.97489.
13. 0.36; 0.91.

SECTION 3.5

1. (a) 1/30 (b) 1/4 (c) 1/5 (d) 1/6 (e) 0 (f) 7/30.
2. (a) 0.6 (b) 0.5 (c) 0.6 (d) 0.57 (e) 0.4 (f) 0.6.
3. 349,188,840 (or $21!/(6!\,7!\,8!)$).
4. There are exactly 13! ways that the 13 volumes can be replaced. If we consider the three volumes of *Tristram Shandy* as one, there are 11! ways of replacing the books with those 3 volumes in a particular order. There are 3! or 6 orders of arranging these, so the probability that they are all together is $6 \cdot (11!/13!)$ or $(6/13) \cdot 12$, which is 1/26; the probability that they are together in the proper order is 1/156 since only 1 of the 6 possible orders is the correct one.
5. There are 25 different sample points; the probability that both will be the same color is 1/5 or 0.2.
6. 125/216.

7. (a) 0.92 (b) 0.92 (c) about 0.098 (d) about 0.444.

8. (a) 8 (b) *hhh, hht, hth, thh* (c) *hhh, hht, hth, htt*
(d) $N = \{hhh, hht, thh, tht\}$, $D' = \{ttt, tth, tht, thh\}$, so N *and* $D' = \{thh, tht\}$.

9. (a) No one plays any of the instruments.
(b) Now $P(P) = 0.5$, $P(V) = 0.8$, $P(C) = 0.5$;
$P(P|V) = 0.5 = P(P)$ so P and V are independent;
$P(P|C) = 0.2 \neq P(P)$, so P and C are not independent;
$P(V|C) = 0.8 = C(V)$, so V and C are independent.
(c) $P(V|P) = 0.8$; $P(C|P) = 0.2$; $P(C \text{ and } V|P) = 0.2$.

10. 0.10.

11. 0.40.

12. $P(W) = 9/20$; $P(R) = 11/20$; $P(A \mid R) = 8/33$.

13. For two engines, the flight will be a success unless both engines fail. The probability that both engines fail is $(0.01)^2 = 0.0001$. Thus the probability of successful completion is $1 - 0.0001 = 0.9999$.
For three engines, the flight will fail if three engines fail $[p = (0.01)^3]$ or if two engines fail. There are $\binom{3}{2}$ or 3 ways two engines can fail, so the probability two engines fail is $3(0.99)(0.01)^2 = 0.000297$. Then the probability of success is $1 - (0.000297 + 0.000001) = 0.999702$.
For four engines the probability of four failures is $(0.01)^4$; the probability of three failures is $4(0.99)(0.01)^3 = 0.00000396$. For any other outcome the flight will succeed. The probability of a successful flight, then is 0.99999603.

14. The probability that a particular machine and its "back-up" both will fail is $(0.01)^2$ or 0.0001. Since there are five machines *and* the events are mutually exclusive—if a pair breaks down, no other pair has a chance to—the probability that the assembly line will shut down is 0.0005.

15. 0.00168.

16. Under the assumption that guinea pigs mate more or less at random, so that hair length in a set of parents is independent, the probability can be shown to be equal to 1/4.

17. 14/19.

18. (a) 2 to 3; 3 to 2 (b) 37 to 1; 10 to 9; 10 to 9 (c) 4 to 1.

19. 0.2875 (not 0.4125).

20. 1/225; 1/225; 11/312 $\doteq$ 0.035.

21. $P(K) = 0.37$.

CHAPTER 4

SECTION 4.1

1. (a) yes (b) no, since the sum of the probabilities is not one (c) yes (d) yes (e) yes, although 0 should be eliminated as its presence is of no value (f) yes.

2. $P(1) = 2/3$; $P(2) = 4/15$; $P(3) = 1/15$.

3. $P(2) = 2/5$; $P(3) = 2/5$; $P(4) = 1/5$.

4. $P(1) = 2/n$, $P(2) = 5/n$, $P(3) = 10/n$, $P(4) = 17/n$; the sum of the probabilities is $34/n$. Since it is a probability distribution, the sum is one, so $n = 34$.

5.

x	$P(x)$
1	0.7
2	0.0
3	0.147
4	0.0882
5	0.0648

6.

x	$P(x)$
0	1/4
1	3/8
2	1/4
3	1/8

7.

x	0	1	2	3	4
$P(x)$	16/81	32/81	24/81	8/81	1/81

SECTION 4.2

1. about 67¢.

2. 0.15; 15.

3. $7,100.00.

4. The probability of winning is 1/38, of losing is 37/38. If you win, you gain $35.00, if you lose, you lose $1.00, so $E(x) = (1/38)(35) + (37/38) \times (-1) = -1/19$. Another way of looking at it is to say that if you win you *get* $36.00, if you lose you get nothing; from this expected receipt you

subtract your cost of \$1.00, so $E(x) = (1/38)(36) + (37/38) \times (0) - 1 = -1/19$. Either way your expected value is $-1/19$ dollar or an average of \$1.00 lost for every \$19.00 bet.

5. $-1/19$ dollars.

6. 5.50; 1.79.

7. 4/3. (See problem 5, section 5.5.)

8. \$5.00.

9. 82.2 women, 75.8 men.

10. 20.

11. 667.

SECTION 4.3

1. $\mu = 0.4$; $\sigma^2 \doteq 1.24$; $\sigma \doteq 1.11$.

2. $\mu = 10.4$; $\sigma^2 = 1.24$; $\sigma \doteq 1.11$.

4. $\mu = 5.9$; $\sigma^2 = 0.89$; $\sigma \doteq 0.94$.

5. $\mu = 1.2$; $\sigma^2 = 11.16 = 9(1.24)$; $\sigma \doteq 3.34 \doteq 3(1.11)$.

7. $\mu = 7$; $\sigma \doteq 5.83$.

8. (a) $\mu = 18.5$; $\sigma \doteq 3.93$ (b) $\mu = 119.46$; $\sigma \doteq 9.58$
   (c) $\mu = 16.20$; $\sigma \doteq 8.86$ (d) $\mu = 1{,}200$; $\sigma = 100$.

9. 1.8178; 1.7498.

10. 1.25; 0.97.

11. 4/3; 8/9.

SECTION 4.4

1. $\mu = 8$, $\sigma \doteq 4.87$.

2. (a) $n = 50$ (c) 4.20 (d) 0.92.

3. (a) (b) $\mu = 4.5$; $\sigma = 1.5$.

x	$P(x)$
2	3/28
3	5/28
4	3/14
5	3/14
6	5/28
7	3/28

4. (a)

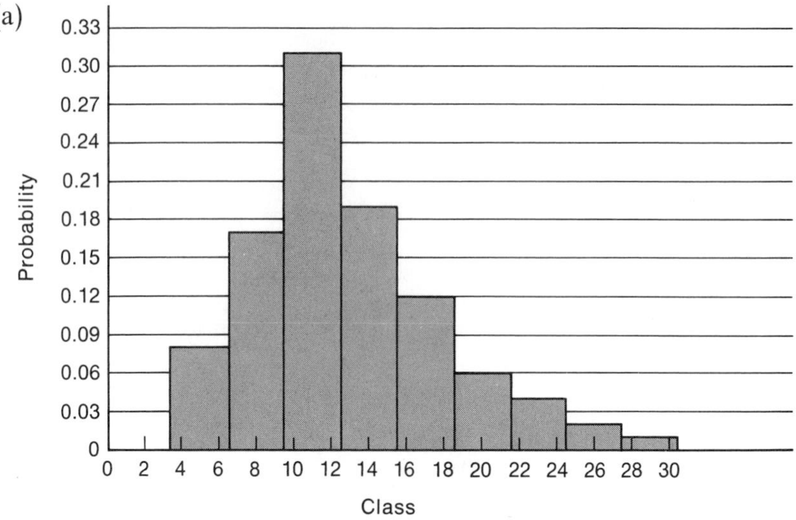

(b)

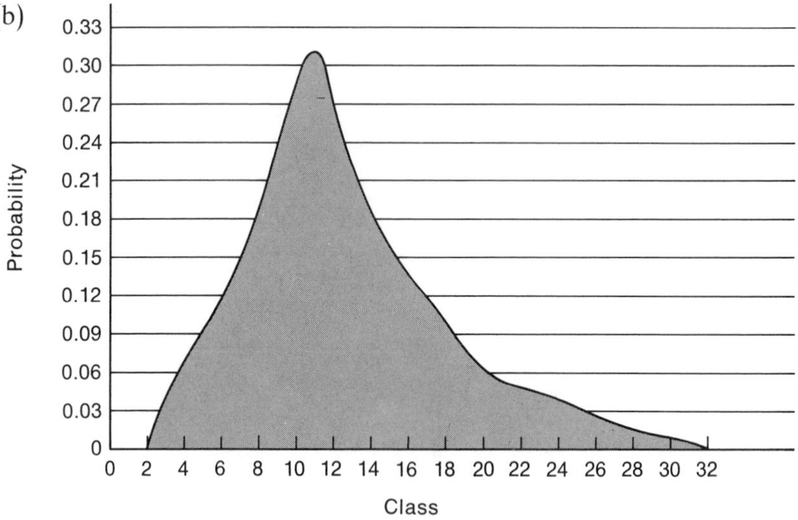

(c) $\mu = 12.8$, $s^2 = 176.4 - (12.8)^2 = 12.56$, so $s \doteq 3.54$.

5. \$240.00.

6. 15/8; 15/16.

7.

x	$P(x)$
0	1/6
1	2/3
2	1/6
$\mu = 1$,	$\sigma^2 = 1/3$

8. 940 per cc.

CHAPTER 5

SECTION 5.1

1. $P(7) = 6/36$, $P(11) = 2/36$, $P(7 \text{ or } 11) = 8/36$, or $p = 2/9$; $n = 4$, $1 - p = 7/9$, so we have, for x successes,

x	$P(x)$
0	$\binom{4}{0}(2/9)^0(7/9)^4 = 2{,}401/6{,}561 \doteq 0.366$
1	$\binom{4}{1}(2/9)^1(7/9)^3 = 2{,}744/6{,}561 \doteq 0.418$
2	$\binom{4}{2}(2/9)^2(7/9)^2 = 1{,}176/6{,}561 \doteq 0.179$
3	$\binom{4}{3}(2/9)^3(7/9)^1 = 224/6{,}561 \doteq 0.034$
4	$\binom{4}{4}(2/9)^4(7/9)^0 = 16/6{,}561 \doteq 0.002$

Then $\mu = 4(2/9) = 8/9$; $\sigma^2 = 4(2/9)(7/9) = 56/81$.

2. $n = 5$, $p = 0.6$, $1 - p = 0.4$. For x blue balls,

x	$P(x)$
0	0.010
1	0.077
2	0.230
3	0.346
4	0.259
5	0.078

$\mu = 5(0.6) = 3$, $\sigma^2 = 5(0.6)(0.4) = 1.2$.

3. If x is the number of games won by the home team, $P(0) \doteq 0.091$; $P(1) \doteq 0.334$; $P(2) \doteq 0.408$; $P(3) = 0.166$; $\mu = 1.65$, $\sigma^2 = 0.7425$.

x	0	1	2	3	4	5
$P(x)$	0.001	0.010	0.044	0.117	0.205	0.246

x	6	7	8	9	10
$P(x)$	0.205	0.117	0.044	0.010	0.001

4. $\mu = 5$, $\sigma^2 = 2.5$.

5. $p = 33/221$, so $\mu \doteq 29.86$, $\sigma \doteq 5.04$.

6. $\mu = 7{,}500$, $\sigma \doteq 43.30$.

7. $P(3) \doteq 0.320$;
   $P(2) \doteq 0.305$;
   $P(3) \doteq 0.145$;
   $\mu \doteq 2.56$.

8. $240/729 \doteq 0.329$.

9. (a) $12/125 = 0.096$.
   (b) $36/125 = 0.288$.

10.

x	$P(x)$
0	0.6561
1	0.2916
2	0.0486
3	0.0036
4	0.0001

$\mu = 0.4$;
$\sigma = 0.6$.

SECTION 5.2

1. 0.317.

2. 0.655.

3. Now the probability that he will be allowed into no more than two houses in the block is the sum of $P(0)$, $P(1)$, and $P(2)$, where $P(x)$ is the probability of being allowed into exactly x houses. Then $n = 6$, $p = 1/3$

so $P(0) = (2/3)^6 = 64/729$; $P(1) = 6(1/3)(2/3)^5 = 192/729$; $P(2) = 15(1/3)^2(2/3)^4 = 240/729$, so $P(x \leq 2) = 496/729 \doteq 0.68$. Now the probability of making a sale is 1/12, since he must be let in (conditional probability) to make a sale. The probability of making no sale at all in 6 tries is $(11/12)^6$ or 1,771,561/2,985,984 or about 0.59. Thus the probability of making at least one sale is 1 − 0.59 or 0.41.

4. 25/32.

5. 5/32.

6. 0.0297.

7. 0.633; 0.656.

8. (a) 0.992
   (b) 0.936.

9. 0.993.

10. 0.967.

SECTION 5.3

1. $P(x \geq 11) = 0.952$; $P(x \geq 12) = 0.887$, so $P(11) = 0.065$.

2. $P(x \leq 5) = 1 - P(x \geq 6) = 1 - 0.019 = 0.981$; $P(5) = 0.054$.

3. $P(x > 18) = P(x \geq 19) = 0.332$; $P(20) = 0.106$.

4. $P(x > 14) = 0.034$; $P(x < 14) = 0.922$.

5. $P(x \geq 7) \doteq 0.543$; $P(x \geq 8) \doteq 0.356$, so $P(7) \doteq 0.187$.

6. $P(x < 12) = 1 - P(x \geq 12) \doteq 0.003$; $P(12) \doteq 0.005$.

7. $p = 0.60$.

8. $r = 13$.

9. $r = 14$.

10. $r = 11$.

11. 0.624; 0.201.

12. 0.983.

13. 0.197. Note: this is a case where the tables do not give a satisfactory answer. Direct calculation yields 0.243. This is because the curve for $n = 20$, $r = 3$, peaks near $p = 0.15$.

14. 0.162.

15. 0.599; 0.389.

16. 0.050.

SECTION 5.4

1. 21 or more.

2. (a) 3 or fewer (b) 3 or fewer (c) 3 or fewer (d) It cannot be concluded that the drug is effective. $\beta = 0.234$. (If $p = 0.20$, $P(x \geq 4) = 0.234$.)

3. $x \geq 3$.

SECTION 5.5

1. The multinomial distribution is used. Using a, b, c for the classifications, $P(a) = 5/15 = 1/3$, $P(b) = 4/15$, $P(c) = 6/15 = 2/5$. Thus we have

a	b	c	*Probability*		
3	0	0	$1(1/3)^3$	$= 1/27$	$\doteq 0.037$
0	3	0	$1(4/15)^3$	$= 64/3{,}375$	$\doteq 0.019$
0	0	3	$1(2/5)^3$	$= 8/125$	$\doteq 0.064$
2	1	0	$3(1/3)^2(4/15)$	$= 4/45$	$\doteq 0.089$
2	0	1	$3(1/3)^2(2/5)$	$= 2/15$	$\doteq 0.133$
1	2	0	$3(1/3)(4/15)^2$	$= 16/225$	$\doteq 0.071$
1	0	2	$3(1/3)(2/5)^2$	$= 4/25$	$= 0.164$
0	1	2	$3(4/15)(2/5)^2$	$= 16/125$	$= 0.128$
0	2	1	$3(4/15)^2(2/5)$	$= 32/375$	$\doteq 0.085$
1	1	1	$6(1/3)(4/15)(2/5)$	$= 16/75$	$\doteq 0.213$

2. There are 20 students altogether, so we have $p_1 = 0.4$, $p_2 = 0.25$, $p_3 = 0.2$, $p_4 = 0.15$. To meet the conditions of the problem we must have (1, 1, 1, 1), (1, 1, 2, 0), (1, 2, 1, 0), or (2, 1, 1, 0). The probabilities are easily found and their sum is 0.276.

3. $P(2) \doteq 0.04$; $P(1) \doteq 0.30$; $P(0) \doteq 0.66$.

4. If x is the number of defectives sold to the customer,

$$P(x) = \binom{4}{x}\binom{16}{5-x}\bigg/\binom{20}{5}$$

x	0	1	2	3	4
$P(x)$	0.282	0.470	0.217	0.031	0.001

5. $P(0) = 1/56$, $P(1) = 15/56$. $P(2) = 30/56$ or $15/28$, $P(3) = 10/56$ or $5/28$; $\mu = 1.875$; $\sigma = .71$.

6. This is a geometric distribution with $p = 1/2$. Then $P(x) = (1/2)^{x-1}(1/2) = (1/2)^x$.

x	1	2	3	4	5
$P(x)$	1/2	1/4	1/8	1/16	1/32

x	6	7	8	9	10
$P(x)$	1/64	1/128	1/256	1/512	1/1,024

By the formula,

$$\mu = \frac{1}{1/2} = 2, \sigma = \left(\frac{1}{1/2}\right)\sqrt{1 - 1/2} = \sqrt{2} \doteq 1.414$$

7. 0.134; 0.393; 0.034.

8. $n = 10$, $p = 0.60$, $\mu = 6$. Then by the Poisson distribution, $P(x \geq 7) = 0.393$. Using the binomial distribution, we have $P(x \geq 7) = 0.382$. For $n = 25$, $p = 0.20$, $\mu = 5$. By the Poisson distribution, $P(x \leq 4) = 0.440$. Using the binomial distribution, we have $P(x \leq 4) = 0.421$.

9. 0.332; 0.329.

10. $1{,}575{,}000/3^{15} \doteq 0.110$.

11. 0.591.

SECTION 5.6

1. A success would be (6, 3, 1), (6, 2, 2), (5, 4, 1), (5, 3, 2), (4, 4, 2), (4, 3, 3) or (6, 4, 1), (6, 3, 2), (5, 5, 1), (5, 4, 2), (5, 3, 3), (4, 4, 3) in some order. Each point has a probability of $(1/6)^3$ or 1/216. The number of ways to obtain each set of 3 is 3! or 6 if all the numbers are different, or 3 if 2 are alike. Thus

$$p = \frac{1}{216}(6 + 3 + 6 + 6 + 3 + 3 + 6 + 6 + 3 + 6 + 3 + 3) = \frac{54}{216}$$
$$= 0.25$$

Then (a) $\mu = 2.5$, $\sigma \doteq 1.37$ (b) 0.94 (c) 0.53.

2. Note that $p = 1/4$ so $P(x \geq 1) = 1 - P(0) = 1 - (3/4)^n$. All that is necessary, then, is to determine when $(3/4)^n$ is less than 1/5. Since $(3/4)^5 \doteq 0.24$ and $(3/5)^6 \doteq 0.18$, then $n = 6$.

3. If $p = 0.6$, then the probability that 9 or more of the patients would be cured is only 0.046. Thus it appears that an investigation should be made.

4. (a) $P(9) = \binom{15}{9}(0.7)^9(0.3)^6 \doteq 0.147$; $P(10) \doteq 0.206$; $P(11) \doteq 0.219$; $P(12) \doteq 0.170$, so $P(9 \leq x \leq 12) \doteq 0.742$.
   (b) For $n = 15$, $p = 0.7$, $P(x \geq 9) = 0.869$ and $P(x \geq 13) = 0.127$. Thus $P(9 \leq x \leq 12) = 0.869 - 0.127 = 0.742$.

5. (a) $P(9) \doteq 0.191$; $P(10) \doteq 0.212$; $P(11) \doteq 0.179$; $P(12) \doteq 0.111$, so $P(9 \leq x \leq 12) \doteq 0.693$.
   (b) For $n = 15$, $p = 0.65$, $P(x \geq 9) \doteq 0.740$; $P(x \geq 13) \doteq 0.077$, so $P(9 \leq x \leq 12) \doteq 0.663$.

6. $\mu = 70$, $\sigma \doteq 4.58$; $P(59.5 < x < 75.5) = P(60 \leq x \leq 75)$. For $x = 59.5$, $z = -2.29$; for $x = 75.5$, $z = 1.20$. Thus $P(60 \leq x \leq 75) = 0.4890 + 0.3849 = 0.8739$.

7. 0.522; 0.150; no, the probability is 0.026.

8. 0.002; yes.

9. Marked degree of ESP on mind reading; some clairvoyant ability.

10. 0.004; 0.114.

11. About 0.85.

12. About 0.16; about 0.825.

13. About 0.48.

14. 0.323.

CHAPTER 6

SECTION 6.2

1. (a) -1.60 (b) 1.30 (c) -1.74 (d) -0.29 (e) 0.63 (f) 2.50.

2. (a) 29.75 (b) 34.83 (c) -15.09 (d) 16.30 (e) 8.87 (f) 23.85.

3. (a) 0.4854 (b) 0.3508 + 0.4382 = 0.7890 (c) 0.4983 − 0.4406 = 0.0577
 (d) 0.1879 − 0.0478 = 0.1401 (e) 0.4988 + 0.4484 = 0.9472
 (f) 0.1664 + 0.4817 = 0.8481.

4. (a) 0.5000 − 0.4236 = 0.0764 (b) 0.5000 − 0.3485 = 0.1515
 (c) 0.5000 + 0.2794 = 0.7794 (d) 0.5000 + 0.4778 = 0.9778.

5. (a) 0.6826 (b) 0.9544 (c) 0.9974 (d) 0.7994 (e) 0.8990 (f) 0.9500
 (g) 0.9802 (h) 0.9902.

6. If 0.1230 of the area under the normal curve is to the right of z, 0.8770 is to the left of z, and 0.3770 is between 0 and z. Therefore $z = 1.16$.

7. (a) 0.1075 (b) 0.3085 (c) 0.8389 (d) 0.6844 (e) 0.2435 (f) 0.8321
 (g) 0.2025
 (h) for $z = 0.48$, 0.1884 lies between 193.4 and 200, so 0.5000 − 0.1884 or 0.3156 lies above 200; for $z = -0.98$, 0.3365 lies between 180 and 193.4 so 0.1635 lies below 180; therefore 0.3156 + 0.1635 or 0.4791 lies above 200 or below 180 (and 0.5209 lies between 180 and 200).

8. If 0.1446 of the area of a normal distribution lies above a score, then 0.3554 lies between the score and the mean. Thus $z = 1.06$. Since $1.06 = (121 - 100)/\sigma$, then $\sigma = 21/1.06 \doteq 19.8$.

9. $\mu = 971.5$.

10. 27.

SECTION 6.3

1. (a) Now $z = (72 - 72.6)/0.4 = -1.50$, so the probability that a given can will weigh less than 72 oz is 0.0668. Thus the expected number of cans which will weigh less than 72 oz is (0.0668)(100,000) = 6,680.
 (b) If the expected number of cans is 300, the probability is 300/100,000 or 0.0030. Thus 0.4970 of the area under the normal curve will lie between 72 and μ. Since 72 is less than μ, z is negative, and $z = -2.75$. Then

$$-2.75 = \frac{72 - \mu}{0.4} \quad \text{or} \quad \mu = 72 + (0.4)(2.75) = 73.1.$$

2. 0.0044.

3. 0.2005; 0.2389.

4. 0.9569; 0.1210.

5. For company B, the probability that dowel rods cannot be used can be determined by finding $z = (4.020 - 4.000)/0.012 \doteq 1.67$, so about 0.0475 of

the lot will have diameters greater than 4.020. Similarly about 0.0475 will have diameters less than 3.080, so about 950 of 10,000 will not be usable. The cost per usable dowel rod, then, is \$460.00/9,050 or about 5.08 cents per rod. For company *A*, a similar procedure gives us $z = (4.020 - 4.000)/0.015 \doteq 1.33$, so about 0.0918 will have diameters too great, and a similar number too small. Thus there will be about 8,164 usable dowel rods at a cost of \$400.00/8,164 or 4.90 cents apiece. Thus company *A* should get the bid.

6. 0.0034.

7. 0.0548.

8. $\mu = 152$, $\sigma = 18$.

SECTION 6.4

1. 0.0532; 0.0537.

3. Now $p = 1/5$, $n = 225$, so $\mu = 45$, $\sigma = 6$. The probability of obtaining more than 60 correct answers is $P(x > 60)$, so the interval containing 60 (59.5 to 60.5) is not included. Thus we wish $P(x > 60.5)$, if we use the normal approximation. Then $z = (60.5 - 45)/6 \doteq 2.58$. Thus 0.0049 is the probability desired.

4. 0.8790.

5. 0.0179.

6. 0.5279.

7. Here $p = 1/2$, $n = 60$; 40 or more, includes the interval 39.5 to 40.5, so we want $P(x > 39.5)$. Now $\mu = 30$, $\sigma = \sqrt{15} \doteq 3.87$, so $z = (39.5 - 30)/3.87$ or about 2.45. The probability is about 0.0071.

8. 0.8584.

9. 0.0985.

10. 0.1335.

11. 0.0359.

SECTION 6.5

1. For $p = 0.7$, $\mu = np = 114.1$, $\sigma = \sqrt{np(1-p)} \doteq 5.85$. For $x = 89.5$, $z = (89.5 - 114.1)/5.85 \doteq -4.20$, so the value of β is very nearly zero.

2. 59; 61.

3. Null hypothesis is $p = 0.40$, where p is the proportion of viewers watching the program. Alternate hypothesis is $p > 0.40$, since the network expects *at least* 40% to watch it. The rejection region is $x \geq 177$. Since $x = 189$, we can conclude that at least 40% of the viewers are watching the premiere show.

4. $\mu = 4$; the probability of ten or more surviving is about 0.01 with the usual treatment, so we can conclude that the new treatment is indeed more successful than previous ones.

SECTION 6.6

1. (a) 0.6268 (b) 0.0301 (c) 0.1736 (d) 110.8 (e) 117.9 (f) 86.1 (g) 0.3944 (h) 0.7324.

2. 5.7.

3. 76.27.

4. 46.75; 46.99 oz.

5. 0.0287.

6. 0.0062.

7. (a) 0.0108 (b) 0.0179 (c) 0.9515.

8. Use $p = 0.6$. Here $n = 150$. For $P(x < 75)$, since 75 is not included, this is equivalent to $P(x < 74.5)$ when using the normal approximation. Then $\mu = 90$, $\sigma = 6$, so $z = (74.5 - 90)/6 \doteq -2.58$, so the probability is only about 0.0049. $P(x \geq 100)$ is equivalent to $P(x > 99.5)$ when using the normal approximation, so $z = (99.5 - 90)/6 \doteq 1.58$, so this probability is 0.0571. Now $P(x > 100.5)$ has $z = (100.5 - 90)/6 = 1.75$; $0.4599 - 0.4429 = 0.0170$, so $P(100) = 0.0170$.

9. $\mu = 60$, $\sigma \doteq 6.48$. For $z = -1.64$, we have $x \leq 49$ as the rejection region. If $x = 53$, we cannot conclude that fewer than 30% will buy the product. If $p = 0.25$, $P(x > 49)$ is about 0.53, yielding a z of -0.08 (using a continuity correction).

10. 0.9372.

11. 0.5714; solution involves a quadratic equation, 93.

12. Probability of chance variation is 0.2358; cannot say that a significant ($\alpha = 0.05$) bias is shown.

CHAPTER 7

SECTION 7.2

1. $\bar{x} = 25$, $s = 4$.

2. Using 25 as assumed mean ($\hat{x}$), we have the following table:

x	$x - 25$	$(x - 25)^2$
28	3	9
25	0	0
20	−5	25
33	8	64
27	2	4
29	4	16
23	−2	4
21	−4	16
24	−1	1
18	−7	49
30	5	25
25	0	0
Total	3	213

Then $\bar{x} = 25 + 3/12 = 25.25$ and $s = \sqrt{[213 - 12(.25)^2]/11} \doteq 4.39$.

3. $\bar{x} = 4.31$, $s \doteq 3.14$.

4. $\bar{x} \doteq 172.51$, $s \doteq 14.71$ (Use 170 or 172 for $\hat{x}$).

5. $\bar{x} = 171.45$; $s \doteq 14.44$.

6. $\bar{x} \doteq 23.9$; $s \doteq 2.20$.

SECTION 7.3

1. (a) 10 (b) 3.16 (c) 1 (d) 16.67 (e) 8.33 (f) 12.5 (g) 8.84 (h) 3.125.

2. $\bar{x} \doteq 3.62$, $s \doteq 2.29$, $s_{\bar{x}} \doteq 0.64$.

3. (a) 0.1660 (b) 0.1660.

4. (a) 0.0853 (b) 0.0749.

5. 1,560.

6. $n = 16$.

7. 0.0013.

8. In this case the question can be restated as "What is the probability that the mean lies between 71.5 and 72.5?" For the part above the mean, $z = 0.5/0.8 = 0.625$ (since the standard error is $8.0/\sqrt{100}$). Using $z = 0.62$, we find that the probability is 0.2324 that the difference is less than 0.50 above the mean. The same procedure applies below the mean, so the probability that the difference between the sample mean and the population mean is less than 0.50 is 0.4648. Note that interpolation and use of $z = 0.625$ in each case will give us about 0.4681; thus the answers should always be regarded as, at best, approximately correct.

9. 2.74.

10. 0.93.

SECTION 7.4

2. 0.0038.

3. (a) $\bar{x} \doteq 2.002$, $s = 0.013$ (b) $\bar{x} = 65$, $s = 6.16$ (c) $\bar{x} = 77.47$, $s = 11.14$.

4. 0.0094.

6. $\bar{x} \doteq 21.4$; $s_{\bar{x}} \doteq 0.24$.

7. 0.9958.

8. (a) 0.8090 (b) 0.9310.

CHAPTER 8

SECTION 8.1

1. (a) 2.63 (b) 23.68 (c) 0.0055 (d) 0.81 (e) 3.96.

2. (a) 0.8414 (b) 0.5098 (c) 0.9990 (d) 0.4972 (e) 0.9652.

3. 0.48.

4. 0.9936.

5. $737.00; reduces it to $588.00.

6. Referring to Table 1, we find that the value of z for which 0.5000 of the area under the curve lies between z and $-z$ is about 0.675. If the population has a standard deviation of 10.84, the standard error of the mean is

$10.84/\sqrt{64}$ or 1.355 for samples of size 64. Then the probable error of estimate is 1.355(0.675) or about 0.91.

7. About 3.34.

8. (a) 0.9356 (b) 0.9250 (c) 0.8858 (d) 0.9342.

9. (a) 148 (b) 211 (c) 298 (d) 365.

10. About 4.12.

SECTION 8.2

1. 2.074 to 2.086; 2.072 to 2.088.

2. 9.72 to 9.94; 9.67 to 9.99.

3. 1,545 to 1,615.

4. No.

5. A good way to handle this problem is to assume a mean, say 10 or 11, then do the calculations. Then $\bar{x} \doteq 10.92$, $s \doteq 1.38$, $s_{\bar{x}} \doteq 0.20$. Then the confidence intervals are 10.53 to 11.31 and 10.40 to 11.44.

6. (a) 11,213 to 12,075 (b) 11,223 to 12,065. The differences are very slight; only about 1/12 of one percent.

7. The probability is 0.98 that the error will be less than 0.42.

8. \$10,706.00 to \$12,180.00; \$10,855.00 to \$12,031.00.

9. 2.9 to 3.5 days.

10. (a) 9.2 to 15.1; 8.6 to 16.8 (b) 115.1 to 171.1; 109.4 to 185.4 (c) 0.011 to 0.014; 0.010 to 0.014 (d) 8.587 to 12.056; 8.213 to 12.879.

SECTION 8.3

1. Since there are nine pieces of data, there are 8 degrees of freedom. The value of t for which 0.025 of the area is above t is 2.306 (from Table 2). Then the 0.95 confidence limits will be $0.630 + ts_{\bar{x}}$ and $0.630 - ts_{\bar{x}}$. Now $s_{\bar{x}} = 0.081/3 = 0.027$, $(2.306)(0.027) \doteq 0.062$, so the confidence limits are $0.630 - 0.062 = 0.568$ and $0.630 + 0.062 = 0.692$. Thus the 0.95 confidence interval is 0.568 to 0.692. Similarly, the value of t for which 0.005 of the area is above t is 3.555. The 0.99 confidence interval is 0.539 to 0.721.

2. 0.76 to 0.88; 0.72 to 0.92.

3. 0.10 to 0.13.

4. 2.82 to 3.46.

5. 4.45 to 5.87.

6. 31.58 to 45.30; 31.02 to 45.86.

SECTION 8.4

1. 2,305; (use $P = 0.60$ as a reasonable estimate.)

2. 0.73 to 0.93.

3. 0.32 to 0.48; 0.30 to 0.50.

4. Now only 130 of the respondents could be used since we want confidence intervals for the proportion of coffee drinkers only. Then $p = 78/130 = 0.60$, $s_p = \sqrt{(0.60)(0.40)/130} \doteq 0.043$; $(1.96)(0.043) \doteq 0.08$, so the 0.95 confidence interval is 0.52 to 0.68; $(2.58)(0.043) \doteq 0.11$, so the 0.99 confidence interval is 0.49 to 0.71.

SECTION 8.5

1. (a) 1,062.7 to 1,101.3; 1,056.5 to 1,107.4
   (b) 95.9 to 123.6; 92.6 to 129.6.

2. (a) 500 (b) 666.

3. 28,128 to 29,156; 27,945 to 29,339.

4. $s_{\bar{x}} = 1.5$; the required intervals are 131.06 to 136.94 and 130.13 to 137.87.

5. Here $n = 4$, $\bar{x} = 174.75$, $s \doteq 30.94$, $s_{\bar{x}} \doteq 15.47$, $t = 3.182$, $15.47(3.182) \doteq 49.23$, so the 0.95 confidence interval is 125.52 to 223.98.

6. 97.

7. 15.65 to 16.39; 15.58 to 16.46. If you can't get this answer, read problem 6, section 8.3.

8. $s_{\bar{x}} = 0.8$, $\bar{x} = 11.55$; critical values for t are 2.131 (for 0.95) and 2.947 (for 0.99). The 0.95 confidence interval is 9.8452 minutes to 13.2548 minutes (or 9 min 51 sec to 13 min 15 sec); the 0.99 confidence interval is 9.1924 minutes to 13.9076 minutes (or 9 min 12 sec to 13 min 54 sec).

9. 3.1 to 3.3.

10. 0.60 to 0.80; 0.57 to 0.83.

11. The probability is 0.95 that the error will be less than 0.048.

12. 0.32 to 0.53; 0.29 to 0.57.

13. 0.075 to 0.165; 0.06 to 0.18.

CHAPTER 9

SECTION 9.1

1. A type II error is made by accepting a false hypothesis, so he has concluded that fertilizer *A* will do a better job than fertilizer *B*. Since this is in error, fertilizer *B* will do as good a job, or a better job, than fertilizer *A*.

2. Assuming that "business is going to be better" means an increase in orders, his hypothesis is that he will have an increase in orders. If he makes a type I error, he has rejected his hypothesis and concluded that business will not be better; since this is in error, it really will be better.

3. The researcher's conclusion is that the implements are not less than 24,000 years old, which is false. A type I error is made by rejecting a true hypothesis, so his hypothesis must have been that the implements were less than 24,000 years old—which he rejected. If he made a type II error he must have accepted the hypothesis which is identical with his conclusion.

4. Actually there are several possible hypotheses, but in the context of the problem the two to be tested are "The additive increases gasoline mileage" and "The additive does not increase gasoline mileage." In the first case, a type I error would be made by concluding that the additive does not increase gasoline mileage, if it does; a type II error would be made by concluding that the additive increases gasoline mileage, if it does not. In the second case, a type I error would be made by concluding that the additive increases gasoline mileage, if it does not; a type II error would be made by concluding that it does not, if it does.

SECTION 9.2

1. In words, we have the following:
   H_0: The average number of children in families in the apartment complex is μ_0. H_1: The average number of children in families in the apartment

complex is not μ_0. In symbols, if μ represents the average number of children in families in the apartment complex, we have
H_0: $\mu = \mu_0$
H_1: $\mu \neq \mu_0$.

2. (a) Let μ represent the population mean number of verbs per week learned under the new method. Then H_0: $\mu = 32$; H_1: $\mu < 32$
(b) H_0: $\mu = 32$; H_1: $\mu > 32$.
Note that the alternate hypothesis depends to a large extent where the researcher wishes to place his emphasis. The first teacher is trying to see if the new method is inferior—if not he will use it; the other teacher is trying to see if the new method is superior—if so he will use it. In any case, it is always more difficult to demonstrate that the alternate hypothesis is highly likely than it is not to.

3. (a) If P represents the proportion of orders going into the area we have H_0: $P = 0.05$; H_1: $P > 0.05$. (b) H_0: $P = 0.05$; H_1: $P < 0.05$. The assumption made is that the office will be opened if more than 5% of the orders go into the area. In the first case, the office will be opened only if it can be shown beyond a reasonable doubt that this is true; in the second case, the office will be opened unless it can be shown beyond a reasonable doubt that less than 5% go into the area.

4. (a) H_0: $\mu_1 = \mu_2$; H_1: $\mu_1 \neq \mu_2$
(b) H_0: $\mu_1 = \mu_2$; H_1: $\mu_1 < \mu_2$ (or $\mu_2 > \mu_1$).

SECTION 9.4

1. Let W_C and W_S represent the weights gained on corn and slop, respectively, (other symbolisms are possible—even likely);
(a) 1. H_0: $W_C = W_S$; H_1: $W_C \neq W_S$ 2. $\alpha = 0.05$
(b) 1. H_0: $W_C = W_S$; H_1: $W_C > W_S$ 2. $\alpha = 0.05$
(c) 1. H_0: $W_C = W_S$; H_1: $W_C < W_S$ 2. $\alpha = 0.05$.

2. His null hypothesis must have been H_0: $A = B$. From the statement of the problem "(he) wants to draw conclusions as to which model actually does the better job," it can be assumed that he had H_1: $A \neq B$. It is possible that he may have had H_1: $A > B$, but not possible that he had H_1: $A < B$ since his results would then have been conclusive as the probability of type II error would have been extremely small. The probability of type II error being great means that he cannot reject H_0, so the sample of model A was not significantly better than that of model B. His conclusions were (formal) to fail to reject H_0 and (informal) to reserve judgment.

3. If 1 sample in 15 would fall so low by chance, the probability is greater than 0.05 that it was by chance, so the shipment would be passed. If 1 sample in 25 would fall so low by chance, the probability is less than 0.05 (actually 0.04), so the shipment would be rejected. Any shipment with a mean in excess of 16.42, no matter how far, would be passed—in terms of the stated problem. In practical terms, however, if a shipment had a sample too high, it would be passed, but the machine might be examined to see if it should be modified in some way.

4. Using the experimental technique we have the following format.
   1. H_0: $P = 0.40$
      H_1: $P > 0.40$
   2. $\alpha = 0.05$
   3. Criteria: For $\alpha = 0.05$, we would have $z = 1.64$ for 0.05 of the area under the normal curve in the right tail if $P = 0.40$. Then $x = \mu + z\sigma$ where $\mu = 100(0.40) = 40$, $\sigma = \sqrt{100(0.4)(0.6)} \doteq 4.90$, $z = 1.64$; then $x = 48.03$ so we will reject H_0 if $x > 48$.
   4. Results: $x = 52$
   5. Conclusion: Reject H_0, the microorganism has a bad effect on the antibiotic.

CHAPTER 10

SECTION 10.1

1. $z = 1.60$; inconclusive.
2. Below 227.6 or above 232.4 oz.
3. $z = 1.43$; accept.
4. 1. H_0: $\mu = 9.2$
      H_1: $\mu > 9.2$
   2. $\alpha = 0.05$
   3. Criteria: Reject H_0 if $z > 1.64$.
   4. Results: $s = 1.1$, $\bar{x} = 9.3$; $s_{\bar{x}} = 1.1/\sqrt{50} \doteq 0.156$;
      $z = (9.3 - 9.2)/0.156 = 0.1/0.156 = 0.64$.
   5. Conclusions: Fail to reject H_0. The claim is substantiated.
5. $z = 2.07$; claim is not substantiated.

6. Since the sample mean is less than 70, the hypothesis can be rejected without going through the tests; z will be negative, and the formal conclusion will be "fail to reject H_0: $\mu = 70$ in favor of H_1: $\mu > 70$." The informal conclusion will be that the research hypothesis will be rejected.

7. $z = 2.62$; yes.

SECTION 10.2

1. $t = 1.75$; reserve judgment.

2. 1. H_0: $\mu = 16.2$
   H_1: $\mu \neq 16.2$
   2. $\alpha = 0.01$
   3. Criteria: Reject H_0 if $t > 2.947$ or if $t < -2.947$.
   4. Results: $s_{\bar{x}} = 1.6/\sqrt{16} = 0.4$; $\bar{x} = 14.8$; $z = (14.8 - 16.2)/0.4 = -3.5$.
   5. Conclusions: Reject H_0. The combination is different and has a shorter effervescing time. NOTE: This conclusion is not stated as a fact, but is based on the level of α; that is, it is at least 0.99 probable that this statement is correct.

3. $t = 1.50$; no.

4. $t = -1.56$; we cannot reject the hypothesis that the mean is actually 16 oz. On the other hand, it might be preferable not to accept either. The best conclusion would be to reserve judgment.

5. $t = -2.418$; the lot is acceptable with the given level of significance.

6. Accept lots A, B, C, E; reject lots D and F.

7. $t \doteq 4.348$; yes.

8. $t = -2.20$; weight is less than 16 oz.

9. Yes; $t \doteq 5.277$.

10. Yes; $t \doteq 2.474$.

SECTION 10.3

1. 1. H_0: $P = 0.5$ (Where P is the proportion preferring brand B)
   H_1: $P > 0.5$
   2. $\alpha = 0.05$
   3. Criteria: Reject H_0 if $z > 1.64$.
   4. Results: $n = 163$, $\bar{x} = 101$, $\mu = 81.5$, $\sigma = \sqrt{163(0.5)(0.5)} \doteq 6.38$; $z = (101 - 81.5)/6.38 \doteq 3.05$

5. Conclusions: Reject H_0. The proportion of customers preferring brand B is greater than 0.5 and greater than the proportion preferring brand A. Stock only brand B.

2. Since we are testing the manufacturer's claim, it is up to us to disprove the claim, so we have H_0: $P = 0.80$; H_1: $P < 0.80$; $\alpha = 0.01$. Criteria are to reject H_0 if $z < -2.33$. Results are $p = 0.72$, $\sigma_P = \sqrt{(0.8)(0.2)/200} \doteq 0.028$, so $z = (0.72 - 0.80)/0.028 \doteq -2.86$. Conclusions: Reject H_0. The manufacturer's claim is probably false. There is less than 1 chance in 100 that the claim is true.

3. $z = 0.83$; accept shipment.

4. $z = -2.74$; reject the claim.

5. $z = 3.23$; more than 4 out of 10 persons still prefer bottles.

6. $z = 5.56$; the contention that dice are loaded is substantiated.

7. $z \doteq 3.15$; hypothesis is substantiated.

8. $z \doteq 6.18$; at least one-third of students favor co-ed dorms.

SECTION 10.4

1. $z = 0.62$; the difference is not significant.

2. $t = 1.43$; we cannot say that learning took place.

3. $z = 1.13$; no significant difference.

4. $t = -2.61$; deny.

5. $t = 0.58$; no.

6. 1. H_0: $\mu_A = \mu_B$
   H_1: $\mu_A \neq \mu_B$
   2. $\alpha = 0.05$
   3. Criteria: Reject H_0 if $t > 2.228$ or if $t < -2.228$.
   4. Results: $n_A = 6$, $\bar{x}_A = 510{,}000$, $s_A = 20{,}000$;
   $n_B = 6$, $\bar{x}_B = 490{,}000$, $s_B = 15{,}000$;

   then

$$s_D = \sqrt{\frac{(20{,}000)^2 + (15{,}000)^2}{6}} \doteq 10{,}206 \text{ and}$$

$$t = \frac{510{,}000 - 490{,}000}{10{,}206} \doteq 1.960$$

5. Conclusions: Fail to reject H_0. There is probably no difference between the brands (with a probability which can actually be determined from more extensive tables of the t distribution).

7. $t = 2.684$; cannot reject H_0. However the obtained value is so close to the critical value of 2.776, it would be wise to reserve judgment, and, if possible, repeat the experiment or obtain additional data.

8. H_1 is $\mu_1 > \mu_2 + 2$; $s_D \doteq 0.666$, $z = [4.3 - (1.9 + 2)]/0.666 \doteq 0.60$. The results do not substantiate the hypothesis.

SECTION 10.5

1. 1. H_0: $P_A = P_B$
   H_1: $P_A > P_B$
   2. $\alpha = 0.01$
   3. Criteria: Reject H_0 if $z > 2.33$.
   4. Results: $p_A = 82/90 \doteq 0.91$; $p_B = 96/110 \doteq 0.87$. Then

$$s_{dp} = \sqrt{\frac{(0.91)(0.09)}{90} + \frac{(0.87)(0.13)}{110}} \doteq 0.044$$

   and $z = (0.91 - 0.87)/.044 \doteq 0.91$
   5. Conclusions: We fail to reject H_0. Firm B will get the contract. NOTE: If a pooled sample proportion is used, $p = 0.89$, $s_{dp} \doteq 0.044$ and the identical results are obtained.

2. $z = -2.14$; company B.

3. $z = -0.48$; no.

4. $z = 0.67$; no.

5. $z \doteq 2.79$; differences probably real; should not be pooled.

6. $z \doteq 1.43$; no.

7. $z \doteq 1.53$; claim is not substantiated.

SECTION 10.6

1. $z = 2.54$; yes.

2. $z = -2.97$; yes.

3. $z = 2.93$; yes.

4. $t = 1.507$; the difference is not significant.

5. $z = -2.04$; the drug is effective.

6. $t = 0.658$; confirm.

7. $z = 1.11$; the jockey's hypothesis is not supported.

8. $z = -2.91$; yes.

9. Yes; $z = 2.16$; company *A*.

10. In this case, we need to determine in each instance the proportion of bulbs which will last at least 1,000 hours. For company *A*'s sample—assuming it represents the population—if the population mean is 1,065 with standard deviation 133, the methods of Chapter 6 give us $z = (1{,}000 - 1{,}065)/133 \doteq -0.49$; from Table 1 we find that 0.1879 + 0.5000 or 0.6879 is the proportion of bulbs we can expect to last at least 1,000 hours. For company *B* similar methods give us $z = (1{,}000 - 1{,}047)/56 \doteq -0.84$ so that we can expect 0.7995 of company *B*'s bulbs to last at least 1,000 hours. Thus we have $p_1 \doteq 0.69$ and $p_2 \doteq 0.80$, from which $s_{dp} \doteq 0.035$, $z \doteq -3.14$, and we can conclude that we should choose company *B* using this criterion.

11. $z = -0.75$; no.

12. $t = -2.236$; no.

13. $\bar{x}_d = 1.04$; SE $\doteq 0.64$; $t = 1.62$; the critical value for t is 1.86; no.

14. $t = 5.252$; procedure is effective.

15. $t = 1.418$; procedure is not effective.

16. $z \doteq 4.24$; differences in attitude are significant.

17. $t \doteq 0.54$; differences are not significant.

18. $z \doteq 0.41, 2.22, 1.58, 2.22$, hysteria and schizophrenia.

CHAPTER 11

SECTION 11.1

(Minor differences in obtained values may be due to differences in rounding.)

1. $\chi^2 = 2.558$; difference is attributable to chance.

2. $\chi^2 = 9.551$; observed differences are significant, and there is some relationship between party preference and preferred issue.

3. $\chi^2 = 77.792$; differences are significant.

4. $\chi^2 = 237.365$; distribution does differ according to weather conditions. Most clearly significant difference is in the fourth quarter.

SECTION 11.2

1. $\chi^2 = 10.661$; even with degrees of freedom reduced to 6 and the critical value thus reduced to 12.592, the loss of sensitivity is such that H_0 cannot be rejected. Proper choice of combinations is essential.

2. $\chi^2 = 2.871$; they appear to be independent.

3. $\chi^2 = 9.489$; the data support a contention that income level and home-owning status are not independent. Examination of the data bear out the contention that low income families are more likely to rent than own homes; however, the reasons he cites are not supported. All that is supported is that significantly fewer low income people than higher income people own homes.

4. $\chi^2 = 173.234$; they are not independent.

5. $\chi^2 = 7.20$; not independent.

6. $\chi^2 \doteq 17.558$; not independent.

7. $\chi^2 \doteq 5.410$; no association.

SECTION 11.3

1. $\chi^2 = 29.164$; yes.

2. $\chi^2 = 26.886$; reject the hypothesis.

3. $\chi^2 = 4.680$; the hypothesis is supported.

4. $\chi^2 = 15.864$; reject H_0. Examination of the data leads to the conclusion that a course in college algebra is better preparation for statistics than a course in logic.

5. $\chi^2 \doteq 9.540$; conformance is satisfactory.

6. $\chi^2 \doteq 94.245$; poll was wrong.

SECTION 11.4

1. $\chi^2 = 10.342$; reject H_0; job satisfaction is related to the job.

2. $\chi^2 = 7.024$; the hypothesis is supported.

3. $\chi^2 = 4.533$; the coin does not appear to be biased.

4. $\chi^2 = 641.347$; he can conclude that there is a relationship. The data would lead one to suspect that the B factor (it occurs in AB also) is related.

5. $\chi^2 = 4.476$, df = 4; substantiates hypothesis of no difference.

6. $\chi^2 = 234.39$, df = 4; there appears to be a relationship between the severity level of the two diseases.

CHAPTER 12

SECTION 12.2

1. The critical value of F for 3 and 80 degrees of freedom at the 0.05 level of significance is 2.72. The results are
   (a) $F = 5.81$; differences are significant;
   (b) $F = 2.58$; differences are not significant, but we should probably reserve judgment;
   (c) $F = 1.60$; differences are not significant.

2. $F = 17.95$; they are not representative of the same population.

3. $F = 21.33$; the differences are significant.

SECTION 12.3

1. The critical value for 4 and 47 degrees of freedom is not on the table, but it may be estimated to be about 2.59. Since $F = 2.03$, we conclude that the differences among the groups is not significant.

2. $F = 31.11$; differences are significant.

3. $F = 5.85$; differences are significant.

SECTION 12.4

1. $F = 3.62$; differences are significant.

2. $F = 6.37$; the systems are not equally effective.

3. $F = 3.43$; we cannot reject H_0, and we should probably reserve judgment.

4. $F = 6.45$; there appears to be a relationship between amount of drug injected and stress level at which adrenalin is released.

CHAPTER 13

SECTION 13.1

1. $r = 0.22$; $R = -0.15$.
2. $r = -0.48$; $R = -0.50$.
3. $r = 0.84$; $R = 0.90$.
4. $r = -1$; $R = -1$.
5. $r = 0.08$; $R = -0.12$.
6. $r \doteq 0.462$.

SECTION 13.2

1. $r = -0.78$; significant.
2. $R = 0.75$, $t = 4.536$; correlation is significant.
3. $r = 0.43$; not significant.

SECTION 13.3

1. $$b = \frac{(10)(695) - (67)(100)}{(10)(481) - (67)^2} = \frac{250}{321} \doteq 0.779$$

   $$a = \frac{(100) - (0.779)(67)}{10} \doteq 4.78$$

   Then $y' = 4.78 + 0.779x$, $r \doteq 0.81$ which is highly significant. This tends to add validity to the regression equation.

2. $y' = 36.54 - 0.156x$

3. $x' = 180.21 - 3.926y$

4. $$b = \frac{20(359{,}875) - (2{,}597)(2{,}692)}{20(348{,}825) - (2{,}597)^2} = \frac{206{,}376}{232{,}091} \doteq 0.889$$

   $$a = \frac{2{,}692 - (0.889)(2{,}597)}{20} \doteq 19.163$$

   $$y' = 19.163 + (0.889)x$$

Then we have

x	y'
83	92.95
97	105.40
112	118.73
124	129.40
146	148.96

5. $b = \dfrac{(30)(195{,}166) - (236)(2{,}431)}{(30)(194{,}099) - (236)^2} = \dfrac{95{,}941}{210{,}809} \doteq 0.455$

$a = \dfrac{(2{,}431) - (0.455)(2{,}369)}{30} \doteq 45.10$

Then $y' = 45.10 + 0.455x$.

SECTION 13.4

1. $y' = 2.94 + 1.3x$; $r \doteq 0.91$; significant
2. $R \doteq 0.79$, $t = 3.866$; significant
3. $y' = 21.125 + 0.625x$
   $x' = 18.74 + 1.245y$

   $\sqrt{(0.625)(1.245)} \doteq \sqrt{0.7781} \doteq 0.88$
4. $r = 0.87$; the contention is substantiated.
5. $F = 26.629$; differences are significant.

APPENDIX B

The Signs Test

1. $n = 13$, $p = 0.50$, $N = 10$, $P = 0.046$; course was effective.
2. $n = 10$, $p = 0.50$, $N = 7$, $P = 0.172$; not significant.

The Wilcoxon T-Test

1. $n = 13$, $T = 15$; critical value is 21; $15 < 21$, so the course was effective.
2. $n = 10$, $T = 6.5$; critical value is 11; $6.5 < 11$, so the radiation did increase deterioration rate.
3. $n = 30$, $T = 83.5$; $z = -3.06$ so we can conclude that the diet is effective.

The Mann-Whitney U-Test

1. $R_1 = 216.5$, $R_2 = 248.5$, $U = 96.5$; $z \doteq -0.66$; not significant.
2. $R_1 = 433.5$, $R_2 = 386.5$, $U = 176.5$; $z \doteq -0.64$; not significant.

The Kruskal-Wallis H-Test

1. $R_1 = 264.5$, $R_2 = 430$, $R_3 = 335.5$, $R_4 = 146$, $H = 18.279$; $18.279 > 7.815$, so the differences are significant.
2. $R_1 = 40$, $R_2 = 82$, $R_3 = 49$, $H = 5.719$; $5.719 \not> 5.991$, so we cannot reject H_0. The obtained value of H is very close to the critical value, however, so we might reserve judgement and perform the experiment again.

Summary

1. $N = 6$, $P(x \geq 6) = 0.377$; not significant. $T = 14$, fail to reject H_0.
2. $U = 5$; differences are not significant.
3. $N = 7$, $P(x \geq 7) = 0.172$; thus we cannot say that learning took place. $T = 13$; fail to reject H_0.
4. $N = 4$; $P(x \geq 4) = 0.188$; for a two-sided alternative this gives $\alpha = 0.376$, which is not significant. $T = 1$; not significant for a two-sided alternative.
5. $T = 0$; significant. $U = 6.5$; not significant.
6. $$H = \frac{12}{(30)(31)}\left[\frac{(211)^2}{10} + \frac{(135)^2}{10} + \frac{(119)^2}{10}\right] - 3(31) \doteq 6.235$$

   Since $6.235 > 5.991$, reject H_0
7. $H = 35.582$; differences are significant.
8. $$H = \frac{12}{(28)(29)}\left[\frac{(86.5)^2}{7} + \frac{(134.5)^2}{8} + \frac{(117)^2}{6} + \frac{(68)^2}{7}\right] - 3(29) \doteq 5.693;$$

   differences are not significant.

9. $H = 9.445$; not significant.
10. $H = 7.656$; significant.
11. $H = 9.762$; significant.
12. $H = 7.733$; significant.